Artificial Intelligence Technologies for Smart and Sustainable Urban Transportation

Scrivener Publishing
100 Cummings Center, Suite 541J
Beverly, MA 01915-6106

Publishers at Scrivener
Martin Scrivener (martin@scrivenerpublishing.com)
Phillip Carmical (pcarmical@scrivenerpublishing.com)

Artificial Intelligence Technologies for Smart and Sustainable Urban Transportation

Integrated Platforms and Use Cases

Edited by

Pethuru Raj
Chief Architect & Vice President, Reliance Jio Platforms Ltd. Bangalore, India

Sudesh Yadav
Govt. College, Ateli, Distt-Mahendergarh, Haryana, India

Manas Kumar Mishra
IMS Engineering College, Ghaziabad, Uttar Pradesh, India

Satya Prakash Yadav
Department of Computer Science and Engineering, Madan Mohan Malaviya University of Technology, Gorakhpur, U.P., India

and

Victor Hugo C. de Albuquerque
Department of Teleinformatics Engineering (DETI), Federal University of Ceará, Brazil

Scrivener Publishing

WILEY

This edition first published 2026 by John Wiley & Sons, Inc., 111 River Street, Hoboken, NJ 07030, USA and Scrivener Publishing LLC, 100 Cummings Center, Suite 541J, Beverly, MA 01915, USA

For more information about Scrivener publications please visit www.scrivenerpublishing.com.

Wiley Global Headquarters
111 River Street, Hoboken, NJ 07030, USA

For details of our global editorial offices, customer services, and more information about Wiley products visit us at www.wiley.com.

The manufacturer's authorized representative according to the EU General Product Safety Regulation is Wiley-VCH GmbH, Boschstr. 12, 69469 Weinheim, Germany, e-mail: Product_Safety@wiley.com.

Library of Congress Cataloging-in-Publication Data

ISBN 978-1-394-34674-5

Cover image: Generated with AI using Adobe Firefly
Cover design by Russell Richardson

Set in size of 11pt and Minion Pro by Manila Typesetting Company, Makati, Philippines

Printed in the USA

10 9 8 7 6 5 4 3 2 1

Contents

Preface

The book delves into the latest advancements and transformative trends in urban transportation, emphasizing the role of emerging digital technologies in fostering sustainability. As urbanization continues to accelerate globally, transportation has become a significant contributor to environmental degradation and climate change. The increase in greenhouse gas (GHG) emissions—including carbon dioxide (CO_2), methane, ozone, nitrous oxide, and chlorofluorocarbons—poses a serious threat to air quality and environmental sustainability. To counteract these challenges, nations worldwide are advocating for smart, eco-friendly urban mobility solutions. The integration of cutting-edge technologies such as artificial intelligence (AI), 5G/6G, cybersecurity, the Internet of Things (IoT), blockchain, edge computing, and cloud-native infrastructures has been widely recognized for its potential to enhance intelligent and energy-efficient transportation systems. Among the most prominent recommendations is the large-scale adoption of electric vehicles (EVs) to mitigate environmental deterioration and promote greener urban ecosystems. In addition to electric mobility, the development of smart microgrids and the utilization of renewable energy sources are gaining traction as proven strategies for achieving long-term sustainability. Furthermore, experts and environmental advocates are championing a range of innovative software platforms and solutions that are essential for modernizing urban transportation. Advanced battery management systems for EVs, for instance, offer enhanced energy efficiency, while intelligent infrastructure management for EV charging ensures automation, optimization, and streamlined operations. Artificial intelligence, initially introduced as a data analytics tool, has evolved into a powerful generative technology, unlocking new possibilities for urban mobility. AI, when combined with other well-established digital innovations, has proven instrumental in revolutionizing transportation networks with intelligent, data-driven decision-making. This book not only highlights the foundational technologies driving this transformation but also explores AI-powered platforms and management solutions shaping the future of urban transportation.

The Editors
September 2026

Part 1

ARTIFICIAL INTELLIGENCE IN SOLVING URBAN PLANNING AND DESIGNING CHALLENGES

1

Illustrating the Sustainability, Challenges, and Concerns of Urban Mobility and Smart Cities

Nilesh Bhosle[1], Amandeep Kaur[2], Raman Kumar[3]*, Yashwant Singh Bisht[4] and Laith H. Alzubaidi[5]

[1]Department of Computing Science and Artificial Intelligence, NIMS Institute of Engineering and Technology, NIMS University Rajasthan, Jaipur, India
[2]Department of Computer Science Engineering, Chandigarh Engineering College, Chandigarh Group of Colleges, Jhanjeri, Mohali, Punjab, India
[3]School of Mechanical Engineering, Rayat Bahra University, Mohali, India
[4]Department of Mechanical Engineering, Uttaranchal Institute of Technology, Uttaranchal University, Dehradun, Uttarakhand, India
[5]Department of Computer Engineering, The Islamic University, Najaf, Iraq

Abstract

The purpose of this chapter is to explore the challenges and sustainability issues associated with urban mobility and smart cities, focusing on the role that digital technologies and smart systems play in the modern planning and management of cities. Among the foundational elements of smart cities, smart mobility is explored for its potential to deliver sustainable transportation solutions, including the use of electric vehicles and Mobility as a Service (MaaS). In the discussion, technological innovations like the Internet of Things (IoT) and artificial intelligence are discussed as enabling improvements in energy efficiency, reduced emissions, and overall quality of life through the use of urban infrastructure and services. It remains difficult to integrate diverse systems and gain public acceptance due to critical obstacles such as data privacy. Through strategic digitalization, urban ecosystems can become more sustainable, efficient, and inclusive, fostering the development of intelligent cities.

**Corresponding author*: ramankakkar@gmail.com

Pethuru Raj, Sudesh Yadav, Manas Kumar Mishra, Satya Prakash Yadav and Victor Hugo C. de Albuquerque (eds.) Artificial Intelligence Technologies for Smart and Sustainable Urban Transportation: Integrated Platforms and Use Cases, (3–22) © 2026 Scrivener Publishing LLC

Keywords: Artificial intelligence, sustainable development, intelligent transportation, smart, urban mobility

1.1 Introduction

Smart mobility strategies incorporate the 17 sustainable development goals set by the Urban Mobility (UN) [1] for 2030, aligning them more closely with sustainable world development. In the coming years, Mobility will undergo dramatic changes, as will city planning and logistics of moving goods and people. With cities around the world growing rapidly, new algorithms for route optimisation, better traffic management, and improved logistics processes are needed. In addition to solving these problems, smart Mobility will continue to lead to real innovation in the future [2, 3]. As smart Mobility develops, sustainability, active transportation, renewable fuels, and citizen engagement will play an increasingly important role. Therefore, citizens and government entities will be affected by a variety of dimensions, including sustainability, economy, and living [2].

The development of smart devices has reached a point where existing networks can be used to connect everyday objects. Hundreds of billions of devices are connected through conventional networks, resulting in the IoT. As skill has developed in urban cities (UC), wireless sensor networks (WSN) [4], and machine-to-machine (M2M), the concept of IoT has become more robust [5, 6]. By utilising uniquely identifiable smart devices, UC can be facilitated with little or no human interaction [7, 8]. A smart device is not only capable of sharing information about itself, but it can also access information authorised by others, leading to contextual decisions [9]. Having gained widespread appeal, the Internet of Things concept has spawned a large number of applications, including smart homes and smart cities, smart warehouses, smart health, etc. [10–13]. Several countries, including the United States, are experiencing dramatic urbanisation, which has led to an increase in smart cities in recent decades. Implementing ICT in city operations led to various improvements in efficiency. Smart cities are not fully interpreted when ICT is used to conduct city operations [14]. Due to its abstraction of all urban models, including telicity, information cities, and digital cities, a smart city brings together all the benefits of all of them [15]. Consequently, smart cities inherit the underlying operating mechanisms used in IoT Consequently, smart cities inherit the underlying operating mechanisms used in IoT [16, 17]. Internet of Things makes smart cities possible, including the creation of data, its management, and its application management. It is generally accepted that a smart city uses telecommunications and information technology to improve the quality of citizen services as well as city operations. As a formal concept, smart cities take into account a variety of perspectives and aspects.

Typically, smart cities are characterised by a constellation of spatially distributed instruments that are linked *via* multiple networks to provide continuous information regarding human and material movement. This data is crucial for determining the city's physical and social form. Only intelligent functions will make cities smart, incorporating and synthesising data to improve efficiency, equity, sustainability, and quality of life. It is expected that smart city technology will grow beyond instrumentation, which is provided by large and small ICT companies that create smart city operating systems with hardware and software. Essentially, this instrumentation is enabling dramatic changes in social organisation as a result of this instrumentation.

1.1.1 Characteristics of a Smart City

Under each attribute, there are a few sub-attributes. Under sustainability, there are sub-attributes for infrastructure, governance, pollution, energy, and climate change, as mentioned in Figure 1.1. Sustainable systems are ecosystems that are balanced in all aspects while providing service to the public and managing city operations. Quality of life is improved when

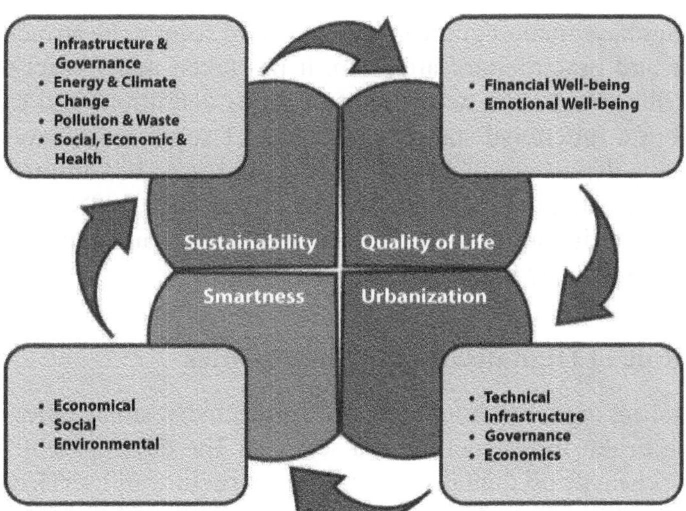

Figure 1.1 An overview of smart city characteristics.

urban citizens are emotionally and financially well-off. Technology, economics, infrastructure, and governance are all part of the urbanisation attribute. As a concept, smartness can be defined as a desire to make cities and their residents more socially, environmentally, and economically prosperous.

A smart city has historically been proposed as a means to improve the quality of life. A commitment to integrating innovative solutions leads to a reduction in social learning barriers and increased social participation, thereby improving citizens' lives. In addition to recognising and utilising skilled citizens, modern city councils implement solid social policies. As a result, the quality of city services is enhanced, while skilled personnel's quality of life and financial status are enhanced. The enhancement of QoL is ultimately determined by employee and citizen financial well-being as well as emotional well-being. The quality of life has been improved in various cities in similar ways. An artists' circle has been established in the city of Yokohama, Japan, which organises exhibitions, workshops, and performances for the benefit of the community [18], thereby improving social values. As part of the campaign in Chicago, underprivileged residents were also provided with healthcare services [19].

1.2 Smart City

1.2.1 An Overview of Smart Cities

Smart cities have made significant progress and seem to be attracting both academic and practitioners' attention. It has taken many years to define and describe smart cities according to various definitions and characteristics. From a functional standpoint, Table 1.1 summarises several studies regarding definitions of smart city. In a smart city, ICT infrastructures, network infrastructures (NIs), physical infrastructures, WSNs, the Internet of Things, and digital infrastructure, as well as human capital and social capital, are all equally important [20, 21].

1.2.2 Role of Digitalisation in Smart Cities

By integrating digital technologies and connectivity, smart cities can revolutionise urban infrastructure and services. The development of smart cities relies heavily on digitalisation for infrastructure and services [30]. In general, there are two primary layers in a city system: infrastructure and services. Infrastructure encompasses the tangible components of a system

Table 1.1 Defined differently by different authors, smart cities.

References	Smart city definition	Indicators
[22]	"Creating and interconnecting human, social, and IT capital will lead to a more sustainable economy and a higher quality of life."	Investing in human, social, and technological capital
[14]	"Using the network infrastructure, smart cities enable social, cultural, and urban development, as well as improving economic efficiency and political efficiency."	-
[23]	In a smart city, public administrations or authorities provide (or aim to provide) new-generation infrastructure and services.	ICT
[20]	Physical, digital and human systems are integrated to create a sustainable, prosperous, and inclusive environment.	Infrastructural components, human capital, and digital components
[24]	WSNs form the basic infrastructure of a CPS, particularly a smart city, and are often described as smart on the Internet of Things.	WSN and IoT
[25]	By incorporating information and communication technology into a smart city, citizens can benefit from a higher quality of life, a stronger local economy, better transportation, better traffic management, better environmental impact, and better communication with governments.	ICT

(*Continued*)

Table 1.1 Defined differently by different authors, smart cities. (*Continued*)

References	Smart city definition	Indicators
[26]	The concept of smart cities is based on the integration of ICT, IoT, and other physical devices connected to the internet in order to optimise city operations, services, and citizen participation.	ICT IoT
[27]	In a smart city, urban problems are solved through technology and governance between stakeholders, as well as improving citizens' quality of life.	Technology revolution
[28]	Utilising digital technologies efficiently so that prioritised services and benefits can be provided to the community. An inability to measure "smart" effectively.	Digital technology
[29]	By using digital solutions, smart cities make traditional networks and services more efficient for their inhabitants and businesses.	Digital solution

and provides the essential services required for the functioning of urban areas [31]. People, buildings, mobility, economics, governance, energy, water, and the environment all depend on city infrastructure, as it serves as a connection between natural resources and the residents [32]. In a city, the service layer provides residents with a range of services. Water distribution, waste management, maintaining the environment, and distributing electricity are a few examples of these activities [32]. To transform a city into a 2.0, a data layer must be incorporated, also referred to as the digital layer.

The development of a smart urban environment and infrastructure has been made possible by the process of digitalisation. It has taken a long time to digitise physical space, energy, information transmission, asset management, and business operations [30]. To achieve the smart city vision, ICT must be used to capture and share information in a timely manner. In urban areas, if information is delivered promptly and accurately, problems can be addressed proactively before they become problematic. A scholarly source states that digital infrastructure can be conceptualised by using a variety of digital layers to provide support.

By gathering, analysing, and utilising real-time data, digitalisation transforms cities and improves services and infrastructure [33]. In addition to IoT devices and sensors [34], there has been an increase in connectivity and communication, and intelligence systems and platforms have been developed. Smart meters and renewable energy sources are being integrated into the energy industry with the help of digital technologies. Sustainability and energy efficiency are achieved as a result. As a result of digitalisation, transportation can be transformed with intelligent traffic management, charging infrastructure for electric vehicles, and ridesharing platforms [29]. Mobilising efficiently and sustainably is made possible by this. As a result of digitalisation, a number of urban services are improving, such as public safety, waste management, and water management [26, 29, 35].

1.2.3 Infrastructural Impacts of Digitalisation in Smart Cities

The digitalisation of smart cities is making them more energy efficient, more sustainable, and more environmentally friendly. It involves a wide range of issues, from waste management to transportation [36]. Digital technologies have had a significant impact on these domains, so we examine how they are integrated into smart cities as part of this section.

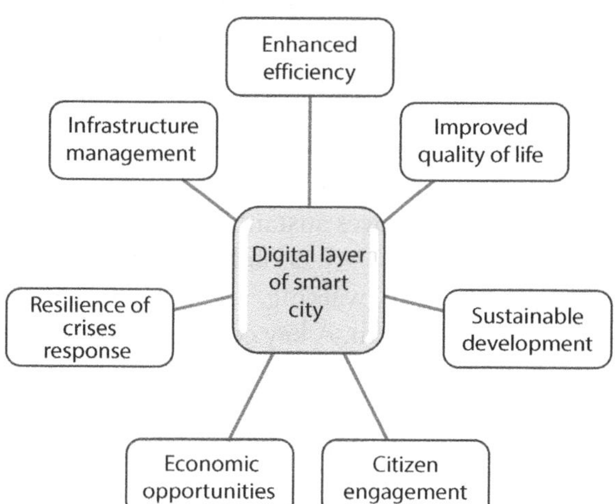

Figure 1.2 A smart city's transformation is influenced by the digital layer.

Smart cities are being affected significantly and extensively by digitalisation. Numerous benefits have been observed as data-driven processes and digital technologies have been integrated into smart cities. Smart cities have been greatly impacted by digitalisation, as shown in Figure 1.2.

When data is utilised in the digitalisation process, decision-making processes are improved, resource allocation is optimised, and public safety, transportation, energy, and waste management are improved [37]. With IoT, AI, and data analytics, smart cities are optimising resources, reducing emissions, and managing the environment [38]. Digitalisation has a significant impact on energy management. By using an energy data and analytics platform in real-time, informed decisions can be made, and energy can be managed more effectively [37, 39]. As part of digitalisation, transportation plays an important role [40, 41]. In addition, waste management is heavily influenced by digitalisation. Monitoring waste levels, optimising collection routes, and improving recycling are possible with smart waste management devices and Internet of Things devices. By using smart waste management and IoT devices, waste levels can be monitored, collection routes can be optimised, and recycling can be improved [42].

1.3 Smart Mobility in Smart Cities

When urban Mobility is incorporated into smart cities, they are smart and sustainable. As nations across the world work to provide smart and affordable transportation to citizens, they heavily invest in transport infrastructure. Smart cities have smart Mobility as one of their key characteristics [43]. Most developing nations face a growing need for cleaner, more environmentally friendly transportation due to rapid urbanisation. As a result of the challenges associated with Mobility in such countries, public transport systems are required to be mass-adopted.

Smart mobility systems provide Mobility as a Service (MaaS), in which intelligent infrastructure improves sustainability, productivity, and intelligence for various stakeholders [44]. Mobility solutions, including electric vehicles, bikes, high-speed rail, walking, and on-demand Mobility, are all part of this transportation system. A key component of quality services for citizens is smart Mobility, which minimises environmental impacts and ensures quality services [45]. The benefits of integrating smart mobility ecosystems are numerous, as shown in Figure 1.3.

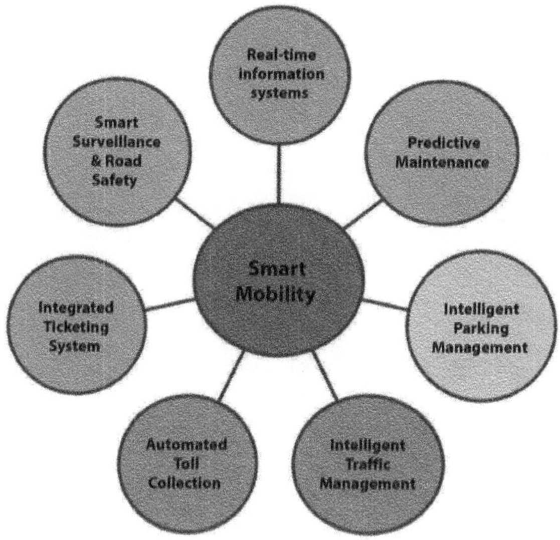

Figure 1.3 Benefits of smart mobility for smart cities.

Systems for real-time information: Real-time Information systems gather data from different connected entities; we need to create efficient and smart information systems by following the following steps.

Maintenance Predictive: Machine learning and artificial intelligence can also be used to predict a wide range of processes. Monitoring and collecting data continuously facilitate this process.

Intelligent Parking Management: Data collected from sensors and connected devices can be used to provide insights into parking availability at different locations through analysis of real-time data.

Intelligent Traffic Management: Utilising effective data analytics can assist in avoiding jams and congestion in real-time. It is also possible to send real-time notifications to connected vehicles regarding parking and routes.

Automated Toll Collection: The automated payment of tolls at toll plazas facilitates hassle-free movements.

Integrated Ticketing Systems: Ticketing systems that use Mobility as a service enable citizens to access various local services.

Smart Surveillance and Road Safety: Cameras and other security devices monitor traffic conditions, improving road safety. A new era of urban planning and management has emerged with the development of connected

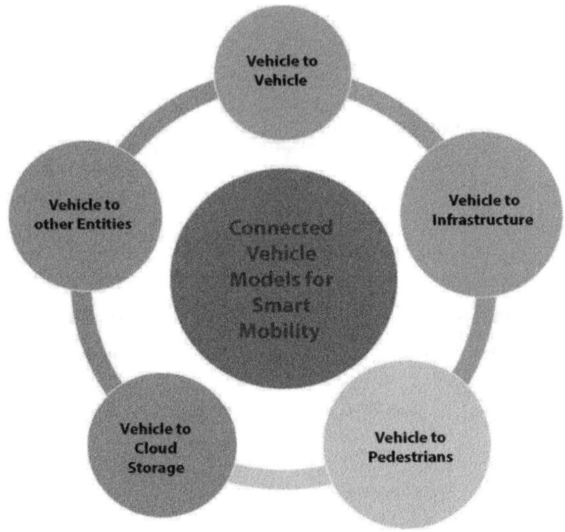

Figure 1.4 Connected vehicle models for smart mobility.

technologies. In addition to the economic costs involved with road mishaps, there is a human cost as well. By implementing intelligent transportation management systems, we are able to achieve smart Mobility, which enables safe and efficient driving and commuting. An example of smart mobility connectivity is shown in Figure 1.4.

Vehicle to Infrastructure: Besides traffic lights, toll booths, parking meters, and pedestrian crossings, the infrastructure of traffic can also be monitored and predicted in real-time.

Vehicle to Vehicle: In addition to traffic lights and tolls, parking meters, and pedestrian crossings, the traffic infrastructure is connected to the vehicles.

Vehicle-to-Cloud Storage: Data sharing, storing, and processing can be done in real-time for vehicle-to-storage connectivity.

Vehicle to Pedestrians: By connecting vehicles with pedestrians *via* phone and wearable devices, as a result of the concept connectivity model, pedestrian safety has been enhanced, and mobility has been made more efficient in real-time.

Vehicles to Other Entities: It is possible to create connectivity by integrating smart Mobility with other components and processes in the smart city.

1.4 Analysis of Security Threats

IoT, WSN, artificial intelligence, and other technologies contribute to the collection, analysis, and sharing of data by autonomous vehicles. In addition to these autonomous vehicles, there are thousands of entities linked together in a larger network. There is a great deal of dependence on the security protocols of the individual devices in connected environments involving heterogeneous devices. Consequently, autonomous vehicles are at risk from a variety of security threats. These attacks can take place on a number of levels, including a network level, an end-user device level, and a software level [46]. At different levels of autonomy, the vulnerability types of autonomous vehicles are shown in Table 1.2.

A number of vulnerabilities raise security and privacy concerns at different levels. Security concerns include those relating to data, networks, vehicles, and finances [47].

A few of the most common security threats are as follows:

Device Hijacking: Autonomous vehicles face a variety of threats, including hijackings. In the event that an attacker modifies the remote vehicle control algorithm, the vehicle system and software can be taken over by the attacker. When a vehicle is hijacked, cybercriminals can alter the navigation, engine, brakes, heating, communication, cameras, and vision systems on the vehicle.

Denial of Service: By exploiting the vulnerable devices on a network, attackers are able to take control of the entire system. Attacks of this type

Table 1.2 Vulnerabilities in autonomous vehicles.

Levels	Target systems	Disrupted services
Sensors	Sensors, including cameras, GPSs, LiDARs, Radars, proximity sensors, and ultrasonics	The ability to park, identify objects, navigate, avoid collisions, identify traffic signals, and cruise control
Device	System of access control	System for preventing theft, keyless entry, jamming of signals, replay attacks
Software	LIN, CAN, FlexRay protocols in vehicles	Communication system

may disrupt communication between vehicles, or between vehicles and infrastructure.

Privacy Infringement: In autonomous vehicles, a continuous stream of data is generated, including the location of the vehicle. Users can track their current location and other whereabouts, as well as their travel history and navigation history, using these data.

Financial Fraud: The use of RFID tags for automatic toll collection in India raises serious concerns about financial security. An insecure network or device may be used to perpetrate banking fraud or other ransomware attacks.

1.4.1 Mobility Trends in Smart Cities in the Future

A smart city develops smart mobility solutions for mobility management. Several innovative approaches are improving smart Mobility's state-of-the-art. Artificial Intelligence has been exploited to create outstanding

Figure 1.5 A generic framework for the design of smart cities.

mobility solutions over the past few years. Visibility can be improved by deep learning-based object detection and identification in conditions of low light, fog, wind, or rain [48–50]. Monitoring road conditions, the environment, and the surrounding area using IoT can help predict future conditions and take appropriate action. With blockchain technology, smart Mobility can benefit from an effective, secure, and privacy-preserving vehicle-to-everything network (V2X) [51–54]. Aside from reducing carbon emissions, alternative fuels, electric vehicles, and improved batteries can also contribute to sustainable development. Communication technologies have been extensively researched for disasters and natural calamities [55, 56]. There is a growing interest in satellite-based communication technology as an option for providing connectivity in extreme and unavoidable circumstances.

Figure 1.5 shows how smart cities are composed of many components. A smart energy system, smart transportation system, and smart health care system are three of the most important components. As a result, smart cities vary in their composition depending on their focus areas. In one smart city, a disaster management system might be incorporated, whereas in another, a waste management system may be integrated. Several smart city components are discussed in the following subheadings.

1.5 Issues and Opportunities Related to Smart Cities

It is important to address existing problems in certain areas to advance smart city concepts further, even with their widespread acceptance. In this section, we discuss the potential advantages of smart cities and how they can realistically be implemented. Research on smart cities has identified numerous challenges through extensive literature reviews. Based on existing research and hands-on experience with smart cities, opportunities in this area were identified as well.

1.5.1 Challenges for Smart Cities

For smart cities to become a reality, they must be designed, implemented, and operated in a complex manner. In designing and operating devices, there are many challenges, including heterogeneity among devices, collecting and analysing huge amounts of data, and ensuring information security and sustainability. As shown in Figure 1.6, a realistic smart city design has several main concerns.

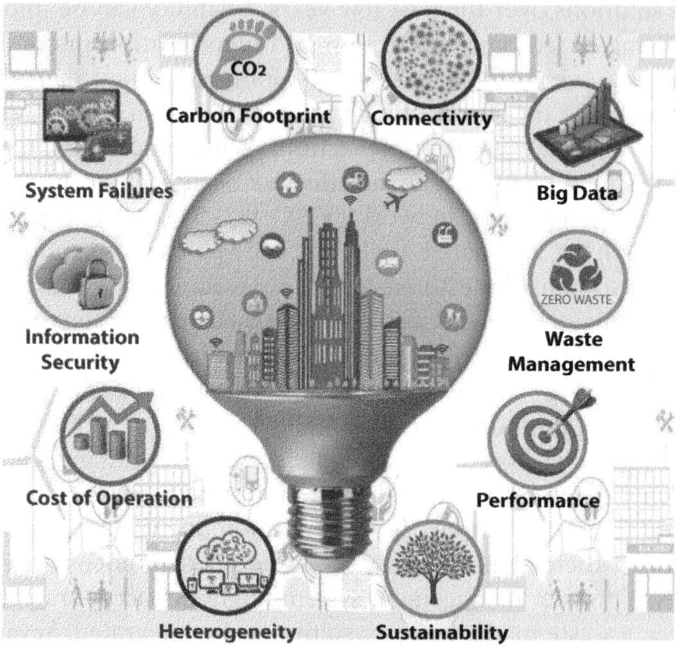

Figure 1.6 Implementing smart cities with a number of challenges.

Smart city architecture also focuses on heterogeneity. Sensors, appliances, and devices present in smart cities can be purchased from multiple vendors and used for a variety of purposes. All these heterogeneous elements must be integrated at the application level in order to realise smart city concepts. A heterogeneous platform, however, limits application integration and interoperability. In smart cities, heterogeneous subsystems must be arranged and accessed universally while hardware and software are chosen and purchased. Despite the additional expense of designing and maintaining high security, smart cities are highly enforced when it comes to infrastructure security and information security. Globally, technological advancements are causing a revolution in major cities. By adapting technology to citizens', visitors', and businesses' needs, cities have improved their quality of life. Despite the rapid growth of technology and malicious threats, smart cities have faced an increased concern about being attacked.

Smart cities are reducing their carbon footprints and maximising resource efficiency to ensure a sustainable future for all. The demand for renewable energy is, therefore, increasing in modern cities as a means of reducing their carbon footprint and ensuring their sustainability. London's power networks were discussed as an important part of the Smart opportunities for London report, focusing on energy utilisation, sustainability, and reducing carbon footprints [57].

Any project aimed at developing a smart city needs to manage failures as well. Natural disasters like floods, earthquakes, and tornadoes can cause failures. An example of a system failure is an infrastructure breakdown or a network outage. It involves immediately recovering from failures and resuming normal operations after a failure has occurred. Yet identifying and implementing a recovery strategy and fault tolerance strategy both in the design phase and in operation results in a higher cost. Implementing failure recovery mechanisms without compromising operational efficiency and costs would be a challenge.

1.5.2 Trends and Opportunities for the Future

It is only within the borders of developed countries that smart cities are being experimented with and implemented. The Singapore government has taken measures to monitor the lives of Singaporeans using citywide sensor data. Through Singapore's Smart Nation Program (SNP), all communities are connected online through existing technological infrastructure. A smart city is beneficial to any urban setting despite the fact that it offers a wide range of benefits. To popularise smart cities throughout the world, further studies should be conducted on cost-efficient design and implementation. As part of ensuring sustainable city operations and managing non-renewable energy scarcity, it is mandatory to incorporate renewable energy sources.

Large amounts of data are generated by smart devices, requiring large amounts of storage. As Big Data grows, traditional data processing techniques will become obsolete. In order to achieve smart city objectives, big data analytics must be explored and incorporated. There have been few studies that address this issue. However, the majority of them have not yet been tested in real-world situations. Therefore, a promising research opportunity is to propose and experiment with Big Data analytics in real smart cities in the future. In connected environments, it is imperative to protect sensitive data.

1.6 Conclusions

With the advent of the Internet of Things, smart cities have become a reality. In urban environments, we can use ICT to promote digital, green, sustainable, and intelligent cities as examples. In comparison to other cities, smart cities possess a holistic vision. Therefore, smart cities are a combination of other strategies for managing urban environments. Smart cities are defined, standardised, and analysed in this paper. There are a number of characteristics and features of smart cities that are explained straightforwardly so readers can grasp their essence. A generic architecture for a smart city is explained in more technical detail after thoroughly examining proposed smart city architectures. Urban residents' quality of life is improved through interoperability between subsystems in a smart city.

Additionally, the article provides examples of smart cities that have been implemented in practice, as well as some current statistics. Due to the vast amount of data processed and the heterogeneous nature of smart devices, smart cities still face a number of serious challenges and issues. The knowledge base was enriched, and future research guidance was provided by identifying several challenges and opportunities for improvement.

References

1. Lee, B.X., Kjaerulf, F., Turner, S., Cohen, L., Donnelly, P. D., Muggah, R., ... & Gilligan, J., Transforming our world: implementing the 2030 agenda through sustainable development goal indicators. *J. Public Health Policy*, 37, 13–31, 2016.
2. Gouveia, J.P., Seixas, J., Giannakidis, G., Smart City Energy Planning: Integrating Data and Tools, in: *Proceedings of the 25th International Conference Companion on World Wide Web - WWW '16 Companion*, ACM Press, Montréal, Québec, Canada, pp. 345–350, 2016, doi: 10.1145/2872518.2888617.
3. Rani, P. and Sharma, R., Intelligent transportation system for internet of vehicles based vehicular networks for smart cities. *Comput. Electr. Eng.*, 105, 108543, 2023.
4. Kumar, N., Rani, P., Kumar, V., Athawale, S.V., Koundal, D., THWSN: Enhanced energy-efficient clustering approach for three-tier heterogeneous wireless sensor networks. *IEEE Sens. J.*, 22, 20, 20053–20062, 2022.
5. Silva, B.N., Khan, M., Han, K., Internet of Things: A Comprehensive Review of Enabling Technologies, Architecture, and Challenges. *IETE Tech. Rev.*, 35, 2, 205–220, Mar. 2018, doi: 10.1080/02564602.2016.1276416.
6. Kumar, N., Rani, P., Kumar, V., Verma, P.K., Koundal, D., Teeech: Three-tier extended energy efficient clustering hierarchy protocol for heterogeneous wireless sensor network. *Expert Syst. Appl.*, 216, 119448, 2023.
7. Gubbi, J., Buyya, R., Marusic, S., Palaniswami, M., Internet of Things (IoT): A vision, architectural elements, and future directions. *Future Gener. Comput. Syst.*, 29, 1645–1660, 7 Sep. 2013, doi: 10.1016/j.future.2013.01.010.

8. Silva, B.N., Khan, M., Han, K., Towards sustainable smart cities: A review of trends, architectures, components, and open challenges in smart cities. *Sustain. Cities Soc.*, 38, 697–713, Apr. 2018, doi: 10.1016/j.scs.2018.01.053.
9. Vermesan, O., *et al.*, Internet of Things beyond the Hype: Research, Innovation and Deployment, in: *Building the Hyperconnected Society-Internet of Things Research and Innovation Value Chains, Ecosystems and Markets*, River Publishers, New York, 2015.
10. Riazul Islam, S.M., Daehan Kwak, M., Kabir, H., Hossain, M., Kwak, K.-S., The Internet of Things for Health Care: A Comprehensive Survey. *IEEE Access*, 3, 678–708, 2015, doi: 10.1109/ACCESS.2015.2437951.
11. Jabbar, S., Khan, M., Silva, B.N., Han, K., A REST-based industrial web of things' framework for smart warehousing. *J. Supercomput.*, 74, 9, 4419–4433, Sep. 2018, doi: 10.1007/s11227-016-1937-y.
12. Jin, J., Gubbi, J., Marusic, S., Palaniswami, M., An Information Framework for Creating a Smart City Through Internet of Things. *IEEE Internet Things J.*, 1, 2, 112–121, Apr. 2014, doi: 10.1109/JIOT.2013.2296516.
13. Khan, M., Silva, B.N., Han, K., Internet of Things Based Energy Aware Smart Home Control System. *IEEE Access*, 4, 7556–7566, 2016, doi: 10.1109/ACCESS.2016.2621752.
14. Hollands, R.G., Will the real smart city please stand up?: Intelligent, progressive or entrepreneurial?, in: *The Routledge Companion to Smart Cities*, Taylor Francis, 2020.
15. Mohanty, S.P., Choppali, U., Kougianos, E., Everything you wanted to know about smart cities: The Internet of things is the backbone. *IEEE Consum. Electron. Mag.*, 5, 3, 60–70, Jul. 2016, doi: 10.1109/MCE.2016.2556879.
16. Silva, B.N., *et al.*, Planning of smart cities: Performance improvement using big data analytics approach in. *Proceedings of the fourth international conference on advances in computing, electronics and communication*, pp. 15–16, 2016, Accessed: Nov. 03, 2024. [Online]. Available: https://www.researchgate.net/profile/Bhagya-Silva/publication/311738615_Planning_of_smart_cities_Performance_improvement_using_big_data_analytics_approach/links/588fd62892851c9794c4a33d/Planning-of-smart-cities-Performance-improvement-using-big-data-analytics-approach.pdf.
17. Bhola, B., *et al.*, Quality-enabled decentralised dynamic IoT platform with scalable resources integration. *IET Commun.*, 1–10, 2022.
18. Sasaki, M., Urban regeneration through cultural creativity and social inclusion: Rethinking creative city theory through a Japanese case study. *Cities*, 27, S3–S9, Jun. 2010, doi: 10.1016/j.cities.2010.03.002.
19. Oakley, D. and Tsao, H., Socioeconomic gains and spillover effects of geographically targeted initiatives to combat economic distress: An examination of Chicago's Empowerment Zone. *Cities*, 24, 1, 43–59, Feb. 2007, doi: 10.1016/j.cities.2006.10.003.
20. S. city framework–Guide, Code of practice for Smart city framework–Guide to establishing strategies for smart cities and communities, 2014, Accessed:

Nov. 03, 2024, [Online], Available: https://www.ospi.es/export/sites/ospi/documents/documentos/BSI_Smart_Cities_Framework.pdf.
21. Rani, P., Hussain, N., Khan, R.A.H., Sharma, Y., Shukla, P.K., Vehicular intelligence system: time-based vehicle next location prediction in software-defined internet of vehicles (SDN-IOV) for the smart cities. *Intell. Things AI-IoT Based Crit.-Appl. Innov.*, 35–54, 2021.
22. Giffinger, R., Fertner, C., Kramar, H., Kalasek, R., Pichler-Milanovic, N., Meijers, E.J., Smart cities, in: *Ranking of European medium-sized cities*, Oct. 2007, Final report, doi: 10.34726/3565.
23. Xie, X., Lu, Q., Herrera, M., Yu, Q., Parlikad, A.K., Schooling, J.M., Does historical data still count? Exploring the applicability of smart building applications in the post-pandemic period. *Sustain. Cities Soc.*, 69, 102804, Jun. 2021, doi: 10.1016/j.scs.2021.102804.
24. Grossi, G. and Pianezzi, D., Smart cities: Utopia or neoliberal ideology? *Cities*, 69, 79–85, Sep. 2017, doi: 10.1016/j.cities.2017.07.012.
25. Ismagilova, E., Hughes, L., Dwivedi, Y.K., Raman, K.R., Smart cities: Advances in research—An information systems perspective. *Int. J. Inf. Manag.*, 47, 88–100, Aug. 2019, doi: 10.1016/j.ijinfomgt.2019.01.004.
26. Toli, A.M. and Murtagh, N., The Concept of Sustainability in Smart City Definitions. *Front. Built Environ.*, 6, 77, Jun. 2020, doi: 10.3389/fbuil.2020.00077.
27. Myeong, S., Park, J., Lee, M., Research Models and Methodologies on the Smart City: A Systematic Literature Review. *Sustainability*, 14, 3, 1687, Feb. 2022, doi: 10.3390/su14031687.
28. Wollman, D., Smart Cities and Communities: A Key Performance Indicators, 2022, Accessed: Nov. 03, 2024, [Online], Available: https://www.academia.edu/download/89260730/nist.sp.pdf.
29. Apanavičienė, R. and Shahrabani, M.M.N., Key Factors Affecting Smart Building Integration into Smart City: Technological Aspects. *Smart Cities*, 6, 4, 1832–1857, Jul. 2023, doi: 10.3390/smartcities6040085.
30. Serrano, W., Digital Systems in Smart City and Infrastructure: Digital as a Service. *Smart Cities*, 1, 1, 134–154, Nov. 2018, doi: 10.3390/smartcities1010008.
31. Rani, P., Verma, S., Yadav, S.P., Rai, B.K., Naruka, M.S., Kumar, D., Simulation of the lightweight blockchain technique based on privacy and security for healthcare data for the cloud system. *Int. J. E-Health Med. Commun. IJEHMC*, 13, 4, 1–15, 2022.
32. Syed, A.S., Sierra-Sosa, D., Kumar, A., Elmaghraby, A., IoT in Smart Cities: A Survey of Technologies, Practices and Challenges. *Smart Cities*, 4, 2, 429–475, Mar. 2021, doi: 10.3390/smartcities4020024.
33. Shahid, K., Hassan, M., Husnain, A., Mukhtar, S., Information and Communication Technologies for New Generation of Sustainable Smart Cities, in: *Sustainable Smart Cities - A Vision for Tomorrow*, A. Almusaed

and A. Almssad (Eds.), IntechOpen, Sweden, 2023, doi: 10.5772/intechopen.107251.

34. Krishnamurthi, R., Kumar, A., Gopinathan, D., Nayyar, A., Qureshi, B., An Overview of IoT Sensor Data Processing, Fusion, and Analysis Techniques. *Sensors*, 20, 21, 6076, Oct. 2020, doi: 10.3390/s20216076.
35. Shahrabani, M.M.N. and Apanavičienė, R., Towards integration of smart and resilient city: literature review. *IOP Conf. Ser. Earth Environ. Sci.*, 1122, 1, 012019, Dec. 2022, doi: 10.1088/1755-1315/1122/1/012019.
36. Alahi, M.E.E., *et al.*, Integration of IoT-Enabled Technologies and Artificial Intelligence (AI) for Smart City Scenario: Recent Advancements and Future Trends. *Sensors*, 23, 11, 5206, May 2023, doi: 10.3390/s23115206.
37. Woetzel, J., *et al.*, Smart cities: Digital solutions for a more livable future. *Exec. Summ.*, 1–28, June, 2018.
38. Mondejar, M.E., *et al.*, Digitalisation to achieve sustainable development goals: Steps towards a Smart Green Planet. *Sci. Total Environ.*, 794, 148539, Nov. 2021, doi: 10.1016/j.scitotenv.2021.148539.
39. Cocchia, A., Smart and Digital City: A Systematic Literature Review, in: *Smart City*, R.P. Dameri and C. Rosenthal-Sabroux (Eds.), pp. 13–43, Springer International Publishing, Progress in IS., Cham, 2014, doi: 10.1007/978-3-319-06160-3_2.
40. Turoń, K., Factors Affecting Car-Sharing Services. *Smart Cities*, 6, 2, 1185–1201, Apr. 2023, doi: 10.3390/smartcities6020057.
41. Olszewski, R., Pałka, P., Turek, A., Solving 'Smart City' Transport Problems by Designing Carpooling Gamification Schemes with Multi-Agent Systems: The Case of the So-Called 'Mordor of Warsaw. *Sensors*, 18, 1, 141, Jan. 2018, doi: 10.3390/s18010141.
42. Vishnu, S., *et al.*, IoT-Enabled Solid Waste Management in Smart Cities. *Smart Cities*, 4, 3, 1004–1017, Jul. 2021, doi: 10.3390/smartcities4030053.
43. Chourabi, H., *et al.*, Understanding Smart Cities: An Integrative Framework, in: *2012 45th Hawaii International Conference on System Sciences*, IEEE, Maui, HI, USA, pp. 2289–2297, Jan. 2012, doi: 10.1109/HICSS.2012.615.
44. Danesin, A. and Linares, P., The relevance of the local context for assessing the welfare effect of transport decarbonisation policies. A study for 5 Spanish metropolitan areas. *Energy Policy*, 118, 41–57, Jul. 2018, doi: 10.1016/j.enpol.2017.12.019.
45. Benevolo, C., Dameri, R.P., D'Auria, B., Smart Mobility in Smart City: Action Taxonomy, ICT Intensity and Public Benefits, in: *Empowering Organizations, vol. 11Lecture Notes in Information Systems and Organisation*, vol. 11, T. Torre, A.M. Braccini, R. Spinelli (Eds.), pp. 13–28, Springer International Publishing, Cham, 2016, doi: 10.1007/978-3-319-23784-8_2.
46. Ren, K., Wang, Q., Wang, C., Qin, Z., Lin, X., The Security of Autonomous Driving: Threats, Defenses, and Future Directions. *Proc. IEEE*, 108, 2, 357–372, Feb. 2020, doi: 10.1109/JPROC.2019.2948775.

47. Parkinson, S., Ward, P., Wilson, K., Miller, J., Cyber Threats Facing Autonomous and Connected Vehicles: Future Challenges. *IEEE Trans. Intell. Transp. Syst.*, 18, 11, 2898–2915, Nov. 2017, doi: 10.1109/TITS.2017.2665968.
48. Pandey, A., Puri, M., Varde, A., Object Detection with Neural Models, Deep Learning and Common Sense to Aid Smart Mobility, in: *2018 IEEE 30th International Conference on Tools with Artificial Intelligence (ICTAI)*, IEEE, Volos, Greece, pp. 859–863, Nov. 2018, doi: 10.1109/ICTAI.2018.00134.
49. Mauri, A., *et al.*, Deep Learning for Real-Time 3D Multi-Object Detection, Localisation, and Tracking: Application to Smart Mobility. *Sensors*, 20, 2, 532, Jan. 2020, doi: 10.3390/s20020532.
50. Karballaeezadeh, N., *et al.*, Intelligent Road Inspection with Advanced Machine Learning; Hybrid Prediction Models for Smart Mobility and Transportation Maintenance Systems. *Energies*, 13, 7, 1718, Apr. 2020, doi: 10.3390/en13071718.
51. Bittencourt, L.F., Diaz-Montes, J., Buyya, R., Rana, O.F., Parashar, M., Mobility-Aware Application Scheduling in Fog Computing. *IEEE Cloud Comput.*, 4, 2, 26–35, Mar. 2017, doi: 10.1109/MCC.2017.27.
52. Vaa, T., Penttinen, M., Spyropoulou, I., Intelligent transport systems and effects on road traffic accidents: state of the art. *IET Intell. Transport Systems*, 1, 2, 81–88, 2007.
53. Lopez, D. and Farooq, B., A blockchain framework for smart mobility, in: *2018 IEEE International Smart Cities Conference (ISC2)*, IEEE, Kansas City, MO, USA, pp. 1–7, Sep. 2018, doi: 10.1109/ISC2.2018.8656927.
54. Sun, J., Yan, J., Zhang, K.Z.K., Blockchain-based sharing services: What blockchain technology can contribute to smart cities. *Financ. Innov.*, 2, 1, 26, Dec. 2016, doi: 10.1186/s40854-016-0040-y.
55. Hank, P., Muller, S., Vermesan, O., Van Den Keybus, J., Automotive Ethernet: In-vehicle Networking and Smart Mobility, in: *Design, Automation & Test in Europe Conference & Exhibition (DATE)*, IEEE Conference Publications, Grenoble, France, pp. 1735–1739, 20132013, doi: 10.7873/DATE.2013.349.
56. Schnieder, M. and West, A., Evaluation of alternative battery charging schemes for one-way electric vehicle smart mobility sharing systems based on real urban trip data, in: *2019 IEEE 5th International forum on Research and Technology for Society and Industry (RTSI)*, IEEE, Florence, Italy, pp. 296–301, Sep. 2019, doi: 10.1109/RTSI.2019.8895556.
57. Authority, G.L., Smart City Opportunities for London. *Lond. Arup-GLA*, 2016.

2

Accentuating Climate Change Adaptation and Vulnerability (CCAV) Challenges

Adil Abbas Alwan[1], Amandeep Kaur[2*], Nilesh Bhosle[3], Sanjeev Kumar Shah[4] and Mohemmed Hussien[5]

[1]*College of Technical Engineering, National University of Science and Technology, Dhi Qar, Iraq*
[2]*Department of Computer Science Engineering, Chandigarh Engineering College, Chandigarh Group of Colleges, Jhanjeri, Mohali, Punjab, India*
[3]*Department of Computing Science and Artificial Intelligence, NIMS Institute of Engineering and Technology, NIMS University Rajasthan, Jaipur, India*
[4]*Department of Electronics & Communication Engineering, Uttaranchal Institute of Technology, Uttaranchal University, Dehradun, India*
[5]*Department of Computer and Information Application, The Islamic University, Najaf, Iraq*

Abstract

This analysis explores the intricate landscape of climate change adaptation, underscoring the necessity of crafting strategies that address vulnerabilities within diverse socio-political contexts. It highlights that vulnerabilities are not solely the result of environmental hazards but are profoundly intertwined with societal and economic structures, adversely affecting marginalised communities. Factors such as adaptation capacity, exposure, and sensitivity significantly influence how vulnerable communities are to climate change impacts. The chapter also examines spatial violence as a crucial framework for understanding how adaptation measures can sometimes cause harm, including forced evictions and urban policies that displace the most vulnerable populations. Addressing climate change adaptation and vulnerability (CCAV) challenges necessitates a multifaceted approach that considers technical, financial, social, and institutional factors. Through case studies in regions like Small Island Developing States, Sub-Saharan Africa, and South Asia, the chapter showcases the unique adaptation challenges in different

**Corresponding author*: amandeep1880.research@cgcjhanjeri.in

Pethuru Raj, Sudesh Yadav, Manas Kumar Mishra, Satya Prakash Yadav and Victor Hugo C. de Albuquerque (eds.) Artificial Intelligence Technologies for Smart and Sustainable Urban Transportation: Integrated Platforms and Use Cases, (23–38) © 2026 Scrivener Publishing LLC

settings, stressing the importance of context-sensitive and inclusive strategies to bolster resilience and promote climate justice.

Keywords: Climate change adaptation, socio-political contexts, spatial violence, climate justice, vulnerability

2.1 Introduction

Identifying vulnerability is the first step towards understanding whether climate adaptation methods are aimed at reducing it. Currently, climate change literature focuses more on vulnerability created by hostile climates than by social and economic systems [1, 2]. Communities' experience of risk is driven by the level of vulnerability they perceive and experience [3]. Three factors make up the level of risk that an individual experiences: vulnerability, exposure, and hazard. An exposed community is located in areas where hazards occur.

In contrast, a vulnerable community has conditions that make it more likely to experience a hazard adversely. Adaptation is now a central component of the global climate change discourse, replacing mitigation as a primary focus. There has been a significant increase in the urgency for effective adaptation strategies as climate impacts have become increasingly inevitable. As part of climate change adaptation, humans and nature adjust to mitigate its adverse effects while taking advantage of any potential benefits that climate change may provide. The process of coping with climate change, however, can be challenging, especially for vulnerable populations. During this chapter, we explore the complexities of climate change adaptation, outline the unique challenges that vulnerability poses, and discuss strategies for navigating these obstacles.

Yangon residents live with violence and conflict daily following the February 2021 military coup, despite not looking exclusively at physical violence. In light of the current and future spatial violence in Myanmar, the context cannot be ignored. As of August 2021, approximately 6,000 political prisoners had been arrested, charged, or sentenced. In recent years, food security and livelihoods have been threatened by rising COVID-19 pandemics [4–6]. Despite Myanmar's shift towards democracy, spatial and structural violence persist, with an increasing number of violent incidents [7, 8].

2.1.1 Adapting to Climate Change Vulnerabilities

The vulnerability of a system to adverse climate impacts is determined by many factors, including its susceptibility to and capacity to cope with these impacts. Three key elements comprise the concept:

Exposure: Physical vulnerability of a system. Flooding and storm surges are among the risks faced by coastal communities as sea levels rise.
Sensitivity: The degree to which a system is affected by climate stressors. Agricultural sectors in drought-prone regions are highly sensitive to changing rainfall patterns, which affect crop yield and food security.
Adaptive Capacity: The ability of a system to cope with extreme weather and climate variations. Financially poor, infrastructure-poor, and socially isolated communities generally have lower adaptive capacities.

Targeted strategies are required to address these components in the context of each population's socio-economic, environmental, and cultural contexts. Adapting to climate impacts is often more challenging for vulnerable groups, such as low-income communities, indigenous populations, and small-island developing states (SIDS) [2]. Examine vulnerability and precarity in relation to each other: In their view, methods can be used to decrease vulnerability and increase precarity by decreasing vulnerability. The term vulnerability refers to the historical exposure to hazards and precarity of livelihoods as well as the 'modern' and produced exposures caused by forced evictions and policies concerning urban planning. By using this diptych, we are able to consider how processes and structures rooted in different times and spaces are reflected in contemporary livelihoods. In light of these findings, researchers have argued that feminist perspectives on gender and climate change need to be added to the literature. A key finding of this research is that social roles, neoliberalism, and politics contribute to gendered insecurities, vulnerabilities, and risks that are linked to climate change [9]. In spite of the fact that climate change is caused by anthropogenic activities everywhere, the impacts and measures required to address it are localised. Also, [10] suggests the 'everyday' gendered effects of climate change policies and planning vary greatly from individual to individual. A second benefit of [11], is that it emphasises the special place women in the global north take in addressing climate change, as opposed to focusing solely on the care and knowledge that women in the global north have about climate change. In spite of this oversimplification, little is done to address women's lack of leadership roles regarding climate policy and practice. Consequently, women are unfairly burdened with taking action. To prevent this, contextualisation and care are vital.

2.2 Related Work

In this literature review, we will examine how spatial violence is related to forced evictions and adaptation to climate change. Following this, a review of the literature on vulnerability will be presented, as well as a look at how vulnerability has evolved and how the enactment of spatial violence can be addressed through policy and planning. It is important to understand that vulnerability does not exist equally across communities. An overview of Feminist Political Ecology is provided in this chapter.

2.2.1 Spatial Violence

Spatial violence is a concept derived from the understanding that violence does not just occur in war zones and physical harm to individuals and communities [10–12]. Spatial violence can be enacted through the process of urbanisation and urban planning in a purposeful and consequential manner, making and perpetuating vulnerability through the process of urbanisation and urban planning. Humans may suffer from different degrees of violence depending on how, where, and where they live because of historical policies and future developments.

It is crucial to understand spatial violence in terms of general definitions of violence as well as its under-theorisation across the social sciences [12]. Their primary concern is urban violence, which oscillates constantly between the visible and invisible, the natural and the social, the institutional and the criminal, making it difficult to define [13]. In the author's view, violence does not follow a linear path and can have detrimental spatial consequences [14]. Thus, spatial violence can be viewed as "war by another means" and is mediated through built environments as opposed to an external force. Many low-income settlements are seen as violent and disorderly because of informality especially compared to large cities. When 'informality' deteriorates and becomes disorderly in low-income settlements in the global south, a narrative of order and control often follows to justify spatial violence. Spatial violence is detrimental and perpetuates itself when it is designed to gain control of policymakers.

In an overarching study titled 'Trajectories of Spatial Violence', South-East Asian cities were compared. During urban development, neoliberal and global agendas were prioritised over residents' needs in favour of neoliberal, globalised agendas [15]. In the long term, spatial violence is the result of entanglement between urbanisation regimes, exclusionary invisibility, disenfranchisement policies, and housing policies [4]. The author suggests that subtler forms of spatial violence still exist in Yangon, as well as a history of forced evictions, such as direct and indirect exclusions. It is becoming increasingly important to understand low-income urban residents' experiences through the lens of spatial violence. In this paper, we examine spatial violence in Yangon resulting from climate adaptation-led development and its close association with vulnerability. Specifically, this paper analyses how spatial violence is legitimised as a necessary step for displaced populations experiencing informal settlement conditions.

The implementation of urban policy and development, forced evictions as well as explicit and covert dispossessions is major forms of spatial violence in Yangon. In urban development, climate change presents a major threat, and this can be used to perpetrate spatial violence. Cities that are ordered and climate-secure are often the victims of dispossession and displacement [16]. The South-East Asia region is notorious for forced evictions due to accumulation by dispossession, as described by [17] when neoliberalism and privatisation are promoting forced evictions without regard for human rights or international law. While forced evictions impact vulnerable communities directly, it is important to consider the importance of home in a broader sense for livelihoods, communities, families, and individuals as well [18]. As part of the Millennium and Sustainable Development Goals, housing quality and tenure security have been identified as key components of development and reducing vulnerability [19].

2.2.2 Response to Climate Change

Climate change adaptation is associated with spatial violence, but physical violence has received relatively little attention [20]. Several studies and reports have examined climate-induced displacement in rural Myanmar. Some recommend that the Myanmar government take action to curb the problem. However, there is a gap in the literature regarding how spatial violence is created and its implications for climate change adaptation in Yangon. Even though climate change has been widely acknowledged, it has not yet had a significant impact on urban planning and policy. The situation is particularly critical in Asian coastal cities [21], where mitigation is insufficient and adaptation to climate change will be necessary to prevent irreversible damage [22]. As an adjustment to climatic changes or their effects, adaptation is the process of reducing harm or exploiting opportunities as a result of actual or expected climatic conditions. As climate change impacts cannot always be easily distinguished from mitigation activities, adaptation must be considered in addition to mitigation [22].

2.3 Key Challenges in Climate Change Adaptation and Vulnerability (CCAV)

Due to a myriad of barriers, adapting to climate change is difficult. In addition to technical challenges, there are also financial, social, and institutional challenges.

2.3.1 Technical Challenges

The limitations of technology can make adaptation difficult. In spite of the increased ability to predict the weather due to advances in climate science, adaptation strategies often require specialised knowledge, tools, and technologies. The installation of flood defences or the implementation of water-saving irrigation techniques in agriculture requires technical expertise that is not always available in developing countries. The maintenance of adaptation technologies can also be costly and complex, especially in rural areas with insufficient skilled labour and inadequate infrastructure.

2.3.2 Financial Constraints

It is often necessary to invest substantial amounts of money in adaptation strategies. Relocating communities, implementing early warning systems, or developing climate-resilient infrastructure all require funds, many of which are lacking in many nations, particularly in the Global South. Many vulnerable countries lack access to climate finance despite increasing initiatives. Additionally, bureaucratic procedures and stringent criteria can make it difficult to get access to available funds.

2.3.3 Social and Cultural Barriers

Social norms, cultural practices, and customs must be considered when adapting measures. Coastal communities in low-lying island nations may, for example, be deeply attached to their land and livelihoods based on fishing or farming. There are logistical challenges as well as cultural disruptions associated with relocating these communities when sea levels rise.

Moreover, social inequalities exacerbate vulnerability, limiting marginalised groups' ability to influence adaptation policies that directly influence them due to their exclusion from decision-making processes.

2.3.4 Institutional and Governance Challenges

Institutions and governance structures that are able to implement and enforce policies are essential for effective adaptation. There are many instances where weak institutional frameworks, insufficient political will, and inadequate coordination between local, regional, and national governments hinder adaptation efforts. Because of this fragmentation, policies are inconsistent, resources are misallocated, and accountability is lacking. The lack of transparency in the allocation of funds may also undermine adaptation efforts, especially in resource-scarce areas.

2.3.5 Multi-Level Governance (MLG) of Climate Change

Researchers studying multi-level governance (MLG) focus on describing, explaining, and evaluating how policy-making capacities dovetail above, below, and beyond the top-down system of government. According to MLG researchers, the EU was a pioneer in transferring power from states to supranational institutions in the first half of the 20th century [23, 24]. The MLG method has been used in recent decades to explore the complex and dynamic relationship between governments and nongovernmental organisations, both within and across territorial boundaries [25]. Unlike traditional political divisions between national, international, and transnational, this new generation of MLG scholarship recognises that policy-making authority flows "upward, downward, and sideways" [26], challenging rigid national distinctions. As a broader definition of MLG, delegating authority is implied, but its dispersion can stem from state disaggregation [27] or private entrepreneurship [28].

MLG literature has largely retained a functionalist optimism despite recent criticism, which has focused primarily on democratic legitimacy, transparency, and accountability [29–31]. Furthermore, MLG's purported efficiency gains are not clearly measured. The need for policy innovation in a rapidly warming world is vital, but MLG cannot automatically promote policy entrepreneurship or multi-level reinforcement. There is a potential for complex MLG arrangements to obscure accountability and to provide special interests with a monopoly on policy-making [32]. MLG is also faced with policy obstructionists", as some factions of society may attempt to undermine or oppose the upscaling or rescaling of policies that threaten them [33]. A strong vested interest has been created due to multi-scale carbon lock-in in climate change [34, 35]. Climate policies often face resistance from private capital, but this should not be ignored. There has been growing evidence that climate action has uneven benefits and costs and that disenfranchised groups might resist certain decarbonisation policies that don't make them more socially inclusive [36].

2.4 Case Studies Highlighting Vulnerability and Adaptation Challenges

Climate change adaptation and vulnerability are illustrated through three case studies that emphasise the unique challenges confronted by vulnerable populations across diverse contexts.

2.4.1 Small Island Developing States (SIDS)

Since SIDS are geographically and economically limited, climate change poses one of the most serious threats. As precipitation patterns change and sea levels rise, their freshwater supply and land resources are negatively affected. Oceans are rising, storms are intensifying, and freshwater supplies are being disrupted by changing precipitation patterns. SIDS are particularly challenged by climate change because their economies depend heavily on tourism and fishing, highly vulnerable industries. In addition, relocation as a means of adaptation has deep sociocultural implications for islanders. As SIDS does not have a strong fiscal capacity, large-scale adaptation projects are difficult to finance, often relying on international aid that may have restrictive requirements.

2.4.2 Rural Farming Communities in Sub-Saharan Africa

Climate change has a significant impact on rural farming communities in Sub-Saharan Africa, an agricultural-based economy. Climate change is causing unpredictable rainfall, prolonged droughts, and rising temperatures that negatively affect crop yields, food security, and water availability [37]. There are many challenges in adapting to drought in these communities, such as drought-resistant crops and efficient irrigation. As a result of traditional land ownership and gender dynamics, adaptation resources are often distributed unevenly, excluding many women who play an important role in agriculture but lack formal ownership rights.

2.4.3 Urban Slums in South Asia

In South Asia, rapid urbanisation coupled with climate vulnerability poses significant adaptation challenges. Several cities, including Mumbai, Dhaka, and Karachi, have informal settlements that are prone to flooding, extreme heat, and water scarcity. These areas are especially vulnerable because of poor housing conditions, overcrowding, and limited access to services. Furthermore, these communities are often politically marginalised and lack access to climate adaptation funds or disaster preparedness resources. Relocation or infrastructure improvements are further complicated by the lack of formal land rights, making it more difficult for durable adaptation solutions to be implemented.

2.5 Strategic Frameworks for Addressing CCAV Challenges

Adapting to climate change in a vulnerable way requires a multidisciplinary approach that combines technical, financial, social, and policy interventions.

2.5.1 Through Community-Based Approaches

An important aspect of community-based adaptation (CBA) is involving local populations in the development and application of adaptation plans. As a result of CBAs, communities can incorporate indigenous knowledge into adaptation strategies and align these strategies with cultural norms and values. It is possible for farmers in drought-prone areas to adopt traditional methods together with modern ones, resulting in more effective and culturally acceptable solutions for conserving water.

2.5.2 Mobilising Climate Finance and Reducing Funding Barriers

To address financial barriers, climate funds need to be made more accessible and equitable. Collaboration between international organisations, governments, and private sector actors is necessary to facilitate developing countries' access to funding. Aside from developing innovative financing mechanisms, such as climate bonds and microfinance, local communities can also benefit from investing in adaptation activities. Funds must be managed in a transparent and accountable manner to ensure that they reach those in need.

2.5.3 Strengthening Institutional Capacity and Governance Frameworks

To adapt effectively to climate change, resilient institutions are essential. It involves improving interagency coordination, building local capacity, and incorporating adaptation planning into existing policies. Strong institutions can facilitate the development of adaptation policies that are coherent, equitable, and well-implemented. It is also essential for policymakers to ensure that marginalised groups are included in the decision-making processes and that climate justice is cultivated, as well as to ensure that adaptation strategies address the needs of all groups.

2.5.4 Innovating and Leveraging Technology

Adapting to climate change is made easier with advances in technology such as satellite monitoring, early warning systems, and digital platforms. Climate impact predictions can be made with these technologies, data-driven adaptation planning can be facilitated, and at-risk populations can be informed in real-time. Access to technology is crucial for ensuring equitable development; however, adaptation solutions must be inclusive and accessible. In addition, capacity-building programs that teach local populations how to use new technologies can help increase their adaptability.

2.5.5 Insufficient Funding and Resources

Many countries, especially low-income ones, lack the financial resources to develop effective strategies for adapting to climate change. As a result of limited funding, it is difficult to implement large-scale adaptation programs, technological advancements, and infrastructure improvements required to reduce vulnerability in the future.

2.5.6 Data Gaps and Uncertainty

It is crucial to gather accurate and comprehensive information on climate change impacts in order to make informed decisions. Despite this, climate variability, environmental vulnerability, and socio-economic impacts are poorly documented in many regions, particularly those located in remote and underdeveloped areas. Since climate change is unpredictable, adaptation measures must be carefully designed and implemented.

2.5.7 Insufficient Localised Solutions

Different regions experience climate change differently, which necessitates a tailored adaptation strategy. Existing solutions are either too generalised or do not address the needs and vulnerabilities of specific regions, resulting in ineffective or misaligned adaptations.

2.5.8 Institutional and Policy Challenges

Adapting to climate change requires cooperation between both the public and private sectors; however, institutional fragmentation often hinders such cooperation. Misaligned policies, lack of political will, and bureaucracy can slow down the implementation of adaptation strategies.

2.5.9 Social and Economic Inequities

People who are vulnerable to climate change include marginalised communities, women, and indigenous groups. The lack of inclusive adaptation strategies will put these groups at greater risk and limit their access to resources, increasing social and economic inequality.

2.5.10 Awareness and Engagement of the Public Lacking

It is often difficult for the general public to understand how climate change impacts and adaptation strategies are being implemented, especially in rural areas. Due to a lack of awareness, adaptation policies may be resisted, conservation measures may be challenged, and opportunities for community-led adaptation may be missed.

2.5.11 Using Fossil Fuels as a Source of Energy

Energy transitions are complex, costly, and take a long time in many countries. In spite of adaptation strategies focusing on renewable energy, the shift from fossil fuels is challenging due to economic dependency and political pressure.

2.5.12 Limitations

Through the implementation of adaptation technologies such as drought-resistant crops, early warning systems, and green infrastructure, we can reduce vulnerability. In many regions, however, advanced technologies remain out of reach due to high costs, insufficient technical expertise, and inadequate infrastructure.

2.5.13 Maintaining a Balance Between Short-Term and Long-Term Needs

Achieving climate adaptation entails competing with immediate socio-economic priorities, such as poverty reduction, health care, and employment. When resource constraints are present, striking a balance between short-term needs and long-term climate vulnerability goals can be challenging.

2.5.14 Adaptation Challenges Based on Ecosystems

It is imperative to develop climate-resilient ecosystems, such as wetlands, forests, and biodiversity. There are, however, challenges such as land-use conflicts, habitat degradation, and lack of coordination across sectors that make it difficult to implement ecosystem-based adaptation strategies.

2.5.15 Efforts to Monitor and Evaluate Adaptation

The long-term impacts of climate change make assessing the effectiveness of adaptation initiatives difficult. As a result, many regions lack established frameworks and metrics for evaluating progress, limiting their ability to refine strategies and ensure that resources are being used effectively.

2.5.16 Global Coordination and Climate Justice

To effectively adapt to climate change, nations must collaborate globally. However, their capacity, priorities, and responsibilities vary greatly, and disparities between developed and developing countries often hinder the coordination of adaptive efforts. These disparities create tension regarding the equitable distribution of funding and resources.

2.6 Conclusion

The urgent and complex challenge of adapting to climate change is particularly critical for those who are vulnerable and have fewer resources. An integrated approach to adaptation that addresses technical, financial, social, and institutional challenges while simultaneously fostering community resilience is essential. Recognising the diversity of vulnerabilities and prioritising climate justice is crucial. As climate uncertainty grows, building sustainable solutions that protect both people and the planet requires collective action, inclusivity, and contextual sensitivity.

References

1. Eriksen, S.H., Nightingale, A.J., Eakin, H., Reframing adaptation: The political nature of climate change adaptation. *Global Environ. Change*, 35, 523–533, 2015.
2. Rigg, J., Oven, K.J., Basyal, G.K., Lamichhane, R., Between a rock and a hard place: Vulnerability and precarity in rural Nepal. *Geoforum*, 76, 63–74, 2016.
3. Cardona, O.D., *et al.*, Determinants of Risk: Exposure and Vulnerability, in: *Managing the Risks of Extreme Events and Disasters to Advance Climate Change Adaptation*, 1st ed, C.B. Field, V. Barros, T.F. Stocker, Q. Dahe (Eds.), pp. 65–108, Cambridge University Press, 2012, doi: 10.1017/CBO9781139177245.005.
4. Rhoads, E.L., Poine, T.S., Win, C.C., Kyed, H., Myanmar–Urban Housing Diagnostic and COVID-19 Rapid Assessment, 2020, Accessed: Nov. 04, 2024. [Online]. Available: https://policycommons.net/artifacts/1258716/myanmar/1828025/.
5. Verma, S., *et al.*, An automated face mask detection system using transfer learning based neural network to preventing viral infection. *Expert Syst.*, 41, 3, e13507, 2024.
6. McNamara, R., Spatial violence through modes of dispossession: A study of vulnerability and climate change adaptation in Yangon, 2020, Accessed: Nov. 04, 2024, [Online], Available: https://open-education-repository.ucl.ac.uk/816/1/McNamara.pdf.
7. Prasse-Freeman, E. and Kabya, K., Revolutionary responses to the Myanmar coup. *Anthropol. Today*, 37, 3, 1–2, Jun. 2021, doi: 10.1111/1467-8322.12649.
8. Rani, P., *et al.*, Federated Learning-Based Misbehaviour Detection for the 5G-Enabled Internet of Vehicles. *IEEE Trans. Consum. Electron.*, 70, 2, 2023.
9. Wonders, N.A., Climate change, the production of gendered insecurity and slow intimate partner violence, in: *Intimate partner violence, risk and security*, pp. 34–51, 2018, Accessed: Nov. 04, 2024, [Online], Available: https://www.taylorfrancis.com/chapters/edit/10.4324/9781315204765-3/

climate-change-production-gendered-insecurity-slow-intimate-partner-violence-nancy-wonders.

10. Bee, B.A., Rice, J., Trauger, A., A Feminist Approach to Climate Change Governance: Everyday and Intimate Politics. *Geogr. Compass*, 9, 6, 339–350, Jun. 2015, doi: 10.1111/gec3.12218.
11. Arora-Jonsson, S., Forty years of gender research and environmental policy: Where do we stand?, in: *Women's Studies International Forum*, pp. 295–308, Science Direct Elsevier Publisher, 2014, Accessed: Nov. 04, 2024. [Online]. Available: https://www.sciencedirect.com/science/article/pii/S0277539514000326.
12. Pavoni, A. and Tulumello, S., What is urban violence? *Prog. Hum. Geogr.*, 44, 1, 49–76, Feb. 2020, doi: 10.1177/0309132518810432.
13. Boano, C., Beyond violence. Toward the politics of inhabitation. *lo Squaderno.*, 59, 67–71, 2021.
14. Herscher, A. and Iyer Siddiqi, A., Spatial Violence. *Archit. Theory Rev.*, 19, 3, 269–277, Sep. 2014, doi: 10.1080/13264826.2014.1037538.
15. Rhoads, E., Sakuma, S., Ortiz, C., Kouvolou-Kouri, M., Astolfo, G., Trajectories of spatial violence in Southeast Asian cities, 2021, Accessed: Nov. 04, 2024, [Online], Available: https://portal.research.lu.se/en/publications/trajectories-of-spatial-violence-in-southeast-asian-cities.
16. Yadav, S.P., Jindal, M., Rani, P., de Albuquerque, V.H.C., dos Santos Nascimento, C., Kumar, M., An improved deep learning-based optimal object detection system from images. *Multimed. Tools Appl.*, 83, 10, 30045–30072, 2024.
17. Harvey, D., Accumulation by dispossession, in: *The new imperialism*, Oxford University Press, 2003.
18. Desmaison, B., Home as a catalyst for resilience. *DPU Work. Paper*, 2016.
19. Chant, S. and McIlwaine, C., Gendered access to land and housing in cities and slums. *Cities, Slums Gender Global South*, 71–92, 2016.
20. Levy, B.S. and Sidel, V.W., Collective violence caused by climate change and how it threatens health and human rights. *Health Hum. Rts. J.*, 16, 32, 2014.
21. Fuchs, R., Conran, M., Louis, E., Climate Change and Asia's Coastal Urban Cities: Can they Meet the Challenge? *Environ. Urbanization ASIA*, 2, 1, 13–28, Mar. 2011, doi: 10.1177/097542531000200103.
22. Change, I.P.O.C., Climate change 2007: Impacts, adaptation and vulnerability. *Genebra, Suíça*, 2001, Accessed: Nov. 04, 2024, [Online], Available: https://msuweb.montclair.edu/~lebelp/PSC643IntPolEcon/IPCCClimateChange2007.pdf.
23. Marks, G., Structural Policy and Multilevel Governance in the EC. *State of the European Community, II. The Maastricht Debates and Beyond/Longman*, 1993.
24. Scharpf, F.W., Community and autonomy: Multi-level policy-making in the European Union. *J. Eur. Public Policy*, 1, 2, 219–242, Sep. 1994, doi: 10.1080/13501769408406956.

25. Elgar, E., and Michael Zürn, *Handbook on Multi-level Governance*, Berlin, Germany, Sonja Wälti, American University, Washington, DC, US, 2010.
26. Hooghe, L. and Marks, G., A postfunctionalist theory of European integration: From permissive consensus to constraining dissensus. *Br. J. Political Sci.*, 39, 1, 1–23, 2009.
27. Slaughter, A.M., *A New World Order Princeton*, Princeton Univ. Press, NJ, 2004.
28. Green, J.F., *Rethinking Private Authority: Agents and Entrepreneurs in Global Environmental Governance*, Princeton University Press, Princeton, 2013, doi: 10.1515/9781400848669.
29. Benz, A., Compounded representation in EU multi-level governance. *Linking EU Natl. Governance*, 82–110, 2003.
30. Peters, B.G. and Pierre, J., Multi-level governance: a Faustian bargain? *Multi-level Governance*, 75–89, 2004.
31. Papadopoulos, Y., Accountability and multi-level governance: more accountability, less democracy?, in: *Accountability and European Governance*, pp. 102–121, 2014, Accessed: Nov. 04, 2024, [Online], Available: https://www.taylorfrancis.com/chapters/edit/10.4324/9781315879390-6/accountability-multi-level-governance-accountability-less-democracy-yannis-papadopoulos.
32. Curry, D., *Network approaches to multi-level governance: Structures, relations and understanding power between levels*, Palgrave Macmillan, Swansea University, UK, 2015, Accessed: Nov. 04, 2024, [Online], Available: https://cronfa.swan.ac.uk/Record/cronfa23817/Details.
33. Hameiri, S. and Jones, L., The politics and governance of non-traditional security. *Int. Stud. Q.*, 57, 3, 462–473, 2013.
34. Bernstein, S. and Hoffmann, M., Climate politics, metaphors and the fractal carbon trap. *Nat. Clim. Change*, 9, 12, 919–925, 2019.
35. Rani, P., Verma, S., Yadav, S.P., Rai, B.K., Naruka, M.S., Kumar, D., Simulation of the lightweight blockchain technique based on privacy and security for healthcare data for the cloud system. *Int. J. E-Health Med. Commun. (IJEHMC)*, 13, 4, 1–15, 2022.
36. McCauley, D. and Heffron, R., Just transition: Integrating climate, energy and environmental justice. *Energy Policy*, 119, 1–7, 2018.
37. Singh, A., *et al.*, Smart Traffic Monitoring Through Real-Time Moving Vehicle Detection Using Deep Learning *via* Aerial Images for Consumer Application. *IEEE Trans. Consum. Electron.*, 70, 4, 1–1, 2024, doi: 10.1109/TCE.2024.3445728.

3

Delineating the Solution Approaches for Sustainable Urban Mobility

Adil Abbas Alwan[1], Amandeep Kaur[2*], Nilesh Bhosle[3], Rajesh Singh[4] and Mohammed Al-Farouni[5]

[1]*College of Technical Engineering, National University of Science and Technology, Dhi Qar, Iraq*
[2]*Department of Computer Science Engineering, Chandigarh Engineering College, Chandigarh Group of Colleges, Jhanjeri, Mohali, Punjab, India*
[3]*Department of Computing Science and Artificial Intelligence, NIMS Institute of Engineering and Technology, NIMS University Rajasthan, Jaipur, India*
[4]*Department of Electronics & Communication Engineering, Uttaranchal Institute of Technology, Uttaranchal University, Dehradun, India*
[5]*Department of Information Technology, The Islamic University, Najaf, Iraq*

Abstract

A growing urban population presents cities with increasing mobility challenges. As autonomous electric vehicles become more common, they will significantly improve safety, reduce emissions, and make transportation more accessible to everyone. The use of artificial intelligence assists city planners and policymakers in prioritizing key urban mobility aspects to develop effective policy proposals. Providing an open-source interoperable prototype, this chapter aims to help smart cities increase the efficiency, sustainability, and forward-thinking nature of urban mobility through the routine use of such recommendation systems. Helsinki, Amsterdam, Messina, and Bilbao have all demonstrated how engine technology can transform urban mobility policies. Helsinki optimized tunnel traffic, Amsterdam optimized bicycle traffic in a new city quarter, and Bilbao used an additional bus line to optimize closing the city centre. Implementing advanced urban mobility management enables more liveable, resilient urban environments.

**Corresponding author*: amandeep1880.research@cgcjhanjeri.in

Pethuru Raj, Sudesh Yadav, Manas Kumar Mishra, Satya Prakash Yadav and Victor Hugo C. de Albuquerque (eds.) Artificial Intelligence Technologies for Smart and Sustainable Urban Transportation: Integrated Platforms and Use Cases, (39–54) © 2026 Scrivener Publishing LLC

***Keywords*:** Electric vehicles, smart city, sustainable environment, transportation, urban mobility

3.1 Introduction

Sustainable and climate-neutral development requires fundamental transformation rather than incremental change [1, 2]. As part of its commitment to sustainable mobility, the commission places a high value on urban transport planning [2]. As a result of traditional planning practices (partly as a result of major challenges in the sustainable development of urban mobility), the European Commission published its Action Plan on Urban Mobility in the early 2000s. During the implementation of this plan, SUMPs will be developed [3]. Throughout Europe, the SUMP concept has been specified and widely distributed to change urban transport planning practices, as shown in Figure 3.1 [4]. A guideline for developing and implementing a sustainable urban mobility plan [5], was included in the "urban mobility package [6, 7]. In addition to complementary resources [8], these guidelines provide planning guidance and have recently been updated [9]. The SUMP cycle is the core of the concept, which consists of four phases of planning: (i) analyzing stakeholders and challenges, (ii) defining strategies (iii) planning for measures and (iv) monitor and implement [9]. SUMP has today become the de facto concept for urban transportation planning in Europe [9].

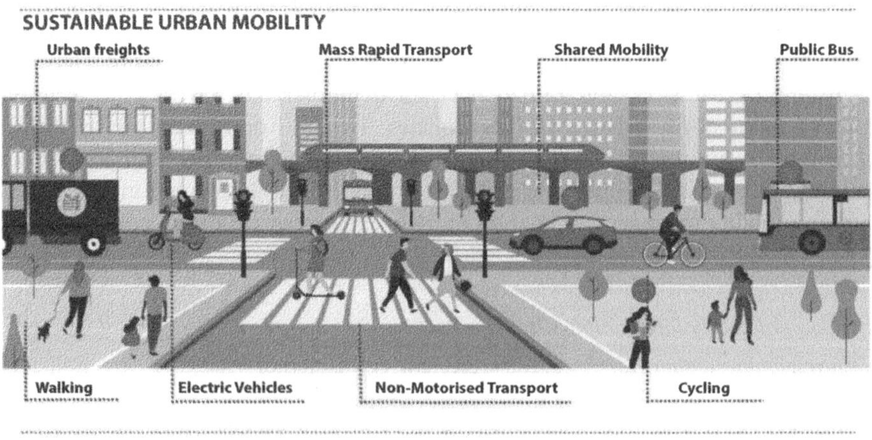

Figure 3.1 Sustainable urban mobility.

SUMP does not support transformations considered essential for sustainability by the European Commission. As well as changing vehicles, infrastructures, mobility options, and travel behaviors, transitioning from one mode of transportation to another requires new policies, regulations, and approaches to planning and governance, as well as changes in social and cultural norms. (For example, desirable images of urban mobility and status symbols are also necessary [10–12]. The transition from urban to rural mobility is interconnected with changes in land use and public health [13–15]. Many variables influence the transition process during its course, including shifting societal preferences, economic developments, and global crises [16–18] and local factors [12, 19, 20]. Multiple actors play a role in transitions, and they have differing roles, responsibilities, interests, and levels of authority, which make them inherently contested, conflictual, and normative (e.g., determining the appropriate goals and measures) [21–23]. Sustainability transitions pose tensions with established planning, prompting a lively debate over whether planning advances or hinders transitions [24–26].

As part of precision agriculture, soil properties are measured by plot sampling, management zones are delineated, crop plans are developed, crops are monitored, and crop protection is provided, among other things [27]. Management zones are subregions of a plot with relatively homogenous soil properties but which require particular input rates from farmers, according to [28]. In agriculture, delineating management zones (DMZs) is also a crucial decision point since soil characteristics-chemical, physical and even electrical-directly influence crop yields [29, 30]. In the crop planning stage (CP for short), specific crops are assigned to each plot, increasing profitability and decreasing the cost of supplying demand [31, 32]. It is usually necessary to perform these stages sequentially because the DMZ output is used as an input for the CP stage. In this study, we apply operations research concepts to integrating rectangular management zones and crop planning problems. The definition of management zones is commonly achieved through the implementation of clustering algorithms. These algorithms may be derived from a variety of sources, such as topographic maps [33], soil samples, yield maps using data from different seasons, or combinations of the above. Farmers are not always satisfied with the irregular shapes obtained by clustering methods when defining rectangular management zones. There is a need to optimize the delineation of rectangular management zones. In the article, the authors develop an optimization problem that minimizes variance in homogeneity when a soil property is taken into consideration as well as non-overlapping management zones [34].

Amsterdam: As Amsterdam, the Dutch capital, grows in population and visitors, traffic and mobility issues are becoming more prevalent. It has a complex traffic pattern with bicycles, cars, and public transportation all merging. It is imperative to develop sustainable mobility solutions based on better data analysis to manage these traffic issues. In the city, cycling is one of the main modes of transportation, so innovations and interventions should take this into account. Amsterdam's municipal government is eager to gain new insights from biking data as it strives to improve its network of bike lanes and cyclist experiences continually.
Bilbao: The Mountains surrounding Bilbao, a northern Spanish industrial port city, are lush and green. Located in the heart of the Basque Country, it houses many skyscrapers. Since the opening of the Guggenheim Museum Bilbao in 1997, the city has seen a revitalization.
Helsinki: Finland's south capital, Helsinki, is located on a peninsula within the Gulf of Finland. Several institutions are flanking the central avenue, Mannerheimintie, including the National Museum, which contains the entire history of Finland. In addition to the Parliament House, Kiasma is an art museum that exhibits contemporary art on Mannerheimintie.
Messina: The Strait of Messina separates Messina from mainland Italy, which is in northeastern Sicily. In the bell tower is a 15th-century astronomical clock, as well as Gothic portals and windows from the Norman Messina Cathedral. Fontana di Orione, for example, has carved inscriptions featuring mythological, while Neptune Fountain has a statue of the sea god on top.

3.2 Related Work

Another application of regionalization is identifying distinct and unique social neighbourhoods based on their demographic characteristics. Census data and multivariate clustering algorithms are used to produce neighborhood cluster types. Geodemographics maps the resulting neighborhood types, but neither spatial compactness nor contiguousness is guaranteed. Consequently, neighborhoods can fragment spatially, which is contrary to the substantive understanding of neighborhoods as organized spaces. To depict discrete neighborhoods more realistically, neighborhood modeling incorporates spatially constrained clustering. Using this approach, for example, to analyze the dynamics of social neighborhoods over time provides an interesting perspective. Regionalization methods pose a number of challenges in neighborhood delineation due to their substantial computational complexity.

It is necessary to assemble spatial units into contiguous regions in a number of social and environmental contexts, including political districts, school districts, police patrol districts, habitat delineations, and several zone aggregations used in modeling [35]. The needs for regionalization have been met through the development of a variety of algorithms. Mixed-integer programming (MIP) models are used here to model typical regionalization problems, which can be solved with software such as GUROBI or GLPK (GNU Linear Programming Kit 2012). The author has developed an algorithm that represents hierarchical clustering with dynamically constrained clustering and partitioning to integrate contiguity constraints into clustering. It gathers spatial units together into compact, contiguous clusters using memory-based randomized greedy and edge reassignment.

Urban passenger mobility can be classified into two major regimes: private cars and public transport. Each has its own set of rules, structures, and actors. Those regimes do not solely rely on technology or technology artefacts such as private cars. To understand land-based personal transportation, we need to consider human agency, social structure, and organizations. Due to their close alignment and interdependence, changes in one socio-technical configuration can result in changes in other aspects as well. To adapt to a new technology, other components within a system must also be adapted. The slow implementation of system-wide changes is due to this reason. For example, autonomous vehicles go beyond introducing new technologies in the field of mobility. A new maintenance network will also be necessary to support autonomous vehicles and adjust regulations, policies, road infrastructure, markets, and practices.

Throughout the illustrations and case studies Geels presents, he strives to demonstrate how technological artefacts have evolved and transitioned over time. For example, sailing ships became steamships, horse-drawn carriages became automobiles, propellers became turbojets, etc. A socio-technical system is considered in this study as a whole as it relates to urban mobility. Therefore, it is more challenging to define the topic of transition. The transition to an SUM is influenced by a multitude of categories that interact and interact with one another. Figure 3.2 shows the configured urban mobility system based on Geels' configuration.

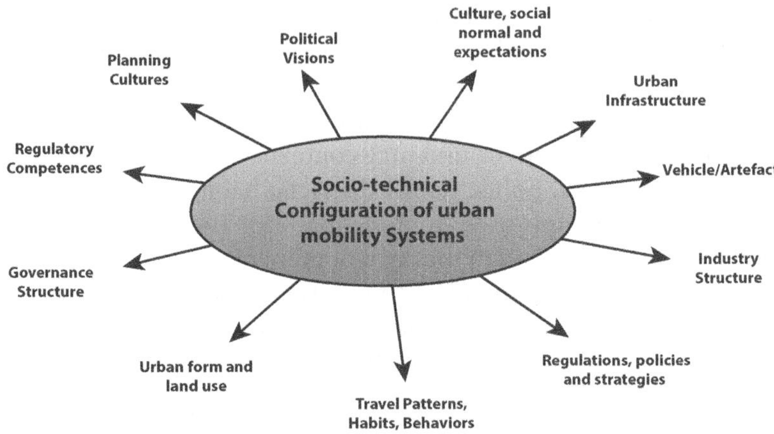

Figure 3.2 System configurations of urban mobility based on socio-technical factors.

Considering the configuration proposed, any transition from a personal vehicle-based regime would involve multiple dimensions, multiple actors, and multiple levels of change.

3.3 Materials and Methods

Data preparation involves collecting and processing a variety of datasets, such as traffic flow, vehicle types, road configurations, and sociodemographic information. To use simulation input data, it must be accurate, comprehensive, and appropriate to realistic scenarios. As a supplement to publicly available data, city teams also provided data on the current conditions of the cities, which were used to develop the baseline scenarios. Since the simulation relied on existing databases and city teams for most of the data, it had a solid foundation. We acquired detailed city maps through OpenStreetMap, and we utilized data from EUSILC to generate synthetic populations where necessary for modeling travel plans. A specialized marginal distribution for each city has been applied to this synthetic population. Models were also enhanced by the incorporation of data from pilot cities whenever available. Due to the large amount of data that large cities manage, we can use this integration to its fullest potential. Study data were taken from both public and city-specific sources and were available to the general public or not. The source code for our software is open source, which is important.

Simulations Data

Street maps are among the free geographic data created by OpenStreetMap. As an international non-for-profit organization, the OpenStreetMap Foundation supports but does not control, the project. The organization is committed to promoting the growth and development of free geospatial data and distributing it to all users and users everywhere. GTFS (General Transit Feed Specification) by Google Developers defines a standard format for sharing information about public transit in different cities and metropolitan areas. Public transport data is contained in GTFS as a series of text files: stop routes, trips, and other information related to schedules [36]. Every city and municipality in the world produce and updates the GTFS feed voluntarily online. Through this interoperable platform, over 900 transit agencies share information about public transportation since 2008. Although North American cities are the largest contributors, European and Asian cities are also becoming more involved [37].

3.3.1 Travel Demand Generation

We can create travel demand models to explain where, when, and how people take trips, as well as how future development in transportation networks might impact them. Models of travel demand can be created by using the following techniques:

Activity-Based: Using template activity patterns, the activity-based technique creates activity and travel schedules through a modular design
Bayesian Networks: Using this technique, you can synthesize daily activity and trip chains based on demographics, travel, and temporal variables. Combining data from various sources is possible with Bayesian networks.
Four-Step Model: Creating travel demand models involves this technique. To model travel demand, practitioners typically take the following steps.

- The study area should be defined and divided into zones.
- Obtaining data on the population, economic activities, and transportation networks within each zone.
- Calculating how many trips each zone generates and attracts using the data.
- Travelling between zones within the study area.
- Different modes are assigned to trips based on traveller preferences.
- It is accomplished by assigning each trip matrix to its corresponding route network.

3.3.2 Traffic Simulation Process

A city's decision-makers resort to a variety of solutions when planning mobility modifications, such as drawing on experience from other solutions, conducting polls, and conducting software simulations. It is possible to simulate city mobility using a variety of free and open-source tools.

MATSim (Multi-Agent Transport Simulation): Agent-based carriage simulations are made possible by the open-source framework MATSim.
Simulation of Urban Mobility (SUMO): Simulating networks in a microscopic, continuous and open-source manner is made possible by SUMO, an open-source, highly portable package. Intermodal simulation is supported, and it can create scenarios with a large number of tools.
TRANSIMS (Transportation Analysis and Simulation System): There are several major modules in the TRANSIMS Project, as well as plans for the project in the near future. In this document, they are summarized. For the purpose of creating a regional transportation system analysis environment, TRANSIMS will utilize advanced computational and analytical techniques. In the simulation environment, a population of individuals and freight loads will be represented with travel activities and plans, and their interactions with the transportation system will be simulated, along with their impact on the environment.

KPI Metric Calculation
For measuring urban mobility, here are some key performance indicators (KPIs) and metrics:

On-time delivery: To achieve a good delivery rate on time, it is necessary to organize the delivery process, optimize routes, and accurately schedule deliveries.
Freight rates: During a specified period, the total cost of freight shipping is divided by the number of units shipped.
Capacity utilization: It refers to how much capacity is actually being utilized versus what capacity is actually being used during operations.
Delivery time: In other words, the time between a shipment leaving a warehouse and a customer receiving it.
Order accuracy: An order fulfilment metric measures how well the product, quantity, and shipping details match up with the order.

3.4 Results Analysis and Discussion

3.4.1 Amsterdam

Between 2006 and 2015, PT became the dominant mode of transport within motorized trips in Amsterdam, with a market share of 23%. From 38% to nearly 50% of costs were recovered during the same period. An EMTAs report published in 2018 and 2009 reported the cost of a monthly pass at €50.50 in 2016 and estimated the area's annual gross domestic product per capita at €34,700. The government provides the majority of PT subsidies by earmarking funds for PTAs; these funds are used for operational costs and small infrastructure projects, while the government funds larger projects directly. As PT subsidies do not rise with inflation, they have created important budget pressures in recent years. Single nationwide smartcards are valid for all PT modes and operators to ensure seamless ticket integration. To manage the smartcard, all national operators form a joint venture. As early as the 1990s, PT was provided with a national travel information system.

At present, there are two main initiatives on the ground in Amsterdam that are driving MaaS development: Amstelland-Meerlanden concessions and Zuidas pilot projects. The two initiatives are metropolitan in scope, despite their seemingly territorial limitations, since they involve travellers and transportation networks that cross concessions and neighborhoods. The majority of Amsterdam's workforce works in service, with only a

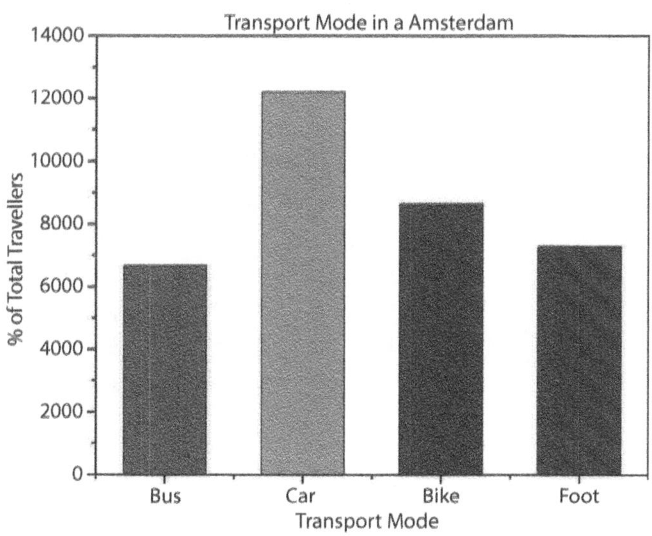

Figure 3.3 Amsterdam modes of transport.

tenth in manufacturing. Telecommunications, information and medical technologies, and business services make up the dominant service sector. Approximately one-fifth of the metropolis's employment has been supported by international trade and transport over the past seven centuries. Around one-eighth of Amsterdam's jobs are in the banking and insurance industry. A further 6% work in the health, cultural, or social services sectors. An estimated 10% of the city's jobs are related to tourism. Despite a thriving service sector, the country faced a high unemployment rate, and about one-eighth of the labor force lacked marketable skills.

Even though Amsterdam's industrial share is decreasing, the remaining industries are shipbuilding, heavy engineering, petrochemicals, food processing (including brewing) and diamond polishing. To reduce unemployment, the city provides training and temporary workers to employers as part of its active economic policy. It resulted in the city's government subsidizing thousands of jobs.

Despite its declining industrial economy, Amsterdam's remaining industries range from shipbuilding to heavy engineering to food processing and brewing to diamond polishing. To reduce unemployment in the city, the city has implemented a proactive economic policy that improves infrastructure connections with surrounding regions and provides employers with training and temporary workers as shown in Figure 3.3. Through subsidized programs implemented by the city government, thousands of jobs were created by the end of the 20th century.

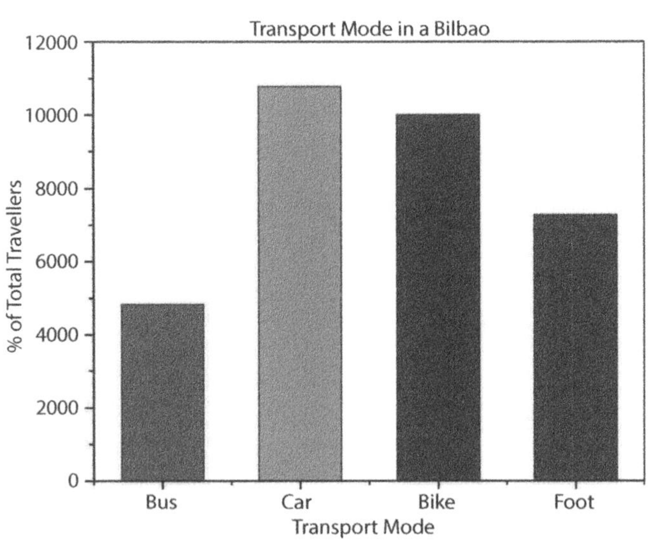

Figure 3.4 Bilbao modes of transport.

Bilbao

As a result of the Guggenheim Museum opening in Bilbao, Spain, the Bilbao Effect has a broad economic and social impact. In 1997, when the museum was built, the city, which previously struggled with unemployment and poverty, was transformed. Since the museum became successful, cities around the world have built iconic structures to attract tourists. Closing the city center for a certain period must be done in such a way that accessibility is maintained and traffic-related issues are reduced so that residents and visitors can enjoy their lives more. Traffic flow can be improved by partially or completely closing Moya Square to private vehicles, as shown in Figure 3.4. These are some of the key findings.

3.4.2 Helsinki

Since Helsinki became Finland's capital in 1810, its population has grown rapidly, going from 4,000 in 1810 to 60,000 by 1890. In December, the city of Helsinki was occupied by leftist rebel units (known as the Reds) shortly after Finland declared independence from Russia in 1917. In 1919, Finland's first president was elected by the Helsinki parliament, stabilizing conditions. World War II interrupts only a century-long development of Helsinki as a trade, industry, and cultural center as shown in Figure 3.5.

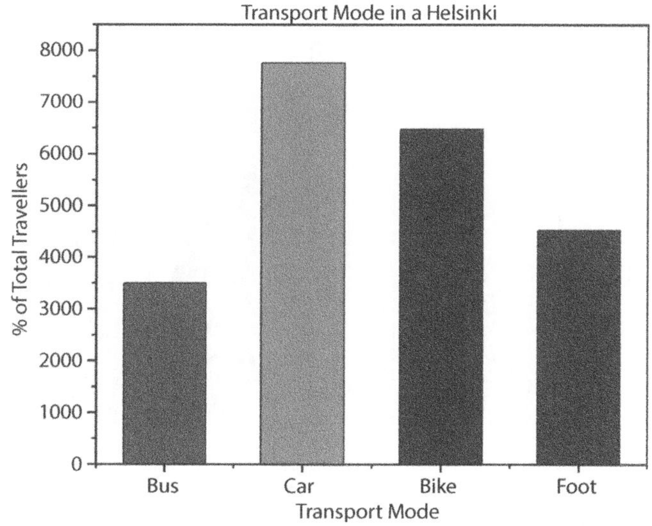

Figure 3.5 Helsinki modes of transport.

With excellent harbor facilities and good rail and road connections to the interior, Helsinki's economy has flourished for many years. As a result, over half of Finland's total imports pass through Helsinki. Helsinki is only a little more than half of Finnish exports since the largest port is somewhere else along the coast. Food, metal, chemical, textile, clothing, apparel, and electrical equipment manufacturing are Helsinki's major industries. Porcelain wares produced by the Arabia porcelain factory, one of Europe's largest, are very well known internationally.

Messina

With its unique geographical constraints and bustling urban life, Messina, a Mediterranean city characterized by its unique geographic constraints, faces significant challenges when it comes to improving its public transportation system. In this case study, a new bus line is introduced to improve accessibility and connectivity between the city's different neighborhoods. There will be a study conducted to compare the impact of the new bus route on traffic flow, public transportation use, and urban mobility as shown in Figure 3.6. There is more than one match for Messina's problem description, including the earthquake and tsunami that struck Messina in 1908 and the Messina algorithm that follows:

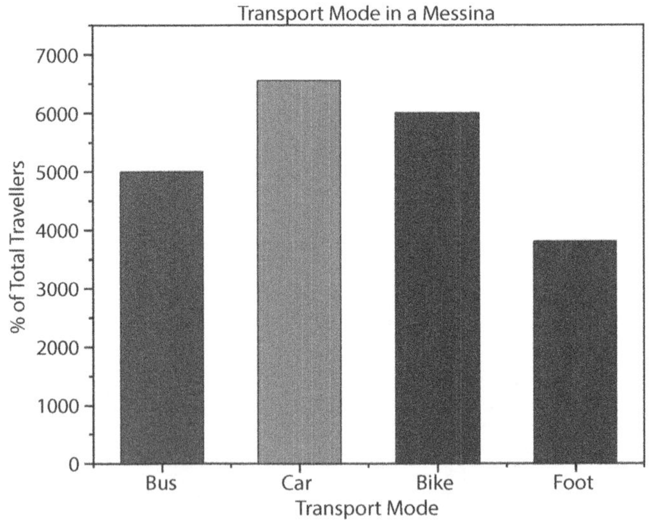

Figure 3.6 Messina modes of transport.

3.5 Conclusion

Through the integration of diverse, city-specific criteria and the provision of a structured approach for policy evaluation, this proposed artificial intelligence-based recommendation system proves effective in dealing with the complexities of urban mobility. Based on case studies, the system's adaptability to different urban contexts demonstrates its potential for fostering sustainable mobility. Due to the framework's open-source nature, it is widely adaptable and refined across smart cities throughout the world. A key element of this initiative is supporting cities in achieving ambitious sustainability goals aligned with the European Green Deal and achieving climate-neutral urban transportation.

References

1. European Commission, Sustainable mobility: The European green deal, 2019, Accessed: Nov. 04, 2024, [Online], Available: https://op.europa.eu/en/publication-detail/-/publication/1045f83c-1d88-11ea-95ab-01aa75ed71a1/language-en.
2. European Commission, Sustainable and smart mobility strategy – putting European transport on track for the future. *Brussels.*, 2020a, Accessed: Nov. 04, 2024, [Online], Available: https://ec.europa.eu/commission/presscorner/detail/en/ip_20_2329.
3. European Commission, White paper - roadmap to a single European transport area - towards a competitive and resource efficient transport system, 2011, Accessed: Nov. 04, 2024, [Online], Available: https://eur-lex.europa.eu/legal-content/EN/TXT/?uri%E2%80%89=%E2%80%89CELEX%3A52011DC0144.
4. Werland, S., Diffusing Sustainable Urban Mobility Planning in the EU. *Sustainability*, 12, 20, 8436, Oct. 2020, doi: 10.3390/su12208436.
5. Wefering, F., Rupprecht, S., Bührmann, S., Böhler-Baedeker, S., Guidelines. developing and implementing a sustainable urban mobility plan. *Workshop*, 2013, Accessed: Nov. 04, 2024, [Online], Available: http://www.trt.it/documenti/guidelines-developing-and-implementing-a-sump_final_web_jan2014b.pdf.
6. European Commission, Annex: A concept for sustainable urban mobility plans to the communication from the commission to the European Parliament, the council, the European economic and social committee and the committee of the regions - Together towards competetive and resource-efficient urban mobility. *Brussels. Eur. Commission.*, 2013, Accessed: Nov. 04, 2024, [Online], Available: https://eur-lex.europa.eu/resource.

html?uri%E2%80%89=%E2%80%89cellar:82155e82-67ca-11e3-a7e4-01aa75ed71a1.0011.02/DOC_4%26format%E2%80%89=%E2%80%89PDF.

7. Rani, P., Comparative studied based on attack resilient and efficient protocol with intrusion detection system based on deep neural network for vehicular system security, in: *Distributed Artificial Intelligence*, pp. 217–236, CRC Press, Boca Raton, 2020.
8. Eltis, The urban mobility observatory - topic guides. *Eltis*, 2021, Accessed: Nov. 04, 2024, [Online], Available: https://www.eltis.org/mobility-plans/topic-guides.
9. Rupprecht Consult, *Guidelines for developing and implementing a sustainable urban mobility plan*, 2nd ed, 2019, Accessed: Nov. 04, 2024, [Online], Available: https://www.eltis.org/sites/default/files/sump_guidelines_2019_interactive_document_1.pdf.
10. Canitez, F., Pathways to sustainable urban mobility in developing megacities: A socio-technical transition perspective. *Technol. Forecast. Soc. Change*, 141, 319–329, Apr. 2019, doi: 10.1016/j.techfore.2019.01.008.
11. Geels, F.W., Low-carbon transition via system reconfiguration? A socio-technical whole system analysis of passenger mobility in Great Britain (1990–2016). *Energy Res. Soc. Sci.*, 46, 86–102, Dec. 2018, doi: 10.1016/j.erss.2018.07.008.
12. Schwanen, T., The Bumpy Road toward Low-Energy Urban Mobility: Case Studies from Two UK Cities. *Sustainability*, 7, 6, 7086–7111, Jun. 2015, doi: 10.3390/su7067086.
13. Banister, D., The sustainable mobility paradigm. *Transp. Policy*, 15, 2, 73–80, Mar. 2008, doi: 10.1016/j.tranpol.2007.10.005.
14. Koszowski, C., Gerike, R., Hubrich, S., Götschi, T., Wittwer, R., Active Mobility: Bringing Together Transport Planning, Urban Planning, and Public Health, in: *Towards User-Centric Transport in Europe*, B. Müller and G. Meyer (Eds.), pp. 149–171, Springer International Publishing, Cham, 2019, Lecture Notes in Mobility, doi: 10.1007/978-3-319-99756-8_11.
15. Rani, P., Verma, S., Yadav, S.P., Rai, B.K., Naruka, M.S., Kumar, D., Simulation of the lightweight blockchain technique based on privacy and security for healthcare data for the cloud system. *Int. J. E-Health Med. Commun. IJEHMC*, 13, 4, 1–15, 2022.
16. CREATE, D4.3 WP4 comparative analysis report, 2018a, Accessed: Nov. 04, 2024, [Online], Available: http://create-mobility.eu/create/Publications/Project-deliverables.
17. Lyons, G. and Davidson, C., Guidance for transport planning and policymaking in the face of an uncertain future. *Transp. Res. Part Policy Pract.*, 88, 104–116, Jun. 2016, doi: 10.1016/j.tra.2016.03.012.
18. Marchau, V.A.W.J., Walker, W.E., Van Wee, G.P., Dynamic adaptive transport policies for handling deep uncertainty. *Technol. Forecast. Soc. Change*, 77, 6, 940–950, Jul. 2010, doi: 10.1016/j.techfore.2010.04.006.

19. Buehler, R., Pucher, J., Gerike, R., Götschi, T., Reducing car dependence in the heart of Europe: lessons from Germany, Austria, and Switzerland. *Transp. Rev.*, 37, 1, 4–28, Jan. 2017, doi: 10.1080/01441647.2016.1177799.
20. Schippl, J. and Truffer, B., Directionality of transitions in space: Diverging trajectories of electric mobility and autonomous driving in urban and rural settlement structures. *Environ. Innov. Soc. Transit.*, 37, 345–360, Dec. 2020, doi: 10.1016/j.eist.2020.10.007.
21. Avelino, F., Power in Sustainability Transitions: Analysing power and (dis) empowerment in transformative change towards sustainability. *Environ. Policy Gov.*, 27, 6, 505–520, Nov. 2017, doi: 10.1002/eet.1777.
22. Hodson, M. and Marvin, S., Can cities shape socio-technical transitions and how would we know if they were? *Res. Policy*, 39, 4, 477–485, May 2010, doi: 10.1016/j.respol.2010.01.020.
23. Marletto, G., Car and the city: Socio-technical transition pathways to 2030. *Technol. Forecast. Soc. Change*, 87, 164–178, Sep. 2014, doi: 10.1016/j.techfore.2013.12.013.
24. Carroli, L., Planning roles in infrastructure system transitions: A review of research bridging socio-technical transitions and planning. *Environ. Innov. Soc. Transit.*, 29, 81–89, Dec. 2018, doi: 10.1016/j.eist.2018.06.001.
25. Malekpour, S., Brown, R.R., De Haan, F.J., Strategic planning of urban infrastructure for environmental sustainability: Understanding the past to intervene for the future. *Cities*, 46, 67–75, Aug. 2015, doi: 10.1016/j.cities.2015.05.003.
26. Wolfram, M., Urban Planning and Transition Management: Rationalities, Instruments and Dialectics, in Co-creating Sustainable Urban Futures, in: *Future City*, vol. 11, N. Frantzeskaki, K. Hölscher, M. Bach, F. Avelino (Eds.), pp. 103–125, Springer International Publishing, Cham, 2018, doi: 10.1007/978-3-319-69273-9_5.
27. Albornoz, V.M., Sáez, J.L., Véliz, M., II, Delineation of Rectangular Management Zones and Crop Planning Under Uncertainty in the Soil Properties. in Operations Research and Enterprise Systems, in: *Communications in Computer and Information Science*, vol. 695, B. Vitoriano and G.H. Parlier (Eds.), pp. 117–131, Springer International Publishing, Cham, 2017, doi: 10.1007/978-3-319-53982-9_7.
28. Roudier, P., Tisseyre, B., Poilvé, H., Roger, J.-M., Management zone delineation using a modified watershed algorithm. *Precis. Agric.*, 9, 5, 233–250, Oct. 2008, doi: 10.1007/s11119-008-9067-z.
29. Moorberg, C.J., *Soil and water conservation: An annotated bibliography*, New Prairie Press, 2019, Accessed: Nov. 04, 2024, [Online], Available: https://newprairiepress.org/ebooks/30/.
30. Peralta, N.R. and Costa, J.L., Delineation of management zones with soil apparent electrical conductivity to improve nutrient management. *Comput. Electron. Agric.*, 99, 218–226, 2013.

31. You, P.-S. and Hsieh, Y.-C., A computational approach for crop production of organic vegetables. *Comput. Electron. Agric.*, 134, 33–42, 2017.
32. Prabakaran, G., Vaithiyanathan, D., Ganesan, M., Fuzzy decision support system for improving the crop productivity and efficient use of fertilizers. *Comput. Electron. Agric.*, 150, 88–97, Jul. 2018, doi: 10.1016/j.compag.2018.03.030.
33. Carr, P.M., Carlson, G.R., Jacobsen, J.S., Nielsen, G.A., Skogley, E.O., Farming Soils, Not Fields: A Strategy for Increasing Fertilizer Profitability. *J. Prod. Agric.*, 4, 1, 57–61, Jan. 1991, doi: 10.2134/jpa1991.0057.
34. Yadav, S.P., Jindal, M., Rani, P., de Albuquerque, V.H.C., C.S.N., Kumar, M., An improved deep learning-based optimal object detection system from images. *Multimed. Tools Appl.*, 83, 10, 30045–30072, 2024.
35. Hussain, N., Rani, P., Chouhan, H., Gaur, U.S., Cyber security and privacy of connected and automated vehicles (CAVs)-based federated learning: challenges, opportunities, and open issues. *Fed. Learn. IoT Appl.*, 169–183, 2022.
36. Google Static Transit, Accessed: Nov. 05, 2024, [Online], Available: https://developers.google.com/transit/gtfs/%20(accessed%20on%2020%20November%202015).
37. GTFS Data Exchange, Accessed: Nov. 05, 2024, [Online], Available: http://www.gtfs-data-exchange.com/%20(accessed%20on%2012%20October%202015).

4

About the Growing Power of Artificial Intelligence (AI) and Blockchain for Fleet Management and Sustainable Societies

Jayant Jagtap[1], Raman Kumar[2*], Kunal Gagneja[3], Anita Gehlot[4] and M. Muhsen Hassan[5]

[1]*Department of Computing Science and Artificial Intelligence, NIMS Institute of Engineering and Technology, NIMS University Rajasthan, Jaipur, India*
[2]*School of Mechanical Engineering, Rayat Bahra University, Mohali, India*
[3]*Department of Computer Science Engineering, Chandigarh Engineering College, Chandigarh Group of Colleges, Jhanjeri, Mohali, Punjab, India*
[4]*Uttaranchal Institute of Technology, Uttaranchal University, Dehradun, Uttarakhand, India*
[5]*Department of Computer Science, The Islamic University, Najaf, Iraq*

Abstract

Artificial Intelligence (AI) and Blockchain technology are revolutionizing fleet management and smart city planning. This chapter examines how these technologies facilitate sustainable societies through their security, decentralization, and scalability attributes. We investigate how blockchain technology facilitates trust and privacy through its decentralized ledgers and immutable data structures. A key feature of AI is that it can make autonomous decisions and process data efficiently, both of which complement blockchain technology. A study examines the synergies between these technologies, addressing issues such as privacy, security, and scalability. The frameworks propose to emphasize the application of artificial intelligence and blockchain technology to intelligent transport systems and energy-efficient infrastructures in sustainable smart cities. A discussion is concluded on the environmental and economic implications of fleet management and urban planning based on AI-blockchain technology.

**Corresponding author*: ramankakkar@gmail.com

Pethuru Raj, Sudesh Yadav, Manas Kumar Mishra, Satya Prakash Yadav and Victor Hugo C. de Albuquerque (eds.) Artificial Intelligence Technologies for Smart and Sustainable Urban Transportation: Integrated Platforms and Use Cases, (55–70) © 2026 Scrivener Publishing LLC

Keywords: Artificial intelligence, blockchain, decision making, privacy, sustainable societies, smart city

4.1 Introduction

There are many impressive attributes of artificial intelligence and blockchain technology, including security, fairness, decentralization, and automation, which are driving their popularity. The technology and possibilities of Blockchain have increased attention since 2008; they have become a dominant subject due to their technology and potential [1, 2]. It is possible to use Blockchain as a decentralized, distributed register of all communication. All mathematically related transactions use private and public keys, which are traceable and recognizable by the participants, instead of using record preservation. Due to the assumption that every computer on the network is a node, blockchain technology offers many advantages, including distributed databases [3]. Such distributed databases allow the download of transaction copies from anywhere. As a result, trust and privacy are raised. In this chapter, we will examine these issues and their solutions.

With blockchain technology, you can also access data anonymously, securely, and with integrity without needing a third party to do so. Smart cities present some limitations and challenges that this technology can address because of these features. Smart cities are increasingly concerned

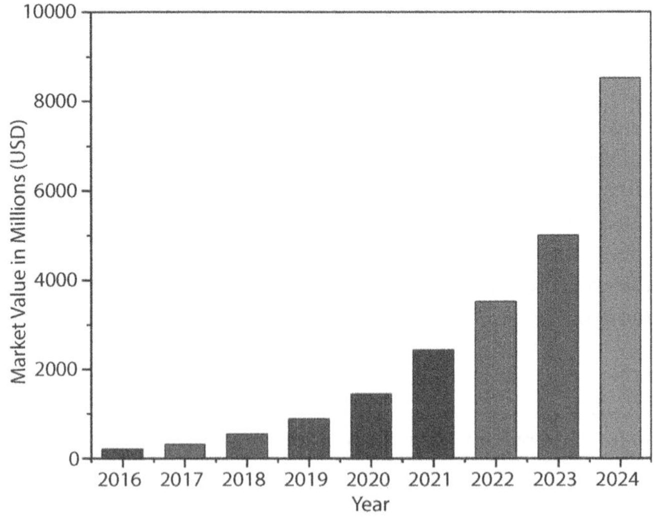

Figure 4.1 Blockchain market from 2016 to 2024.

with public safety and security, especially in dealing with terrorism, traditional crime, and cybercrime. Figure 4.1 shows how data and security issues related to IoT accounted for 750 million dollars in 2016 (IIoT-world hearsay) and how such expenses are anticipated to increase to 4.5 billion dollars by 2021 (Statista, 2021). As smart cities develop, many of the same problems will exist but will not occur because of their interconnected nature.

Smart cities pose privacy and security issues for investors, so they should solve the entire scenario to ensure these issues [2]. As a means of achieving this giant goal, smart city planners and professionals should pull related projects and encourage the development of the required smart systems. With recent attempts and considerations, we can ensure the future of smart cities. Due to its reliability and security, blockchain can be used to develop intelligent transport systems [4].

4.1.1 Artificial Intelligence and Blockchain

We propose that blockchains can not only be used to maintain datasets on chains for AI but also to host AIs that are advanced enough to work with data and advance knowledge independently. Moreover, we explore how such an implementation may advance in the coming years. It is a controversial proposal, especially in a decentralized environment where everyone can access and utilize its computing capabilities. Within a decade, give or take, we expect this proposal to be realized. Blockchains are distributed databases that are shared among computers. In the early days of cryptocurrency, blockchains were recognized for their significance in creating a digital database [5]. Through the implementation of blockchain technology, various systems can be made more secure, reliable, and trustworthy. Since its invention in 1991, blockchain technology has evolved into cryptocurrencies, decentralized finance, non-fungible tokens, and smart contracts [6]. Several countries are exploring blockchain technology, whether cautiously or enthusiastically. Despite this, blockchain technology has received considerable interest from the public and private sectors. North American blockchain technology markets are predicted to grow rapidly. Even though crypto had a bad year in 2018, blockchain technology expenses increased in the United States (U.S.) [7]. In 2025, blockchain revenues in the U.S. will surpass $37 billion (at a 44.5% annual growth rate).

Artificial intelligence (AI), which differs from AI, is designed to create intelligent machines that can perform a range of tasks autonomously, which humans normally handle [8]. The process by which machines become intelligent is not necessarily explained by making them intelligent.

According to Norvig and Russell, Turing's research is complemented by a four-part AI principle. Humanly thinking, rationally thinking, acting humanly, and acting are the principles. Aside from smart assistants (such as Siri and Alexa), there are autonomous cars, robotic advisors, bots, spam filters, and Netflix recommendations [8]. The AI market offers a variety of applications, such as machine learning (ML), natural language processing (NLP), automation, and complex decision-making algorithms [9]. According to Figure 4.2, the global market for AI-based software is predicted to grow at a significant rate and reach 126.00 billion USD by 2025 from 51.27 billion USD in 2022 [10]. AI-based techniques are increasingly being adopted in various fields based on this pattern.

There have been several blockchain and artificial intelligence technologies that have emerged throughout history, but their applications have become increasingly popular across a wide range of industries [11–15]. Blockchains and artificial intelligence have been around for several centuries, but their applications have become increasingly popular across a wide range of industries as time has progressed. Blockchains provide infrastructure, while artificial intelligence establishes engagement rules [16]. Data security and privacy may be issues with vehicular network applications since attackers could exploit the central server's model updates to access confidential data. The blockchain can validate each transaction, and local

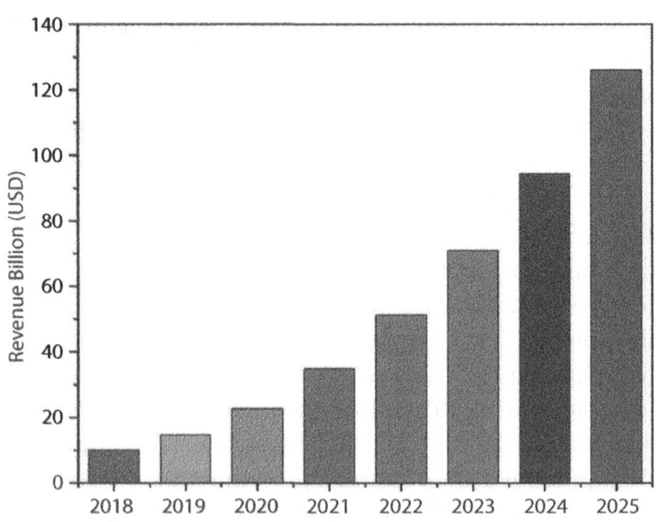

Figure 4.2 Revenue from the AI software market worldwide from 2018 to 2025 (billion USD).

training data regarding the completed transactions can be stored locally [17]. Transportation infrastructure is faced with major challenges due to data breaches. Using blockchains, data can be encrypted properly, breaches can be minimized, and vehicle networks can be optimized. Blockchains can also counter some of AI's other challenges, including high-speed servers for collecting updates and high bandwidth requirements, as a result of their decentralized storage [17]. There can be a lot of data exchange among vehicular networks, which may cause problems for complex decentralized blockchain systems. Utilizing AI-based algorithms could solve the scalability issue with blockchains by processing big data effectively. By applying AI to transportation systems, energy-intensive transport systems can also be made more energy-efficient.

4.1.2 Sustainable Smart City Society

Creating a sustainable city that is secure, scalable, efficient, and powered by artificial intelligence necessitates meeting several key requirements.

- **Resilient environment:** With urbanization and globalization combined, cities face unprecedented challenges, requiring resilient and sustainable infrastructure.
- **Interoperable and flexible:** Interoperability is essential for enabling physical objects to be interconnected through communication technologies. The Internet of Things can be used to control smart home appliances remotely and to transmit home data to the cloud, for example [18].
- **Decision support system:** The development of a smart city requires the coordination of a variety of complex and diverse systems to balance industrialization, economic growth, and environmental appreciation. The decision-making process for optimal solutions must, therefore, be highly adaptive.
- **Behavior monitoring:** Computer simulations are used in smart cities to simulate adaptive behavior. Because the web of going generates pattern behavior data, a reciprocal relationship cycle generates data describing economic, social, and physical interactions in a network of networks.
- **Energy sources and distribution:** It is imperative to meet energy quality and renewable resource requirements. Therefore, distributed technologies should be integrated into the power grid and other systems.

- **Intelligent infrastructure:** To build a sustainable smart city, everyone needs to collaborate and use intelligent technologies.
- **Scalability:** A smart city's infrastructure allows citizens to collect and process data in real-time using a number of hardware and software components. The city needs a distributed IoT-blockchain cloud infrastructure to become sustainable.
- **Smart healthcare:** In the healthcare industry, ICT can provide benefits through the integration of big data and smart systems, as well as cutting-edge critical analysis.
- **Secure infrastructure:** Data and networks must be monitored for malicious activity, authentication protocols set up, entities authorized, and network control policies developed to ensure security.

4.2 Literature Survey and Contribution

Unreliable environments and networks make it more challenging to verify the integrity of messages sent to users. The most forthright communicator cannot be identified, either. The academic community responded by devising Smart Coin, a cryptocurrency incentive scheme based on blockchain technology. There was a belief that improvising in-network would improve safety and efficiency. As a result of the proposed plan, mobility is improved, traffic is reduced, road accidents are reduced, and scam and deception messaging are eliminated. I believe that vehicles would benefit from being able to rate their communication sources based on their reliability. Blockchain accounts are credited with Intelligent Coins in proportion to how well drivers decode messages.

As AI, Blockchain, and the Internet of Things (IoT) become more prevalent in the tourism sector, they have significant potential to elevate sustainability while offering immersive experiences. While this amalgamation presents a myriad of challenges, it must be carefully considered and resolved strategically.

4.2.1 Privacy and Security Concerns

Integrating these technologies involves collecting and processing a great deal of sensitive information [19–22]. In today's world, privacy and security are critical concerns in a number of areas, from personal information to blockchain transaction data [23–25]. Using AI and IoT to collect real-time data and analyze user behavior, data breaches are becoming more prevalent. Although Blockchain is hailed as a secure technology, the entire ecosystem is susceptible to vulnerabilities, which makes security a challenge. It is imperative to devise robust encryption methods and implement stringent security measures to mitigate these risks.

Previously, blockchain security has been studied primarily from a specific perspective instead of from a holistic perspective. Based on its unique advantage of empowering computing systems, blockchain technology has been identified in the literature for a wide variety of potential applications, including smart cities [26–28], smart homes [29], fog computing [30], IoT wireless sensor networks [27], cloud computing [28, 29] and mobile computing. In these cases, blockchain technology can be used to improve efficiency. As well as economies of scale, blockchain acceptance is heavily influenced by them. The focus of researchers has been on mitigating security risks rather than exploring the possibilities of blockchain technology despite the many advantages that it offers. To make blockchain applications safe and secure across industries, we must address security issues in blockchain environments.

To achieve sustainable smart cities, we present a comprehensive taxonomy of security in blockchain technology. This chapter aims to provide a comprehensive discussion of several key factors that will lead to a sustainable, smart society through the convergence of blockchains and artificial intelligence. Using blockchain and artificial intelligence, this research examines how smart cities can develop sustainably.

As a result of this study, the following scientific contributions were made:

- The purpose of this paper is to present background information about Blockchain, its convergence with artificial intelligence, and an architecture for smart cities based on blockchain and artificial intelligence. It is intended to provide readers with a basic understanding of blockchain technology.
- An overview of IoT smart city security challenges, vulnerabilities and limitations is presented and discussed in this survey. We also discuss some technological and business challenges associated with blockchain and artificial intelligence.
- This survey provides extensive details about intelligent infrastructure for smart cities that are sustainable and

environmentally friendly. A number of ongoing projects are also integrating blockchain technology and artificial intelligence.

- Security solutions for blockchain-AI technology are presented in the context of sustainable smart cities. This paper finalizes its assessment of the interrelationship between urban issues at present and sustainability, as well as possible directions for the future based on our survey.

4.3 Blockchain to Support Smart Cities' Operations

Nodes, which are participants in the Blockchain, can confirm and reject transactions and view them. Businesses use them to track asset movements and record transactions. There are tangible and intangible assets, such as land, cars, and patents. Transaction data is stored in block-based structures using cryptographic hashes and block-based structures. Blockchain technology is decentralized and distributed, so every block of data that is added to it cannot be changed. A blockchain is a secure, immutable, and cryptographically secure distributed ledger because of its immutability and cryptographic security. Currently, it is append-only, which means consensus or peer-to-peer agreement is required to update it. Distributed ledger technology (DLT) is a broader concept that includes blockchain [31]. Since blockchain technology utilizes a distributed ledger, it is resilient because it does not have a single point of vulnerability. Digital signatures establish a unique connection between each block and the previous block. To alter a record, you must disrupt earlier records, which makes the information unchangeable. Using blockchain technology, participants can transfer assets over the Internet without involving a third party centered on a central server.

The Blockchain technology behind Bitcoin emerged recently as a result of the subprime crisis in 2008, causing the existing financial system to lose its trust [32]. Satoshi Nakamoto explains how algorithms are computed through the use of a distributed ledger and consensus in his white paper on the Bitcoin protocol. Its purpose is to eliminate traditional financial intermediaries and promote direct peer-to-peer transactions [33].

Virtual currencies have been attempted many times, but they have failed due to double spending. Trust intermediaries, such as banks, are currently being introduced to eliminate double-spending. Blockchain technology makes transferring assets, like virtual currencies, secure over the Internet without the need for trusted intermediaries. In addition to currencies, blockchain technology could benefit many other areas, making it one of the most promising technologies ever.

Blockchain technology enables businesses to share updated ledgers *via* peer-to-peer replication whenever a transaction takes place, for example. The network's participants act as publishers and subscribers, and data is synchronized across participants' computers. As a result of the blockchain network, duplication of effort is eliminated, and intermediaries are eliminated, resulting in a more efficient and economical process. Additionally, consensus models reduce network vulnerability by validating transaction information. There is security, authentication, and verifiability in transactions.

4.4 Blockchain Benefits

Unlike traditional databases, blockchains store data in tamper-proof formats that can only be accessed by valid users. Figure 4.3 illustrates the primary characteristics of blockchain that make it unique. Understanding these characteristics is crucial to realizing its potential.

Figure 4.3 Blockchain benefits.

Distributed shared ledger: Distributed systems that are append-only eliminate centralized databases' single points of failure by sharing them across business networks.

- **Consensus:** A transaction cannot be committed and appended to the ledger without the consent of all validating parties.
- **Provenance:** Using blockchain technology, an asset's history can be accessed.
- **Immutability:** As long as the shared ledger is indelible, all information can be trusted.
- **Finality:** The blockchain does not allow transactions to be reversed once they have been completed.
- **Smart contracts:** Blockchains contain code that is executed by computers/nodes upon triggering an event. It is essentially possible to automate the execution of an "if this, then that" statement.

Using blockchain technology, you can disrupt any transaction that relies on trusting information. Due to the advent of blockchain technology, all trusted intermediaries have been disrupted in one way or another, and it addresses the issues associated with current information-related transactions. In the blockchain, transactions are verified by peer-to-peer networks, creating a permanent, unalterable copy of all transactions.

4.5 Types of Blockchain Networks

Public or private blockchain networks depend on who owns them. In public blockchain networks, membership is open and unrestricted, but in private blockchain networks, membership is controlled by a central authority. Both networks have the same inherent technology, but their dynamics and utility differ. Nodes are motivated to stay in the network based on their incentives. Public blockchains reward participants for staying on the network, and private blockchains don't need to create this incentive to encourage participation.

An organization or corporate network may not benefit from the democratization of a transparent public registry since the parties know one another and are familiar with the participants. Public blockchains, according to consensus, are capable of handling certain applications, such as cryptocurrency (bitcoin)-based transactions. For blockchain technology to be used as an enterprise solution, a private blockchain ecosystem with a higher level of regulatory control would be needed.

Several blockchain applications are still in development, so the technology is evolving continuously and iteratively. A distributed system composed of multiple private blockchains connected by a public network may help resolve the issue of public vs. private blockchains. Blockchains in that shared ecosystem interact with each other in a symbiotic way, just like private networks do with the Internet.

Blockchain technology is being used in a variety of smart city applications, such as healthcare, transportation, supply chain management, education, manufacturing, and construction. Blockchain applications in several areas and the services they provide [34–36].

4.6 Blockchain Suitability

The tool should only be used when there is a need to share data between multiple parties, and a common understanding of the data is needed. Blockchain is about much more than just sharing data.

There are several situations in which a blockchain solution can be deemed viable and efficient:

- Multiple intermediaries complicate, increase the cost, and increase the time of a transaction.
- Businesses benefit tremendously when transactions are sped up, and delays are reduced.
- A business participant's transactions are dependent on one another.
- Data validation should be updated for actions undertaken by more than one person.
- In order for a business to succeed, participants must build trust between them. As a result, blockchain technology may not be able to solve all the problems associated with transactions.

Fleet Management: By 2030, DHL intends to electrify 60% of its "last mile" vehicle fleet, providing a starting point for this study. Throughout this paper, direct CO2 emissions from operations, also known as Tank-to-Wheel emissions (TtW), are formulated as a 60% reduction by 2030 over 2022.

A large number of other parcel delivery companies [37], are also exploring ways to reduce their fleet's emissions. Since 2010, Aramex has reduced its fleet emissions by 20%, and a long-term electric vehicle conversion program is in place [38]. The goals of FedEx have never been higher. A 50% electric vehicle fleet will be acquired by FedEx by 2025, and in 2030, the company will use only electric vehicles [39]. By 2050, UPS aims to achieve CO2 neutrality. Currently, there are no specific goals for electric vehicles. Instead, it has been announced that electric vehicles will be part of a fleet of alternative fuel vehicles and advanced technologies to improve sustainability [40].

As a result, parcel service providers' vehicle fleets could be procured economically optimally in 2030, lowering CO2 emissions by 60%. Various acquisition strategies, as well as scenarios for key parameters, should be investigated for electric vehicles designed for parcel delivery services. CO2 emissions throughout the period over which the target savings will occur are not only taken into account but also the total savings by 2030. A decision criterion is calculated using the DIPO tool, a professional fleet management tool with economic efficiency and sustainability capabilities. A collection of Excel spreadsheets was created for teaching and research purposes at Ludwigshafen University of Applied Sciences [41]; in the expansion stage, LCA values were calculated more accurately. Using the ten diesel vehicles in 2022, the fleet emissions target for 2030 will be calculated based on 25 tons of carbon dioxide (tCO2). According to the 2030 vehicle fleet emission target, CO2 emissions will reach 9.8 tons by 2030. Due to the fact that TtW emissions are the only emissions that matter, it is easy to determine the vehicle fleet constellations for the target year. There must be six electric vehicles in the fleet, or 60% of the fleet. The 2030 target value will be reached through three different strategies.

4.7 Conclusion

This study emphasizes the transformative potential of AI and blockchain integration in addressing challenges faced by modern urban societies. By leveraging the decentralized and secure nature of blockchain alongside AI's decision-making capabilities, these technologies can create sustainable solutions for smart city development. The proposed framework addresses critical aspects like privacy, security, and scalability, providing a pathway for energy-efficient transport systems and intelligent infrastructure. As the adoption of these technologies continues to grow, future research should focus on overcoming scalability and interoperability challenges, ensuring widespread implementation in diverse sectors. Through strategic investments and collaborative innovation, AI and Blockchain can significantly contribute to sustainable urban development and enhanced fleet management strategies.

References

1. Zhang, L., Cai, Z., Wang, X., Fakemask: A novel privacy preserving approach for smartphones. *IEEE Trans. Netw. Serv. Manag.*, 13, 2, 335–348, 2016.
2. Rani, P., Verma, S., Yadav, S.P., Rai, B.K., Naruka, M.S., Kumar, D., Simulation of the lightweight blockchain technique based on privacy and security for healthcare data for the cloud system. *Int. J. E-Health Med. Commun. IJEHMC*, 13, 4, 1–15, 2022.
3. Rani, P., Hussain, N., Khan, R.A.H., Sharma, Y., Shukla, P.K., Vehicular intelligence system: time-based vehicle next location prediction in software-defined internet of vehicles (SDN-IOV) for the smart cities. *Intell. Things AI-IoT Based Crit.-Appl. Innov.*, 4, 35–54, 2021.
4. Rani, P. and Sharma, R., Intelligent Transportation System Performance Analysis of Indoor and Outdoor Internet of Vehicle (IoV) Applications Towards 5G. *Tsinghua Sci. Technol.*, 29, 6, 1785–1795, Dec. 2024, doi: 10.26599/TST.2023.9010119.
5. Rani, P., Singh, P.N., Verma, S., Ali, N., Shukla, P.K., Alhassan, M., An implementation of modified blowfish technique with honey bee behavior optimization for load balancing in cloud system environment. *Wirel. Commun. Mob. Comput.*, 2022, 1–14, 2022.
6. Blockchain, Accessed: Nov. 04, 2024, [Online], Available: https://www.investopedia.com/terms/b/blockchain.asp.
7. Hashcash, Accessed: Nov. 04, 2024, [Online], Available: https://www.hashcashconsultants.com/media/the-growth-and-prospect-of-the-global-blockchain-industry/.
8. Bluetin, Accessed: Nov. 04, 2024, [Online], Available: https://builtin.com/artificial-intelligence.
9. Rani, P. and Sharma, R., An experimental study of IEEE 802.11 n devices for vehicular networks with various propagation loss models, in: *International Conference on Signal Processing and Integrated Networks*, Springer, pp. 125–135, 2022.
10. Artificial intelligence software market revenue worldwide 2018-2025, Accessed: Nov. 04, 2024, [Online], Available: https://www.statista.com/statistics/607716/worldwide-artificial-intelligence-market-revenues/.
11. Miglani, A. and Kumar, N., Deep learning models for traffic flow prediction in autonomous vehicles: A review, solutions, and challenges. *Veh. Commun.*, 20, 100184, 2019.

12. Inedjaren, Y., Maachaoui, M., Zeddini, B., Barbot, J.-P., Blockchain-based distributed management system for trust in VANET. *Veh. Commun.*, 30, 100350, 2021.
13. Gupta, R., Tanwar, S., Kumar, N., B-IoMV: Blockchain-based onion routing protocol for D2D communication in an IoMV environment beyond 5G. *Veh. Commun.*, 33, 100401, 2022.
14. Parvaresh, N., Kulhandjian, M., Kulhandjian, H., D'Amours, C., Kantarci, B., A tutorial on AI-powered 3D deployment of drone base stations: State of the art, applications and challenges. *Veh. Commun.*, 36, 100474, 2022.
15. Mchergui, A., Moulahi, T., Zeadally, S., Survey on artificial intelligence (AI) techniques for vehicular ad-hoc networks (VANETs). *Veh. Commun.*, 34, 100403, 2022.
16. Sandner, P., Gross, J., Richter, R., Convergence of blockchain, IoT, and AI. *Front. Blockchain*, 3, 522600, 2020.
17. Javed, A.R., *et al.*, Integration of blockchain technology and federated learning in vehicular (iot) networks: A comprehensive survey. *Sensors*, 22, 12, 4394, 2022.
18. Iqbal, A., *et al.*, Interoperable Internet-of-Things platform for smart home system using Web-of-Objects and cloud. *Sustain. Cities Soc.*, 38, 636–646, 2018.
19. Mhlanga, D. and Ndhlovu, E., Digital Technology Adoption in the Agriculture Sector: Challenges and Complexities in Africa. *Hum. Behav. Emerg. Technol.*, 2023, 1–10, Sep. 2023, doi: 10.1155/2023/6951879.
20. Pradhan, N.R., Sharma, H.K., Choudhury, T., Mor, A., Mohanty, S., A Blockchain Enabled Medical Tourism Ecosystem. in Intelligent Systems and Machine Learning, in: *Lecture Notes of the Institute for Computer Sciences, Social Informatics and Telecommunications Engineering*, vol. 471, Nandan Mohanty, S., Garcia Diaz, V. and Satish Kumar, G. A. E. (Eds.), pp. 158–169, Springer Nature Switzerland, Cham, 2023, doi: 10.1007/978-3-031-35081-8_13.
21. Panpeng, V., Fang, M., Phaphuangwittayakul, A., Rattanadamrongaksorn, T., Preliminary Study and Implementation of Chiang Mai Tourism Platform Based on DOSA. in *Proceedings of Fifth International Congress on Information and Communication Technology*, in: *Advances in Intelligent Systems and Computing*, vol. 1184, X.-S. Yang, S. Sherratt, N. Dey, A. Joshi (Eds.), pp. 511–521, Springer Singapore, Singapore, 2021, doi: 10.1007/978-981-15-5859-7_51.
22. Arcuri, M.C., Gai, L., Ielasi, F., Ventisette, E., Cyber attacks on hospitality sector: Stock market reaction. *J. Hosp. Tour. Technol.*, 11, 2, 277–290, 2020.
23. Prashyanusorn, V., Kaviya, S., Yupapin, P.P., Surveillance system for sustainable tourism with safety and privacy protection. *Procedia Soc. Behav. Sci.*, 2, 1, 74–78, 2010, doi: 10.1016/j.sbspro.2010.01.016.
24. Talwar, S., Dhir, A., Kaur, P., Mäntymäki, M., Why do people purchase from online travel agencies (OTAs)? A consumption values perspective. *Int. J. Hosp. Manag.*, 88, 102534, Jul. 2020, doi: 10.1016/j.ijhm.2020.102534.

25. Prashyanusorn, V., Fujii, Y., Kaviya, S., Mitatha, S., Yupapin, P.P., Sustainable Tourism Using Security Cameras with Privacy Protecting Ability. *J. Inf. Secur.*, 01, 02, 68–73, 2010, doi: 10.4236/jis.2010.12008.
26. Sharma, P.K. and Park, J.H., Blockchain based hybrid network architecture for the smart city. *Future Gener. Comput. Syst.*, 86, 650–655, 2018.
27. Sharma, P.K., Moon, S.Y., Park, J.H., Block-VN: A distributed blockchain based vehicular network architecture in smart city. *J. Inf. Process. Syst.*, 13, 1, 184–195, 2017.
28. Sharma, P.K., Rathore, S., Park, J.H., DistArch-SCNet: blockchain-based distributed architecture with li-fi communication for a scalable smart city network. *IEEE Consum. Electron. Mag.*, 7, 4, 55–64, 2018.
29. Singh, S., Sharma, P.K., Park, J.H., SH-SecNet: an enhanced secure network architecture for the diagnosis of security threats in a smart home. *Sustainability*, 9, 4, 513, 2017.
30. Sharma, P.K., Chen, M.-Y., Park, J.H., A software defined fog node based distributed blockchain cloud architecture for IoT. *IEEE Access*, 6, 115–124, 2017.
31. Rauchs, M., *et al.*, Distributed Ledger Technology Systems: A Conceptual Framework. *SSRN Electron. J.*, 2018, doi: 10.2139/ssrn.3230013.
32. Sanders, A., The subprime crisis and its role in the financial crisis. *J. Hous. Econ.*, 17, 4, 254–261, Dec. 2008, doi: 10.1016/j.jhe.2008.10.001.
33. Nakamoto, S., Bitcoin: A peer-to-peer electronic cash system. Decentralized Business Review. 21260, 2008, Available online, Accessed: Nov. 05, 2024, [Online], Available: https://www.debr.io/article/21260.pdf.
34. Xie, J., *et al.*, A Survey of Blockchain Technology Applied to Smart Cities: Research Issues and Challenges. *IEEE Commun. Surv. Tutor.*, 21, 3, 2794–2830, 2019, doi: 10.1109/COMST.2019.2899617.
35. Upadhyay, A., Mukhuty, S., Kumar, V., Kazancoglu, Y., Blockchain technology and the circular economy: Implications for sustainability and social responsibility. *J. Clean. Prod.*, 293, 126130, Apr. 2021, doi: 10.1016/j.jclepro.2021.126130.
36. Teisserenc, B. and Sepasgozar, S., Adoption of blockchain technology through digital twins in the construction industry 4.0: A PES&xxx2121;S approach. *Buildings*, 11, 12, 670, 2021.
37. Imarc, Top companies in the courier, 2023. Accessed: Nov. 04, 2024, [Online], Available: http://express,https//www.imarcgroup.com/courier-express-parcel-companies.
38. Aramex, Aramex begins fully electric vehicles pilot in KSA, 2020, January 15. Accessed: Nov. 04, 2024, [Online], Available: https://www.aramex.com/us/en/media/in-the-news/details?contentId=77327688-b3f2-659d-9310-%20ff0100e7fe0c&module=stories#:~:text=The%20company%20is%20committed%20to,negative%20impact%20on%20%20the%20environment%20BAFA.

39. Werwitzke, C., FedEx beschafft ab 2025 zu 50 Prozent E-Lieferwagen, 2021. Accessed: Nov. 04, 2024, [Online], Available: https://www.electrive.net/2021/03/03/.
40. FedEx, Electrifying our future, 2022, Accessed: Nov. 04, 2024, [Online], Available: https://about.ups.com/us/en/social-impact/environment/sustainableservices/electric-vehicles—about-ups.html.
41. Ludwigshafen University of Business and Society (LUBS), Bongard, S., Friesenhahn, A., Does it Pay Off? Tool-supported Profitability Analysis of Alternative Drive Technologies, in: *6th FEB International Scientific Conference 2022*, Ludwigshafen University of Business and Society (LUBS), Ludwigshafen, Germany, J. Wolff, and Ludwigshafen University of Business and Society (LUBS), Ludwigshafen, Germany, Ludwigshafen, Germany, University of Maribor Press, Ludwigshafen, Germany, pp. 365–374, 2022, doi: 10.18690/um.epf.5.2022.35.

5

Testifying the Criticality of the Internet of Things (IoT), 5G and AI: A Perfect Combination for Battery Management

Preeti Rani[1]*, Raman Kumar[2], Amrita Singh[3], Jayant Jagtap[4] and Muntather Almusawi[5]

[1]*College of Computing Sciences & IT (CCSIT), Teerthanker Mahaveer University, Moradabad (U.P.), India*
[2]*School of Mechanical Engineering, Rayat Bahra University, Mohali, India*
[3]*Department of Computer Science Engineering, Chandigarh Engineering College, Chandigarh Group of Colleges, Jhanjeri, Mohali, Punjab, India*
[4]*Department of Computing Science and Artificial Intelligence, NIMS Institute of Engineering and Technology, NIMS University Rajasthan, Jaipur, India*
[5]*Department of Computer Science, The Islamic University, Najaf, Iraq*

Abstract

In this study, we examine how IoT, 5G, and artificial intelligence (AI) can be integrated into battery management systems (BMSs) to transform electric vehicles and large-scale energy storage systems. With IoT, 5G, and AI-driven predictive analytics, these technologies collectively enable smarter, more efficient, and proactive battery management. The present study examines how these three factors can improve energy efficiency, safety, and device longevity while addressing environmental concerns. Wireless communication frameworks may perform better than traditional wired systems when applied to specific BMS applications, as our findings demonstrate. In this study, a real-time energy management strategy using neural networks and fuzzy logic provides an evaluation of the feasibility and performance of wireless networks within electric vehicle battery packs, offering a new perspective on the design of BMS in the future.

**Corresponding author*: preetiresearcher1@gmail.com

Pethuru Raj, Sudesh Yadav, Manas Kumar Mishra, Satya Prakash Yadav and Victor Hugo C. de Albuquerque (eds.) Artificial Intelligence Technologies for Smart and Sustainable Urban Transportation: Integrated Platforms and Use Cases, (71–88) © 2026 Scrivener Publishing LLC

Keywords: Artificial intelligence, battery management, energy efficiency, internet of things, smart city, vehicle

5.1 Introduction

All industries need battery management systems (BMS) to manage energy efficiency and performance, including electric vehicles (EVs), consumer electronics, smart cities, and renewable energy storage. To manage batteries effectively, three revolutionary technologies must be combined: IoT, 5G, and artificial intelligence (AI). As a result, traditional BMS can be transformed into proactive self-correcting systems by combining these technologies [1]. As IoT, 5G networks, and artificial intelligence (AI) are converging, technology is transforming various industries. A triad of powerful technologies is revolutionizing battery management for devices in multiple sectors, including electric vehicles, consumer electronics, and large-scale energy storage systems [2]. Batteries must be managed efficiently not only to extend their life and optimize performance but also to address environmental concerns associated with waste and energy consumption. A new paradigm for smart, reliable, and efficient battery management will emerge as a result of the convergence of IoT in real-time monitoring, 5G in rapid data transfer, and AI in predictive analytics [3].

With the Internet of Things, connected devices are continuously collecting data, including battery usage patterns, temperatures, and charge cycles. Adding 5G networks improves this framework by allowing devices to communicate faster and more reliably, enabling near-instantaneous response times needed for time-critical applications. A combination of AI and big data provides advanced algorithms that are capable of analyzing large datasets, predicting battery behaviour, and optimizing charging. In conjunction with 5G, IoT, and AI, a synergy of these technologies holds great promise for reducing battery consumption and increasing device lifespans [3].

Technology is constantly evolving at a rate that has never been seen before. The convergence of the IoT, 5G, and AI is reshaping industries, economies, and daily life at an unprecedented rate. Technology is driving innovation because it enhances connectivity, automates processes, and allows for real-time data-driven decisions. As billions of IoT devices are connected, we will be able to communicate seamlessly and exchange data, and 5G will ensure the ultra-fast and reliable connectivity we need to support these interactions. The potential of this vast data is simultaneously unlocked by artificial intelligence, enabling intelligent analysis as well as autonomy in decision-making. In addition to improving current systems, the cooperation of IoT, 5G, and AI is fundamentally changing business, the way cities function, and how people interact. In this introduction, we explore the critical role that these technologies play in advancing society's digital transformation and the reasons for their importance.

For BMS internal communications and buses to operate properly, high quality of service must be ensured. Besides ensuring safety, BMSs also determine how much power is left in Lithium-Ion batteries. Wireless buses may have advantages such as weight reduction, ease of manufacturing, and weight reduction compared with wired buses when communicating between BMS sub-modules. In this study, a wireless medium has been used to communicate between BMS sub-modules. Typically, IoT networks are built on top of physical layers with 2.4GHz frequencies and MAC protocols. For networks that use the Internet of Things (IoT), these protocols provide a reasonable amount of bandwidth on this frequency band. TSCH (Time Slotted Channel Hopping) (IEEE Std 802.15.4-2015) is the standard that has been developed with industrial application requirements in mind [4, 5]. There are several reasons why this protocol is popular: it permits simultaneous transmissions on multiple channels. As a result of dedicated timeslots, collisions can be greatly reduced, and channel hopping makes external interference equally distributed between scheduled timeslots. As a result, we use them in this work as the physical and MAC layer protocols. A wireless application of this nature was considered using other wireless technologies. There were several options, including Wi-Fi, but it would have consumed too much power. Despite its attractiveness, Bluetooth Low Energy (BLE) lacks the capabilities of IEEE Standard multi-hop routing. The 802.15.14 standard influenced our decision. Vehicles require this feature to function properly. Communication technology has to be evaluated in the EV environment in order to be used effectively for battery management since the metallic parts of the car may negatively impact the link quality. To evaluate pack network conditions, we conducted experiments. To achieve a high level of quality, we developed a linear programming (LP) strategy based on the results of the above analysis.

Several of these papers have already been published at conferences [6]. It is presented here in an extended form. As a first addition, the added content contains data about the link quality of each radio channel, which has been incorporated into the network performance evaluation. A second set of experiments was conducted to measure IoT networks' response to Wi-Fi traffic generated in the environment. In this final part of our research, we propose a new topology for managing batteries with wireless communications, which is based on a simple algorithm called simple descent (SD) that allows faster processing times than linear programming (LP).

In this study, we are not attempting to determine which type of communication is best for building automation systems. The purpose of the study is not to examine the feasibility of wireless communication within BMS but to find out how such a network functions within the battery pack environment and to determine what performance levels can be reached.

5.1.1 Energy Management Strategy Description

This paper studies an energy storage system combining batteries and supercapacitors. The strategy is to use neural networks offline training and implement them in real-time. Part I of this paper decomposes the load power demand into different frequency components using wavelet transforms. Afterwards, a neural network is trained offline using the datasets obtained. A combined neural network model is presented in the second part.

An energy management control strategy proposed in this study suppresses battery power peaks while reducing battery power variations at the same time. The strategy employs wavelet transforms, neural networks, and fuzzy logic to achieve its goals. We first offline decompose the battery and supercapacitor frequency bandwidths into two separate frequency components using the wavelet transform. In the second step, the frequency data obtained above is used to build and train a neural network model. As inputs, the neural network model calculates load power demand and its low-frequency component, and as outputs, the battery's power demand. Therefore, the neural network predicts battery power demand based on driving cycle variations. A supervisory controller using fuzzy logic is employed to manage supercapacitor voltage within a certain fluctuating range. A real-time hardware platform for validating the proposed energy management control strategy is developed using 72 V batteries and 96 V supercapacitors as storage mediums.

5.2 Literature Review

Batteries are increasingly being integrated with IoT, 5G, and artificial intelligence, with studies highlighting their individual and combined potentials.

5.2.1 Managing an EV Battery Pack

Batteries can be wired in series or parallel within a battery pack. A pack's capacity is increased by adding more cells, but its voltage output is determined by how many cells are in series. By adding cells simultaneously, the most charged cells discharge into the least charged cells. Whenever items are wired in series, the BMS must balance them because they may become unbalanced over time. It is also necessary to monitor every cell individually in a series. A whole pack should not be used if any one cell exceeds the safety window [7]. There are several popular topologies in commercial electric vehicles, including 96S1P and 96S2P.

As electric cars are divided into modules, they typically have six to sixteen cells, so the wire harness is smaller. CSUs (also known as Cell Supervision Units) supervise the modules and communicate with the Master Control Units (MCU) *via* I2C, SPI, and UART [8, 9]. BMSs in EVs also have external interfaces for sharing data with other systems, like engine controllers and displaying information on the dashboard. Today's EVs use CAN, a common network interface in vehicles [10].

5.2.2 IoT in Battery Management

In battery management, IoT is used to monitor and manage battery systems in real-time. Sensors connected to the IoT can monitor battery temperatures, voltage, current, and state of charge (SOC) continuously. IoT-enabled battery monitoring systems provide more accurate state-of-health (SOH) estimations, allowing battery degradation to be predictably predicted and premature failures to be avoided [11, 12]. Electric vehicles (EVs) and renewable energy systems, which rely heavily on battery health, benefit greatly from such insights.

5.2.3 Wireless BMS Incentive Program

A BMS application does not typically use wireless communications. Simple and inexpensive, copper wire is a good choice. A solution to the quality of service constraints is to replace wired communication links with specific microcontrollers. Although wireless communication has its downsides, there are also some upsides. Assembling a device without wires simplifies the procedure while reducing weight and space. CSUs with microcontrollers can also distribute the processing of cell data. A local copy of the information can be saved and re-used later, even if the cells are re-used elsewhere. In this case, the battery is said to be in its second life [13].

5.2.4 5G as a Catalyst for Rapid Data Transmission

By enhancing speed, latency, and connectivity, 5G networks will enhance IoT systems significantly. As a result of 5G's high data transmission rates, IoT devices can communicate between themselves in real-time with minimal delay, which is crucial for battery systems that need to react quickly when environmental conditions change. An IoT system powered by 5G can, for example, send data instantly, allowing a centralized system to adjust battery usage immediately, improving reliability.

5.2.5 AI and Predictive Analytics in Battery Optimization

The use of AI algorithms provides sophisticated tools for analyzing battery performance data, allowing for the identification of trends and predictive failures before they occur. In a new study, machine learning models are applied to data generated by the Internet of Things to predict optimal charging times, helping to prevent overcharging and extending battery life. Additionally, AI models can adapt to usage patterns in real-time to enable predictive maintenance and dynamic optimization. It is especially beneficial to use these techniques in electric vehicles since improper battery management can result in performance problems and poor range.

5.2.6 Synergy of IoT, 5G, and AI in Battery Management

5G networks, combining IoT, 5G, and AI, enable rapid processing and action on data from IoT devices, with AI optimizing results. A 5G IoT-AI framework, for instance, can provide real-time insights into smart grid energy efficiency and battery longevity, as shown in [14]. Consumer electronics and industrial applications can benefit from intelligent energy management through interconnected systems that automatically adjust based on predictive algorithms.

As a result, IoT, 5G, and AI have the potential to transform battery management, each contributing unique advantages that together promote a sustainable and efficient energy supply. As a result of the combination of these technologies, a range of innovations across industries can be achieved in the area of battery management. A data-driven, optimized battery management system may be possible as a result of further research and development in this integrated approach.

5.3 The Internet of Things (IoT) in Battery Management

Data collection, sharing, and action are possible through IoT, which is an interconnected network of devices. Battery management uses IoT devices for real-time data collection, such as voltage, temperature, and current levels, that are then processed for predictive maintenance [15]. The Internet of Things can provide real-time battery health monitoring, allowing early detection of potential problems and preventing costly failures.

5.3.1 Real-Time Monitoring and Predictive Maintenance

Using IoT sensors and predictive analytics, companies can predict and prevent battery failures through continuous battery monitoring. As a result of IoT-enabled sensors, stakeholders can receive alerts about potential problems, such as overheating or cell imbalance. In the case of critical thresholds breached, the system may be shut down [16].

For instance, electric vehicles use IoT-based sensors to monitor battery health, charging cycles, and temperature. As a result of this continuous data collection, the vehicle's BMS is able to optimize energy consumption and extend battery life by making real-time adjustments [17].

5.3.2 Data Collection and Data-Driven Insights

Battery behavior under various conditions can be uncovered by IoT devices collecting and transmitting a great deal of data. By analyzing these insights, batteries can be designed better, manufacturing processes improved, and materials improved [18]. As IoT-driven data improves battery management, 5G and AI will be able to refine it further.

5.4 5G Connectivity: Enabling High-Speed, Low-Latency Data Exchange

The benefits of 5G go well beyond speed; they include a wide range of bandwidths, ultra-low latency, and massive connectivity, which are essential for supporting the vast network of battery management IoT devices. Data transmission and reception over 5G ensure seamless communication between devices, cloud servers, and AI algorithms in real-time [19].

5.4.1 Enhancing Real-Time Decision Making

Any delay in real-time decision-making can have a detrimental impact on the functionality or safety of EVs, drones, or autonomous robots. IoT sensors generate vast amounts of data, which 5G allows for almost instantaneous processing and action. As a result, dynamic battery management is vital since it may be necessary to make adjustments in real-time to charge, discharge, or cool batteries to maintain optimal performance [20].

5.4.2 Scalability of IoT Networks

There are many advantages to 5G, including the ability to connect multiple devices simultaneously. Scalability is especially important in battery management systems for smart cities or industrial applications, where thousands of devices must be monitored and controlled in real-time. By leveraging 5G, IoT networks can be built and managed more reliably and efficiently, improving data-driven insight and decision-making [21].

5.5 Artificial Intelligence (AI): The Brain Behind Smart Battery Management

There are many advantages to 5G, including the ability to connect multiple devices simultaneously. Artificial Intelligence is able to predict battery degradation patterns and estimate battery life, as well as recommend actions aimed at extending battery life through machine learning algorithms [22]. AI models can predict the future state of batteries based on historical and real-time data gathered from IoT sensors, preventing failures and optimizing performance.

- **Automated Decision-Making:** In addition to monitoring, AI enables BMSs to operate autonomously. An EV, for instance, can use AI to adjust battery usage dynamically based on the driver's habits and environment. In addition to enhancing safety, automated decision-making can mitigate risks like overcharging and rapid discharging at high loads in real-time.

- **Adaptive Charging and Discharging:** A battery's life can be extended and efficiency improved with AI-driven algorithms that optimize charging and discharging processes. As an example, in renewable energy systems that use battery storage, AI can optimize energy storage during peak production times and regulate discharge based on forecasts of consumption. Batteries can be made more efficient, energy is saved, and the lifespan of the system is extended through adaptive control of these processes.
- **Edge Computing and Distributed AI:** As data is processed closer to the source, latency is reduced, and the response time is improved through edge computing. AI algorithms deployed at the edge, such as on IoT devices, allow immediate adjustments without requiring centralized servers. Remote and mission-critical applications that rely on battery health and availability can benefit greatly from distributed AI.
- **Real-Time, Comprehensive Battery Insights:** With IoT, 5G, and AI, we can collect, process, and analyze data in real-time. As IoT-enabled sensors collect data and 5G ensures rapid transmission, AI can interpret this data in real-time, optimizing battery performance through critical adjustments [23].
- **Proactive and Predictive Maintenance:** In conjunction with AI-driven predictive analytics and IoT-based continuous monitoring, BMS can proactively detect signs of battery deterioration and make recommendations before any serious problems occur. The 5G network's high-speed connectivity enables accurate prediction at all levels, from individual devices to grid-scale systems [24].
- **Prolonged Battery Life and Reduced Costs:** With AI, IoT, and 5G, battery life can be prolonged, replacement costs can be reduced, and consequently, costs can be reduced. As a result of this trifecta of technologies, battery life cycles are likely to be extended across an array of applications, reducing the environmental impact and making systems more economical [25].

- **Enhanced Security and Reliability:** By incorporating security protocols into 5G and AI, cyber threats are mitigated. Additionally, IoT systems can be integrated with AI-driven security measures that detect anomalies in real-time, benefiting sectors like healthcare, defense, and energy with enhanced battery protection [26].

These types of studies analyze what is needed, what is lacking, and what must be done to develop a secure and privacy-conscious 5G network in the near future by taking into account emerging paradigms such as the Internet of Things, fog computing, and blockchain technology. With fog computing, some tasks are moved to the edge of a network system, thus improving overall network performance [27]. In spite of the fact that blockchain could ensure adequate robustness to information transmission, it has its limitations [28]. In addition to distributed and peer-to-peer applications, blockchain technology does not require any central authority or trusted intermediary.

Nevertheless, due to their constant connection to the network, devices of this type are more likely to be tracked down and vulnerable to numerous types of attacks, such as eavesdropping, impersonation, man-in-the-middle attacks, denial-of-service attacks (DoS), replays, and repudiations [29]. In a 5G network, where a large amount of data is being transferred, maintaining high-quality service and ensuring network reliability simultaneously are a challenging and complex task. The objective is to prevent mobile devices from being breached by data breaches and spreading harmful content. It is, therefore, imperative that new security and privacy solutions be developed for the 5G network, given that 5G technology will be widely adopted in the future.

Electric vehicles, especially pure electric vehicles, have been considered one of the best traffic tools for developing green transportation systems in terms of reducing emissions [30, 31]. Battery performance directly influences the economics and dynamics of pure electric vehicles due to their role as energy storage units. The sole battery system for electric vehicles still faces some challenges despite significant improvements in battery performance in the past decades [32].

It is becoming increasingly difficult for battery management systems to re-use batteries [33]. By 2025, the number of electric vehicles will have grown by nearly 500% [13]. BMS algorithms must be characterized in the laboratory to achieve the highest level of performance [34]. As a result, they are only effective for the first time a battery is used. Due to the unique changes in electrochemical characteristics that occur when a battery is used for the first time, traditional BMS approaches based on empirical modeling will not be sufficient to manage used batteries since batteries of the same size and chemistry will have similar characteristics.

5.6 BMS's Goals and Challenges

The following section provides a brief overview of some of the challenges involved in the design of battery management systems. To find out a battery's state of charge (SOC), count how many coulombs it has [34]. A recursive method of computing SOC. There are, however, several sources of error associated with coulomb counting:

- **Initial SOC error:** Since this is a recursive integration, the bias will persist as long as the initial SOC assumption is incorrect.
- **Current measurement error:** Noise and bias are common issues with current sensors; inexpensive, simple sensors are less accurate.
- **Current integration error:** With coulomb counting methods, a rectangle is used to approximate current integration. When sampling intervals increase, errors are caused by rapid changes in load.
- **Batteries' capacities are unknown [35].** Using coulomb's method, battery capacity depends on temperature, usage patterns, and time (battery age) [36].
- **Timing oscillator error.** A timing oscillator provides a SOC clock. That is, time is measured from a timing oscillator. Using a timing oscillator with a drift or error will impact coulomb measurements.

Figure 5.1(b) illustrates how SOC can be calculated based on voltage. Figure 5.1(c) illustrates how the SOC is estimated using fusion. The parameters of the OCV, battery capacity, ECM, and voltage and current measurement sensors are required for fusion-based methods. The methods presented in References [37] both use neural networks, which include features such as SOC changes, current changes, temperature changes, and parameters used in internal circuits [37]; the voltage curve was used as a feature. Reference [37], considered several factors, including the voltage curve. It continues to be difficult for researchers to estimate SOH in real time for battery management systems.

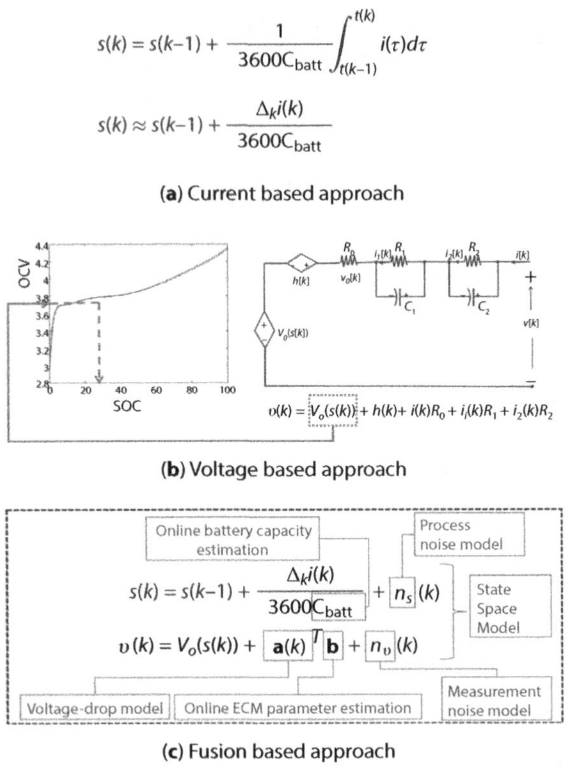

Figure 5.1 Estimating the state of charge.

5.6.1 Optimal Charging

Compared to gasoline refueling, battery charging is primitive: time-consuming, inefficient and unsafe. There has been a great deal of interest in OCAs in recent years. Continuous trickle current-based charging is a common method of charging. It takes around 10 hours to charge due to the low charging current [38]. To reduce charging time, increase the accusing present. The battery's OCV will increase as it charges, and the battery terminal will exceed its safety threshold when this occurs. Hence, the higher current applied at the beginning of the process must be reduced once the voltage reaches a certain threshold. Hence, CC-CV (constant current, constant voltage) [38], is now widely accepted that fast charging can be achieved using this strategy.

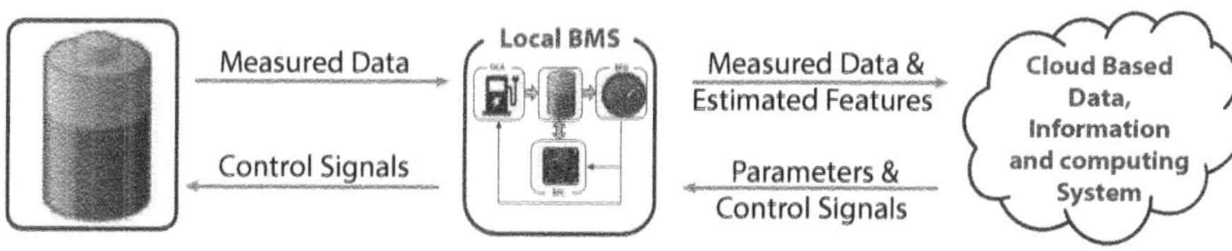

Figure 5.2 A system that assists in managing batteries *via* the cloud.

It is, therefore, necessary to reduce the high current applied at the initial stage when the terminal voltage reaches a certain level. Hence, CC-CV (constant current, constant voltage) (CC-CV) [38], has become widely used to fast-charge mobile devices.

5.6.2 Fast Characterization

In addition to characterizations of SOC and SOH, a BMS requires another offline analysis. "Data was analyzed to determine the state of charge (SOC) and open-circuit voltage (OCV). Current methods of characterizing SOH use the number of cycles, requiring hundreds or even thousands of data cycles. The characterization of batteries can be carried out while they are in use in real-time, which can reduce characterization time. Past proposals for real-time SOC characterization have been made [39].

The local characterization of SOH for a particular battery pack is difficult compared to the characterization of SOC. With today's abundance of connectivity, a real-time characterization of SOCs is possible. Figure 5.2 above illustrates how a cloud-based BMS can estimate vital parameters of battery packs in real time using data from numerous batteries.

Today, the abundance of connectivity enables the measurement of SOC and SOH characteristics in real-time. With IEEE permission, this image was taken from Reference [40].

5.7 Conclusion

As IoT, 5G, and AI combine, battery management systems are able to monitor, process, and predict more accurately in real-time. This study emphasizes how wireless communication can be advantageous in complex environments, such as electric vehicle battery packs, where weight reduction and distributed processing are crucial. Fuzzy logic is combined with neural networks in this energy management strategy to ensure safe and efficient battery management. Various industries can fully benefit from further exploration and optimization of these technologies for energy storage and consumption management.

References

1. Fleischer, C., Sauer, D.U., Barreras, J.V., Schaltz, E., Christensen, A.E., Development of software and strategies for Battery Management System testing on HIL simulator, in: *2016 Eleventh International Conference on Ecological Vehicles and Renewable Energies (EVER)*, IEEE, pp. 1–12, 2016, Accessed: Nov. 11, 2024, [Online], Available: https://ieeexplore.ieee.org/abstract/document/7476438/.
2. Popp, A., Fechtner, H., Schmuelling, B., Kremzow-Tennie, S., Scholz, T., Pautzke, F., Battery management systems topologies: Applications: Implications of different voltage levels, in: *2021 IEEE 4th International Conference on Power and Energy Applications (ICPEA)*, IEEE, pp. 43–50, 2021, Accessed: Nov. 11, 2024, [Online], Available: https://ieeexplore.ieee.org/abstract/document/9639285/.
3. Gabbar, H.A., Othman, M., Abdussami, M.R., Review of Battery Management Systems (BMS) Development and Industrial Standards. Technologies 2021, 9, 28, in: *A Note: MDPI stays neutral with regard to jurisdictional claims in published*, 2021, Accessed: Nov. 11, 2024, [Online], Available: https://www.academia.edu/download/71495733/pdf.pdf.
4. IEEE Standard for Low-Rate Wireless Networks, IEEE Std 802.15.4-2015, 1–709, 2016, doi: 10.1109/IEEESTD.2016.7460875.
5. Pister, K., Thubert, P., Dwars, S., Phinney, T., Industrial Routing Requirements in Low-Power and Lossy Networks. *RFC Editor*, RFC5673, 1–27, Oct. 2009, doi: 10.17487/rfc5673.
6. Le Gall, G., Montavont, N., Papadopoulos, G.Z., Enabling IEEE 802.15.4-2015 TSCH based Wireless Network for Electric Vehicle Battery Management, in: *2020 IEEE Symposium on Computers and Communications (ISCC)*, IEEE, Rennes, France, pp. 1–6, Jul. 2020, doi: 10.1109/ISCC50000.2020.9219551.
7. Bonfiglio, C. and Roessler, W., A cost optimized battery management system with active cell balancing for lithium ion battery stacks, in: *2009 IEEE Vehicle Power and Propulsion Conference*, IEEE, Dearborn, MI, pp. 304–309, Sep. 2009, doi: 10.1109/VPPC.2009.5289837.
8. Chatzakis, J., Kalaitzakis, K., Voulgaris, N.C., Manias, S.N., Designing a new generalized battery management system. *IEEE Trans. Ind. Electron.*, 50, 5, 990–999, Oct. 2003, doi: 10.1109/TIE.2003.817706.
9. Weicker, P., *A systems approach to lithium-ion battery management*. Artech house, USA, 2013, Accessed: Nov. 05, 2024, [Online], Available: https://books.google.com/books?hl=en&lr=&id=pXIiAgAAQBAJ&oi=fnd&pg=PA17&d

q=Weicker,+P.+(2013).+A+systems+approach+to+Lithium-Ion+battery+management+artech+house&ots=OpNgVnunGa&sig=8iATT7IqMKOiXefGbR6K4O5LJgA.

10. Tuohy, S., Glavin, M., Hughes, C., Jones, E., Trivedi, M., Kilmartin, L., Intra-vehicle networks: A review. *IEEE Trans. Intell. Transp. Syst.*, 16, 2, 534–545, 2014.
11. Li, X., Feng, J., Meng, Y., Han, Q., Wu, F., Li, J., A Unified MRC Framework for Named Entity Recognition. *Association for Computational Linguistics*, 5849–5859, Nov. 22, 2022. arXiv: arXiv:1910.11476. Accessed: Nov. 05, 2024, [Online], Available: http://arxiv.org/abs/1910.11476.
12. Wang, F., *et al.*, A luminous quasar at redshift 7.642. *Astrophys. J. Lett.*, 907, 1, L1, 2021.
13. Martinez-Laserna, E., *et al.*, Technical Viability of Battery Second Life: A Study From the Ageing Perspective. *IEEE Trans. Ind. Applicat.*, 54, 3, 2703–2713, May 2018, doi: 10.1109/TIA.2018.2801262.
14. Ma, J., *et al.*, The 2021 battery technology roadmap. *J. Phys. D: Appl. Phys.*, 54, 18, 183001, 2021.
15. Sun, P., Bisschop, R., Niu, H., Huang, X., A Review of Battery Fires in Electric Vehicles. *Fire Technol.*, 56, 4, 1361–1410, Jul. 2020, doi: 10.1007/s10694-019-00944-3.
16. Rani, P. and Sharma, R., Intelligent transportation system for internet of vehicles based vehicular networks for smart cities. *Comput. Electr. Eng.*, 108543, 2023.
17. Rani, P. and Sharma, R., IMFOCA-IOV: Intelligent Moth Flame Optimization based Clustering Algorithm for Internet of Vehicle, in: *2023 14th International Conference on Computing Communication and Networking Technologies (ICCCNT)*, IEEE, pp. 1–6, 2023, Accessed: Nov. 09, 2024, [Online], Available: https://ieeexplore.ieee.org/abstract/document/10307646/.
18. Rani, S., Talwar, R., Malhotra, J., Ahmed, S.H., Sarkar, M., Song, H., A novel scheme for an energy efficient Internet of Things based on wireless sensor networks. *Sensors*, 15, 11, 28603–28626, 2015.
19. Rani, P., Verma, S., Yadav, S.P., Rai, B.K., Naruka, M.S., Kumar, D., Simulation of the lightweight blockchain technique based on privacy and security for healthcare data for the cloud system. *Int. J. E-Health Med. Commun. (IJEHMC)*, 13, 4, 1–15, 2022.
20. Wenger, M., Waller, R., Lorentz, V.R., März, M., Herold, M., Investigation of gas sensing in large lithium-ion battery systems for early fault detection and safety improvement, in: *IECON 2014-40th Annual Conference of the IEEE Industrial Electronics Society*, IEEE, pp. 5654–5659, 2014, Accessed: Nov. 11, 2024, [Online], Available: https://ieeexplore.ieee.org/abstract/document/7049366/.
21. Kong, L., Li, C., Jiang, J., Pecht, M.G., Li-ion battery fire hazards and safety strategies. *Energies*, 11, 9, 2191, 2018.

22. Rani, P., Hussain, N., Khan, R.A.H., Sharma, Y., Shukla, P.K., Vehicular Intelligence System: Time-Based Vehicle Next Location Prediction in Software-Defined Internet of Vehicles (SDN-IOV) for the Smart Cities, in: *Intelligence of Things: AI-IoT Based Critical-Applications and Innovations*, F. Al-Turjman, A. Nayyar, A. Devi, P.K. Shukla (Eds.), pp. 35–54, Springer International Publishing, Cham, 2021, doi: 10.1007/978-3-030-82800-4_2.
23. Zhang, P., Lu, J., Wang, Y., Wang, Q., Cooperative localization in 5G networks: A survey. *ICT Express*, 3, 1, 27–32, 2017.
24. Panwar, N., Sharma, S., Singh, A.K., A survey on 5G: The next generation of mobile communication. *Phys. Commun.*, 18, 64–84, 2016.
25. Devi, R., Jha, R.K., Gupta, A., Jain, S., Kumar, P., Implementation of intrusion detection system using adaptive neuro-fuzzy inference system for 5G wireless communication network. *AEU Int. J. Electron. Commun.*, 74, 94–106, 2017.
26. Zhang, H., Liu, N., Chu, X., Long, K., Aghvami, A.-H., Leung, V.C., Network slicing based 5G and future mobile networks: Mobility, resource management, and challenges. *IEEE Commun. Mag.*, 55, 8, 138–145, 2017.
27. Ferrag, M.A., Maglaras, L., Argyriou, A., Kosmanos, D., Janicke, H., Security for 4G and 5G cellular networks: A survey of existing authentication and privacy-preserving schemes. *J. Netw. Comput. Appl.*, 101, 55–82, 2018.
28. Cao, J., Ma, M., Li, H., Fu, Y., Liu, X., EGHR: Efficient group-based handover authentication protocols for mMTC in 5G wireless networks. *J. Netw. Comput. Appl.*, 102, 1–16, 2018.
29. Li, S., Da Xu, L., Zhao, S., 5G Internet of Things: A survey. *J. Ind. Inf. Integr.*, 10, 1–9, 2018.
30. Manzetti, S. and Mariasiu, F., Electric vehicle battery technologies: From present state to future systems. *Renew. Sustain. Energy Rev.*, 51, 1004–1012, 2015.
31. Hannan, M.A., Azidin, F.A., Mohamed, A., Hybrid electric vehicles and their challenges: A review. *Renew. Sustain. Energy Rev.*, 29, 135–150, 2014.
32. Hannan, M.A., Hoque, M.M., Mohamed, A., Ayob, A., Review of energy storage systems for electric vehicle applications: Issues and challenges. *Renew. Sustain. Energy Rev.*, 69, 771–789, 2017.
33. Ahmadi, L., Fowler, M., Young, S.B., Fraser, R.A., Gaffney, B., Walker, S.B., Energy efficiency of Li-ion battery packs re-used in stationary power applications. *Sustain. Energy Technol. Assessments*, 8, 9–17, Dec. 2014, doi: 10.1016/j.seta.2014.06.006.
34. Plett, G.L., *Battery management systems, Volume I: Battery modelling*, Artech House, USA, 2015, Accessed: Nov. 11, 2024, [Online], Available: https://books.google.com/books?hl=en&lr=&id=suLRCgAAQBAJ&oi=fnd&pg=PR5&dq=Plett,+G.L.+Battery+Management+Systems,+Volume+I:+Battery+Modeling%3B+Artech+House:+Norwood,+MA,+USA,+2015&ots=aCwQGX0wlo&sig=hJdqYG3W9cAG5XZcrfihH-elLp4.

35. Balasingam, B., Avvari, G.V., Pattipati, B., Pattipati, K.R., Bar-Shalom, Y., A robust approach to battery fuel gauging, part II: Real time capacity estimation. *J. Power Sources*, 269, 949–961, 2014.
36. Avvari, G.V., Pattipati, B., Balasingam, B., Pattipati, K.R., Bar-Shalom, Y., Experimental set-up and procedures to test and validate battery fuel gauge algorithms. *Appl. Energy*, 160, 404–418, 2015.
37. Eddahech, A., Briat, O., Bertrand, N., Delétage, J.-Y., Vinassa, J.-M., Behavior and state-of-health monitoring of Li-ion batteries using impedance spectroscopy and recurrent neural networks. *Int. J. Electr. Power Energy Syst.*, 42, 1, 487–494, 2012.
38. Cope, R.C. and Podrazhansky, Y., The art of battery charging, in: *Fourteenth Annual Battery Conference on Applications and Advances. Proceedings of the Conference (Cat. No. 99TH8371)*, IEEE, pp. 233–235, 1999, Accessed: Nov. 11, 2024, [Online], Available: https://ieeexplore.ieee.org/abstract/document/795996/.
39. Waag, W. and Sauer, D.U., Adaptive estimation of the electromotive force of the lithium-ion battery after current interruption for an accurate state-of-charge and capacity determination. *Appl. Energy*, 111, 416–427, 2013.
40. Balasingam, B. and Pattipati, K., Elements of a robust battery-management system: From fast characterization to universality and more. *IEEE Electrif. Mag.*, 6, 3, 34–37, 2018.

6

Using Local Knowledge and Sustainable Transport for Greener Mobility

Jayant Jagtap[1], Amrita Singh[2], Sandeep Singh[3*], Shivani Pant[4] and Haider Mohammed Abbas[5]

[1]Department of Computing Science and Artificial Intelligence, NIMS Institute of Engineering and Technology, NIMS University Rajasthan, Jaipur, India
[2]Department of Computer Science Engineering, Chandigarh Engineering College, Chandigarh Group of Colleges, Jhanjeri, Mohali, Punjab, India
[3]Department of Civil Engineering, School of Engineering and Technology, Bahra University, Waknaghat, Distt. Solan, HP, India
[4]Uttaranchal Institute of Technology, Uttaranchal University, Uttarakhand, India
[5]Department of Computer Science, The Islamic University, Najaf, Iraq

Abstract

Adapting to the rapid urbanization that is harming health, the environment, and mobility, cities need to implement sustainable transportation systems and green mobility frameworks to be resilient and sustainable. According to this chapter, natural-based solutions (NBS) can improve urban connectivity and promote healthy, eco-friendly lifestyles. This study examines sustainable urban mobility solutions, including multimodal transportation, electric vehicles (EVs), and smart city concepts, with a focus on the UN's Sustainable Development Goals, such as clean energy, health, and responsible consumption. In this report, it emphasizes how COVID-19 influenced transport choices by shifting the emphasis from public transport to personal safety. In this chapter, the Greening Mobility Framework (GMF) is proposed, offering guidelines for supporting green mobility across a range of levels. In addition to technological advancements, the GMF promotes user engagement and regulations that reduce emissions, make sustainable practices more accessible, and encourage public acceptance of green technologies.

**Corresponding author*: drsandeep1786@gmail.com

Pethuru Raj, Sudesh Yadav, Manas Kumar Mishra, Satya Prakash Yadav and Victor Hugo C. de Albuquerque (eds.) Artificial Intelligence Technologies for Smart and Sustainable Urban Transportation: Integrated Platforms and Use Cases, (89–104)

***Keywords*:** Electric vehicles, green mobility, smart cities, sustainable development goals, sustainable transportation, urban connectivity, urban mobility

6.1 Introduction

Many cities have suffered significant environmental changes as a result of urbanization, which has been further exacerbated by climate change [1, 2]. Human health has also deteriorated as a result of urbanization processes [3], and sedentary habits have become more prevalent [4]. Automobile dependency [5], desk-bound jobs [6] and reduced physical activity contribute to sedentary lifestyles that are detrimental to citizens' health [7, 8]. As cities become more urbanized, they need to become more resilient [9–11] by integrating nature-based solutions (NBS) into urban green infrastructure projects [12, 13]. Since NBS is relevant to environmental factors, it can contribute to sustainability targets [14, 15]. As a result of NBS, ecosystem services are generated that have a positive impact on human health and resilience [12, 16]. NBS, however, can only be beneficial if green spaces are strategically located and connected within cities [15]. NBS connectivity and access enable residents to move safely and sustainably around cities [17]. Urban green spaces are more capable of providing ecosystem services when they are more connected [18, 19].

As cities and surrounding urban areas develop intelligently and sustainably, the transportation system is an important issue affecting smart and sustainable mobility, which is a topical and vital issue today. As the United Nations estimates, three million people migrate from rural areas to cities every week, doubling urban populations in the near future. Over the next few decades, urban areas will be unable to accommodate rapidly growing populations due to outsized homes, inadequate energy and traffic infrastructure, and deteriorating air quality [20]. Sustainability and smart urban transportation are integral to the attainment of Sustainable Development Goals (SDGs) [21, 22]. As part of Agenda 2030, several Sustainable Development Goals have been outlined to address global sustainability challenges such as poverty, income inequality, and environmental degradation. One hundred ninety-five countries reaffirmed in 2015 their commitment to attain the Sustainable Development Goals, both financially and non-financially [23]. Each goal's progress towards the SDGs can be monitored and evaluated using specific indicators. In the United Nations' view, a good transport system is largely responsible for improving society's quality of life by facilitating access to education, health care, social services, employment, and business opportunities. In addition to reducing pollution, climate change can also be reduced if transport systems and vehicles are designed efficiently and environmentally friendly [23]. However, other goals are closely related to sustainable transportation and mobility even though none of them explicitly mention it, such as Sustainable cities (Goal 11), Responsible consumption (Goal 12), Industry, innovation, and infrastructure (Goal 9), and Decent Work and Economic Growth (Goal 8). A healthy and prosperous society, clean water, and reliable energy are all driven by sustainable transport. As a result, sustainable transportation is incorporated into several of the 2030 Agenda's targets, including food security, health, energy, economic growth, infrastructure, and cities. As well as understanding the regulatory framework set by countries and international organizations within the UN context, it is important to know what support instruments have been proposed for developing smart and sustainable transportation, such as car sharing, electric cars, light vehicles, and other environmentally friendly options. There is no doubt that market development, consumer needs, and new technologies in various industries influence the development of these issues. Almost all action plans developed by governments and international organizations focus on improving public transportation as a potential method to reduce emissions and create other environmental benefits.

In the aftermath of COVID-19, transportation users' behaviours and preferences have changed. A marked change in the 2020 International Consumer Trends Chart can be seen after incorporating the COVID-19 pandemic [24]. A durable and adaptable consumer trend will dominate 2020. Consumers need to be safe, secure, and plan their time effectively to reduce anxiety and worry. Since the COVID-19 pandemic struck, there has been a shift in priorities towards personal safety and hygiene, as well as the desire for comfort, which has led to a shift in consumption preferences from constantly available to experiencing [25, 26]. Prior research concluded that safety and health concerns drove people to distance themselves from public transportation and use individual vehicles instead [27]. According to experts, consumers are taking chances, putting their plans into motion, and seizing opportunities now as opposed to 2020" [25]. Consumers have reconsidered daily conveniences due to the pandemic, preferring private transportation instead of public transportation, which has also contributed to climate change. The development of mobility creates new opportunities to improve the well-being of citizens in areas such as traffic congestion, parking, and air pollution [28]. Sustainability in urban areas refers to emissions levels that are low or zero, smart technologies, and behaviour changes among citizens [28]. Smart technologies are improving the efficiency, monitoring, and safety of urban planning and development. Technology can also improve accessibility, ensure user-centred and efficient mobility, and improve the use of transportation through new platforms and solutions. It is becoming easier and easier to use mobility in urban areas as it becomes autonomous, electric, and connected. Consumer needs are also being taken into account, and air quality is also being protected [29]. The number of cars per person keeps increasing, as does the amount of traffic in cities [30, 31]. The above factors challenge urban transportation systems, which prompts efforts to improve efficiency, intelligence, and sustainability [32]. Taking cities to climate neutrality and implementing the European Green Deal requires urban mobility. Technical solutions must be aligned with the end user and city needs to reach climate neutrality. An innovative mobility and transportation concept must take urban transportation into account and be embraced by all stakeholders: Technology providers, municipalities, and end users must collaborate. A key aspect of this concept development is to optimize performance, ease of use, and maximum throughput. Smart cities pay special attention to this issue because it includes technology accessibility (particularly ICT) as a key element of the concept, as well as the ICT capability of cities.

By developing new technologies, enabling citizens to use and access them, investing in people and training, and putting intangibles into place, smart cities are created [33, 34]. Smart cities are defined as ones that use ICT, technology, and innovation to improve residents' quality of life and help the city grow. Smart mobility is also included in the definition, along with the promotion of economic growth, environmental sustainability, and effective urban planning [35]. In his description of smart cities, the author focuses more on human-centricity than ethnocentrism [36]. Poulos believes that a smart city should have the following attributes: resources, transportation, urban infrastructure, living, government, economy, and cohesion. The creation of exceptional infrastructure and the efficient use of urban mobility systems are essential to achieving sustainable development [37].

6.2 Related Work

In addition to affecting households, multimodality also affects individuals. The assessment of impacts and factors should take into account not only the population but also the attributes of the larger urban environment. The majority of this is based on the size, distances, and urban structure of the city in objective terms [38]. It is possible, however, that multimodal public transportation could lead to people changing modes of transportation, resulting in them being unable to perform their usual activities, leading to even greater use of automobiles [39]. Furthermore, Nobis states that public transportation also plays an important role in determining a municipality's multimodal behavior [40]. Public transportation plays a major role in minimizing the environmental impacts of cars. Often, public transportation is underserved in cities and suburbs because of low population density, which makes it economically unviable [41]. Integration and full utilization of multimodal public transportation are essential for solving transport problems, particularly urban mobility. As a result, it is more efficient, lower-cost, greener, and contributes to the well-being of cities. An urban area dominated by road transport cannot function effectively when it comes to personal transportation. There is little efficiency in this use of transport space, and road transport requires a great deal of space. A sustainable mobility system will consist of multiple transportation technologies with different characteristics regarding usability, infrastructure requirements, and environmental impact [42].

Sustainability-oriented urban mobility plans provide a balance between all important modes of transportation in an attempt to achieve sustainability. With European Mobility Week (EMW), travellers can change their traditional travel patterns. In this campaign, the European Commission raises awareness about sustainable urban mobility. European Municipal Week promotes sustainable transport among residents by increasing, promoting, and implementing it. In this campaign, residents will be introduced to sustainable mobility alternatives, public multimodal transport will be introduced, and the problems that cities face will be clarified. Every year, in addition to raising public awareness, Nitra contributes to improving sustainable mobility through the EMW. Environment-friendly behaviors lead to greater satisfaction in life. Neither individual carbon footprints nor subjective well-being has been shown to be strongly correlated with green behavior [43], for example, work-life balance [44].

Individuals' pro-environmental behaviors are strongly correlated with their mode of transportation. Individuals can alleviate mobility's environmental burden by choosing modes of transport that are less polluting than driving cars. Active modes of mobility, such as public transport, are the most eco-friendly way to travel since they involve zero carbon emissions [45].

6.3 Greening Mobility Necessities

To achieve more sustainable mobility, it is essential to follow a holistic approach to electric vehicle promotion and installation. To meet future global energy challenges, we developed a "Greening Mobility Framework" that incorporates these factors. Among the most important aspects of this are the people involved in making decisions, the potential for greener mobility (taking into account different dimensions) and how the proposed measures/actions will affect the transportation system. As a result of the study [46], and the variables used to rank cities based on their readiness to transition to electric vehicles, we identified Greening Mobility Necessities (GMN). A GNM reflects its environment and supports its development, as shown in Table 6.1. Using the future GMF, specific needs and challenges can be adequately addressed, and guidance can be provided at the EU, national, and local levels. Providing favorable taxes, electricity, and infrastructure to support clean vehicles on a national level can also encourage their use. To implement green mobility on a local level, it is important to ensure that parking spaces are reserved and/or free, enforce emission zones, prioritize clean vehicle lanes, and provide financial support for charging infrastructure. Parking spaces should be reserved and/or free, emission zones should be enforced, lanes for clean vehicle movements should be prioritized, and financial support should be provided to charge infrastructure.

6.3.1 Green Transport Standards

The Global Mobility Report focuses on sustainable challenges and improvements for global transport systems, which addresses issues such as accessibility, efficiency, safety, and environmental challenges. Among the key recommendations are a number of ones relating to sustainable mobility, the integration of intelligent traffic management technologies, traffic safety standards, the promotion of green transportation, and increased investments in sustainable transportation infrastructure. Sustainable development goals include achieving sustainability, mitigating climate change, safeguarding health and well-being, developing infrastructure, industrializing, and pursuing innovation [47].

Sustainable transport standards are closely tied to the Sustainable Development Goals since they are accessible, efficient, safe, and environmentally friendly. As a result of implementing these standards, conveyance systems will be improved worldwide, congestion in urban areas will be reduced, carbon emissions will be reduced, and traffic safety will

Table 6.1 An overview of GMF.

Situation	GMN	Samples	Gains/motives	Recommendations
Fuel Cost Environment	Gas Price	Increase Gas Price	Gain attractiveness for EVs	Create an environment that is conducive to lower energy costs
	Electricity Price	Reduce Electricity Price	Gain attractiveness for EVs	
	Electricity costs for PEVs that are adaptable depending on when they are used	The price of electricity for recharging PEVs overnight is low	Gain Money	
Purchase Incentives	Purchase EV Incentives	Businesses or governments can receive a normal or urban tax credit or rebate if they purchase a PEV	The share of the car sooq occupied by PEVs is increasing, making them more competitive.	Incentives for electric vehicles and electric vehicle charging stations
	Purchase EVSE Incentives	A tax credit or rebate for businesses or governments purchasing EVSEs	Increase PEVs attractiveness and usage	
Fees	Reduced Fees	Fees for registration, vehicle taxes, and sales taxes	Reduce the cost of driving an electric vehicle	In order to make EVs more appealing to the general public, fee-based incentives should be provided
Fleet Programs	National/Urban Flect Programs	Introduce PEV car sharing services, integrate PEVs into flects	By demonstrating the benefits of PEVs, citizens will be more likely to use EVs	Programs focusing on flects are introduced

(*Continued*)

Table 6.1 An overview of GMF. (*Continued*)

Situation	GMN	Samples	Gains/motives	Recommendations
Infrastructure	Charging stations per capita	In what proportion does each city have charging stations?	Reducing range anxiety	A sufficient number of charging stations should be provided per capita in a proper location
Parking	Parking Privileges	Owners of PEVs can park for less	Time and money saved by reducing parking costs	Allow EVs to park on the street
Traffic Management	HOV Lane access	Owners of PEVs are entitled to use the HOV lanes	Reduced travel time	Provide EVs with high-occupancy vehicle lanes
	Emission Zones	Almost all polluted vehicles lack an accessory or are modified in some way	Low-emission zones should be created EVs will become more attractive	Establish low-emission zones and strict emission standards.
Human	Ecodriving	Public education/ awareness campaigns related to Ecodriving	Increasing awareness, increasing revenue, and reducing emissions	Educate/make the general public about eco driving and promote campaigns to encourage it
	Feedback for Surveys and Observations	Improve the level of knowledge, awareness, and feedback of EV drivers and potential drivers	Provide services to the general public by collaborating with them	Get feedback on drivers' mobility behaviour and needs and assess their knowledge and understanding of EVs by observing and interacting with them

be improved. At the same time, related industries and technological innovation will also be stimulated. To achieve the Sustainable Development Goals, the Global Mobility Report provides practical guidelines and action options through its sustainable transport standards framework.

6.4 Principles of the Sustainable Mobility Paradigm

Sustainability in mobility requires consistent policies that incorporate these elements. This article suggests four key elements:

a. Technology must be utilized effectively, which means investing in transportation modes, information systems, and the transportation system itself, as well as directing industry priorities (e.g., alternative fuels and hybrid vehicles). Individuals are generally not opposed to these options since they allow them to maintain their daily routines without making major adjustments. Additionally, it is imperative to change driver behavior by adhering to speed limits, observing ecological principles, and increasing occupancy. Vehicle performance would improve, enabling efficiency to be achieved; however, people might drive more as a result, which would lead to rebound effects.
b. A fuel price increase or a fee charged to road users is necessary to reflect the external cost of transportation in the actual cost of travel. Congestion can be reduced and environmental quality improved through demand management, but it requires public support to be successful. As a result, the number of trips would be reduced, trip distances would be reduced, modal shares would be changed, and better vehicles would be promoted.
c. A comprehensive land-use plan, including regulations and physical restraint measures, needs to be implemented to support shorter travel distances. Distances travelled could be reduced, and modal splits could be altered through improved proximity.
d. It is also vital that personal information be consistently used to raise awareness, make demonstrations, persuade, and market individually to individuals. Despite its importance, acceptance is often overlooked when it comes to sustainable mobility.

As sustainable mobility moves towards an objective-based planning system, a range of policy interventions are being implemented, along with an important other element: stakeholder support. A key part of this discussion is understanding behavior and exploring ways to achieve cooperation and support to achieve actual change. Social welfare must be considered in conjunction with personal utility.

Public acceptance is influenced by a variety of factors, as summarized in Table 6.2. Due to the fact that many cities are already actively implementing sustainable mobility policies, the first four elements are widely used, such as providing information, making sure that effective communication and involvement occur, packaging policy measures, and selling benefits. As a result of its direct benefits, this message should be an effective tool for influencing public opinion and politics. It is essential to gain a better understanding of acceptability to develop an effective, sustainable mobility strategy. There is a need to go beyond automatic responses and continuous evaluations based on individual benefits and embrace reasoned action that examines the entire process from a societal standpoint, taking into consideration all options and implications. There is no doubt that a change in behavior is an important component of the solution (Table 6.2), but the final three elements are the most controversial. In implementing policy packages on sustainable mobility, we use a phased approach to be realistic and consistent, moving flexibly towards more long-term goals.

Table 6.2 Key elements for public acceptance of sustainable mobility.

Detailed information	The key is to educate the public, promote awareness, and use media and social pressure to promote them. Bringing sustainable mobility to the attention of individuals and businesses, emphasizing its positive economic, social, and health benefits
Communicating and being involved	Throughout the process, there must be clear aims and a clear understanding of how the strategy will affect those affected by it. The goal is to get stakeholders' support and understanding of the proposals as well as their buy-in. Make sure expectations and results are consistent.

(*Continued*)

Table 6.2 Key elements for public acceptance of sustainable mobility. (*Continued*)

Wrapping	A combination of push and pull policy measures is necessary for them to work together. In addition to reducing car use or raising its cost, well-publicized initiatives should also promote the availability and attractiveness of alternatives to driving alone, such as carpooling, public transportation, cycling, and walking, all of which are funded by car pricing measures.
Benefit-selling	Publicizing benefits, despite cost, inconvenience, and sacrifice, is essential. Alternative modes of transportation support funding to relieve road congestion. Drivers support the development of alternative modes of transportation to alleviate congestion on the roads. Everyone benefits from cleaner air and safer traffic conditions. To relieve parking shortages, more people should walk, cycle, and use public transportation. The impact of all of these is significant and direct, and all individuals can play a role in promoting them.
Stage-by-stage adoption of controversial policies	A positive outcome and measurable improvement in quality of life are needed to build support. The purpose of politics is to reflect prevailing preferences and form opinions at the same time. Success depends on accepting responsibility and committing to change through actions.
Measures and policies should be consistent.	The pricing of futures is a measure that is common to all futures. The implementation of such measures is urgent, even if they will not have an immediate impact. Manufacturers and other transport suppliers should be encouraged to develop regulations, standards, subsidies, and tax incentives to achieve the best energy efficiency and environmentally friendly technology. Transport emissions should be treated with caution, especially with respect to global warming effects, and longer-term actions should be consistent with this principle. There are many problems in the transportation system that are not caused by the transport industry but by sectors outside of it. To accomplish this, a more holistic perspective must be adopted, one that integrates decision-making across sectors and expands public participation.

(*Continued*)

Table 6.2 Key elements for public acceptance of sustainable mobility. (*Continued*)

Flexibility	"Today's decisions should be made in a way that does not unnecessarily restrict the scope for future decisions to assess individual and agency adaptive behaviors. It is impossible to prescribe or outline the correct procedures. Depending on the circumstances, it is necessary to analyze and implement policies separately and to be flexible when policies don't meet expectations. Sustainable mobility must assess risk and be reversible. It is not acceptable to be adaptable if you do nothing or act weakly. A clear decision-making process, leadership, analysis, and monitoring are necessary to ensure that policy actions are effective.

6.5 Conclusion

Mobility in cities can significantly reduce environmental impacts and improve public health, making it an important component of sustainable city development. Throughout the chapter, NBS implementation, multi-modal transit, and the adoption of EVs as foundational elements of green cities are highlighted as interconnected goals of mobility, urban planning, and sustainability. It incorporates local needs, adapts policies to support green transport options, and supports technological innovation as a way of promoting sustainable mobility. For sustainable urban mobility to flourish, interdisciplinary collaboration, proactive public engagement, and gradual adoption of policies are essential. Through these combined efforts, cities will become climate-neutral, will have a better quality of life, and will be more resilient to natural disasters.

References

1. Kelly, F.J. and Zhu, T., Transport solutions for cleaner air. *Science*, 352, 6288, 934–936, May 2016, doi: 10.1126/science.aaf3420.
2. Kammen, D.M. and Sunter, D.A., City-integrated renewable energy for urban sustainability. *Science*, 352, 6288, 922–928, May 2016, doi: 10.1126/science.aad9302.
3. Zhou, M., *et al.*, The associations between ambient air pollution and adult respiratory mortality in 32 major Chinese cities, 2006–2010. *Environ. Res.*, 137, 278–286, 2015.

4. Normile, D., China Rethinks Cities. *Science*, 352, 6288, 916–918, May 2016, doi: 10.1126/science.352.6288.916.
5. Geller, A.L., Smart Growth: A Prescription for Livable Cities. *Am. J. Public Health*, 93, 9, 1410–1415, Sep. 2003, doi: 10.2105/AJPH.93.9.1410.
6. Popkin, B.M., Urbanization, Lifestyle Changes and the Nutrition Transition. *World Dev.*, 27, 11, 1905–1916, Nov. 1999, doi: 10.1016/S0305-750X(99)00094-7.
7. Hodson, R., Urban health and well-being. *Nature*, 531, 7594, S49–S49, 2016.
8. Deweerdt, S., Mobility: The urban downshift. *Nature*, 531, 7594, S52–S53, 2016.
9. U. N. Habitat, Habitat III: new urban agenda. *Quito Online Httphabitat3 Orgwp-Content Uploads NUA-Engl. Pdf UN Habitat*, 52, 1, 86–91, 2016.
10. Moran, E.F. and Lopez, M.C., Future directions in human-environment research. *Environ. Res.*, 144, 1–7, 2016.
11. Rani, P., Singh, P.N., Verma, S., Ali, N., Shukla, P.K., Alhassan, M., An implementation of modified blowfish technique with honey bee behavior optimization for load balancing in cloud system environment. *Wirel. Commun. Mob. Comput.*, 2022, 1–14, 2022.
12. Lafortezza, R. and Chen, J., The provision of ecosystem services in response to global change: Evidences and applications. *Environ. Res.*, 147, 576–579, May 2016, doi: 10.1016/j.envres.2016.02.018.
13. Kabisch, N., *et al.*, Nature-based solutions to climate change mitigation and adaptation in urban areas: perspectives on indicators, knowledge gaps, barriers, and opportunities for action. *Ecol. Soc.*, 21, 2, 15, 2016, Accessed: Nov. 06, 2024, [Online], Available: https://www.jstor.org/stable/26270403.
14. Badiu, D.L., *et al.*, Is urban green space per capita a valuable target to achieve cities' sustainability goals? Romania as a case study. *Ecol. Indic.*, 70, 53–66, 2016.
15. Ngom, R., Gosselin, P., Blais, C., Reduction of disparities in access to green spaces: Their geographic insertion and recreational functions matter. *Appl. Geogr.*, 66, 35–51, Jan. 2016, doi: 10.1016/j.apgeog.2015.11.008.
16. Gómez-Baggethun, E. and Barton, D.N., Classifying and valuing ecosystem services for urban planning. *Ecol. Econ.*, 86, 235–245, Feb. 2013, doi: 10.1016/j.ecolecon.2012.08.019.
17. Aultman-Hall, L., Hall, F.L., Baetz, B.B., Analysis of Bicycle Commuter Routes Using Geographic Information Systems: Implications for Bicycle Planning. *Transp. Res. Rec. J. Transp. Res. Board*, 1578, 1, 102–110, Jan. 1997, doi: 10.3141/1578-13.
18. Hansen, R. and Pauleit, S., From Multifunctionality to Multiple Ecosystem Services? A Conceptual Framework for Multifunctionality in Green Infrastructure Planning for Urban Areas. *AMBIO*, 43, 4, 516–529, May 2014, doi: 10.1007/s13280-014-0510-2.

19. Rani, P., Hussain, N., Khan, R.A.H., Sharma, Y., Shukla, P.K., Vehicular intelligence system: time-based vehicle next location prediction in software-defined internet of vehicles (SDN-IOV) for the smart cities. *Intell. Things AI-IoT Based Crit.-Appl. Innov.*, 35–54, 2021.
20. U. Nations, New urban agenda, in: *Habitat III–The United Nations Conference on Housing and Sustainable Urban Development*, 2017.
21. Halkos, G. and Gkampoura, E.-C., Where do we stand on the 17 Sustainable Development Goals? An overview on progress. *Econ. Anal. Policy*, 70, 94–122, Jun. 2021, doi: 10.1016/j.eap.2021.02.001.
22. Rani, P. and Sharma, R., Intelligent Transportation System Performance Analysis of Indoor and Outdoor Internet of Vehicle (IoV) Applications Towards 5G. *Tsinghua Sci. Technol.*, 29, 6, 1785–1795, Dec. 2024, doi: 10.26599/TST.2023.9010119.
23. U. Nations, *The Sustainable Development Goals Report 2019*, 2019.
24. Verma, S., *et al.*, An automated face mask detection system using transfer learning based neural network to preventing viral infection. *Expert Syst.*, 41, 3, e13507, 2024.
25. Angus, A. and Westbrook, G., Top 10 global consumer trends. *Euromonitor Int. Lond. UK*, 2019, Accessed: Nov. 06, 2024, [Online], Available: https://is.muni.cz/do/econ/sm/management_rozvoje/media/literatura/TOP_10_Global_Consumer_megatrends_2020_Euromonitor.pdf.
26. Yadav, S.P., Jindal, M., Rani, P., de Albuquerque, V.H.C., dos Santos Nascimento, C., Kumar, M., An improved deep learning-based optimal object detection system from images. *Multimed. Tools Appl.*, 83, 10, 30045–30072, 2024.
27. Mikelsone, E., Atstaja, D., Koval, V., Uvarova, I., Mavlutova, I., Kuzmina, J., Exploring Sustainable Urban Transformation Concepts for Economic Development. *Stud. Appl. Econ.*, 39, 5, May 2021, doi: 10.25115/eea.v39i5.5209.
28. van Well, L., Ballerino, C.C., Johansson, M., Strategies and priorities for cores and peripheries in the Baltic Sea Region: Territorial cohesion and polycentricity. *Eur. XXI*, 14, 191, 2006.
29. Pfeiffer, A., Burgholzer, A., Kanag, D., Coupling of the mobility and energy infrastructures as urban mobility needs evolve, in: *Solving Urban Infrastructure Problems Using Smart City Technologies*, pp. 107–128, Elsevier, Amsterdam, 2021, Accessed: Nov. 06, 2024, [Online], Available: https://www.sciencedirect.com/science/article/pii/B978012816816500005X.
30. Akhmetshin, E.M., Sekerin, V.D., Pavlyuk, A.V., Shichiyakh, R.A., Allanina, L.M., The influence of the car sharing market on the development of ground transport in metropolitan cities. *Theor. Empir. Res. Urban Manag.*, 14, 2, 5–19, 2019.
31. Cohen, K., Human behavior and new mobility trends in the United States, Europe, and China. *Fondazione Eni Enrico Mattei (FEEM)*, 2019, Accessed:

Nov. 06, 2024, [Online], Available: https://www.jstor.org/stable/pdf/resrep21774.pdf.

32. Singh, A., *et al.*, Smart Traffic Monitoring Through Real-Time Moving Vehicle Detection Using Deep Learning *via* Aerial Images for Consumer Application. *IEEE Trans. Consum. Electron.*, 70, 1–1, 2024, doi: 10.1109/TCE.2024.3445728.
33. Alfaro Navarro, J.L., López Ruiz, V.R., Nevado Peña, D., The effect of ICT use and capability on knowledge-based cities. *Cities*, 60, 272–280, Feb. 2017, doi: 10.1016/j.cities.2016.09.010.
34. Tranos, E., Reggiani, A., Nijkamp, P., Accessibility of cities in the digital economy. *Cities*, 30, 59–67, Feb. 2013, doi: 10.1016/j.cities.2012.03.001.
35. Ahmed, D.M. and Anguluri, R., Conceptual understanding of smart cities. *Int. J. Sci. Res.*, 3, 12, 1470–1471, 2014.
36. Dashkevych, O. and Portnov, B.A., Criteria for Smart City Identification: A Systematic Literature Review. *Sustainability*, 14, 8, 4448, Apr. 2022, doi: 10.3390/su14084448.
37. Rani, P., Verma, S., Yadav, S.P., Rai, B.K., Naruka, M.S., Kumar, D., Simulation of the lightweight blockchain technique based on privacy and security for healthcare data for the cloud system. *Int. J. E-Health Med. Commun. IJEHMC*, 13, 4, 1–15, 2022.
38. Klinger, T., Moving from monomodality to multimodality? Changes in mode choice of new residents. *Transp. Res. Part Policy Pract.*, 104, 221–237, Oct. 2017, doi: 10.1016/j.tra.2017.01.008.
39. Krygsman, S., Activity and travel choice(s) in multimodal public transport systems. *Dspace.*, 2004, Accessed: Nov. 06, 2024, [Online], Available: https://dspace.library.uu.nl/bitstream/handle/1874/1332/?sequence=7.
40. Nobis, C., Multimodality: Facets and Causes of Sustainable Mobility Behavior. *Transp. Res. Rec. J. Transp. Res. Board*, 2010, 1, 35–44, Jan. 2007, doi: 10.3141/2010-05.
41. McKenzie, B.S., *Public transportation usage among us workers: 2008 and 2009*, US Department of Commerce, Economics and Statistics Administration, US, 2010.
42. Brunner, H., Hirz, M., Hirschberg, W., Fallast, K., Evaluation of various means of transport for urban areas. *Energy Sustain. Soc.*, 8, 1, 9, Dec. 2018, doi: 10.1186/s13705-018-0149-0.
43. Melo, P.C., Ge, J., Craig, T., Brewer, M.J., Thronicker, I., Does work-life balance affect pro-environmental behaviour? Evidence for the UK using longitudinal microdata. *Ecol. Econ.*, 145, 170–181, 2018.
44. Andersson, D., Nässén, J., Larsson, J., Holmberg, J., Greenhouse gas emissions and subjective well-being: An analysis of Swedish households. *Ecol. Econ.*, 102, 75–82, 2014.
45. Chapman, L., Transport and climate change: a review. *J. Transp. Geogr.*, 15, 5, 354–367, 2007.

46. Clark-Sutton, K., Siddiki, S., Carley, S., Wanner, C., Rupp, J., Graham, J.D., Plug-in electric vehicle readiness: Rating cities in the United States. *Electr. J.*, 29, 1, 30–40, Jan. 2016, doi: 10.1016/j.tej.2015.12.006.
47. Xu, F., Xu, W., Zhang, X., Wang, Y., Wang, F., Analysis of Traffic Characteristics and Distance Optimization Design between Entrances and Exits of Urban Construction Projects and Adjacent Planar Intersections. *Sustainability*, 15, 11, 8656, 2023.

Part 2

GREEN REVOLUTION IN IOV

7

Expounding the Importance of Explainable AI for Greener Transportations

Abhilasha Jadhav[1], Amrita Singh[2*], Adil Abbas Alwan[3], Ruby Pant[4] and Haider Alabdeli[5]

[1]Department of Computing Science and Artificial Intelligence, NIMS Institute of Engineering and Technology, NIMS University Rajasthan, Jaipur, India
[2]Department of Computer Science Engineering, Chandigarh Engineering College, Chandigarh Group of Colleges, Jhanjeri, Mohali, Punjab, India
[3]College of Technical Engineering, National University of Science and Technology, Dhi Qar, Iraq
[4]Department of Mechanical Engineering, Uttaranchal Institute of Technology, Uttaranchal University, Dehradun, Uttarakhand, India
[5]Department of Computer Science, The Islamic University, Najaf, Iraq

Abstract
Intelligent machines are transforming industries, including sustainable transportation, thanks to Artificial Intelligence (AI). Greener transport solutions can be promoted through improved traffic management, reduced emissions, and improved efficiency with AI systems. Due to the AI algorithm's inherent complexity, stakeholders and end users often remain in the dark about the decision-making process. XAI addresses this challenge by making AI systems easier to understand and more trustworthy by improving transparency and interpretability. In this chapter, the benefits of transparency for safety, regulatory compliance, and stakeholder trust are discussed in relation to the critical role XAI plays in promoting sustainable transportation. The chapter explores the implications of XAI for greener and smarter transportation through case studies.

Keywords: Artificial intelligence, greener transport, improved efficiency, sustainable transportation, traffic management, explainable AI (XAI)

**Corresponding author*: Amrita2638.research@cgcjhanjeri.in

Pethuru Raj, Sudesh Yadav, Manas Kumar Mishra, Satya Prakash Yadav and Victor Hugo C. de Albuquerque (eds.) Artificial Intelligence Technologies for Smart and Sustainable Urban Transportation: Integrated Platforms and Use Cases, (107–126) © 2026 Scrivener Publishing LLC

7.1 Introduction

Transport is no exception to AI's power to accelerate innovation across industries. With AI-driven models, environmental impacts can be reduced, and sustainable mobility can be advanced from traffic flow optimization to fleet emissions management. In spite of their complexity, explainability remains one of the biggest challenges. Artificial intelligence decisions should be understandable and trusted by users, regulators, and stakeholders. Explainable AI aims to do this. As a result, explainable AI is critical to ensuring trust and safety, as well as maximizing AI's environmental benefits. Throughout this chapter, we examine how explainable AI can advance sustainable transportation systems. Through the XAI framework, stakeholders are better able to trust, make informed decisions, and comply with regulations. A discussion of explainable AI's role in greener transportation will be presented in this chapter, along with use cases and key challenges.

Global urbanization and a growing fleet of vehicles are posing challenges in transportation, traffic, logistics, and environmental sustainability in major cities [1]. Maintaining sustainable transportation systems relies heavily on technology. 5G networks, for example, enable intelligent transport systems (ITS) in smart cities [2, 3]. Fuel-efficient and road-capacity-efficient vehicles have been developed in recent years to meet sustainable transportation demands. Traffic control, route planning, and trip planning are all achievable with 5G-enabled single-circuit board user interfaces [4]. In this way, products and services can be delivered effectively and efficiently both within and outside the country. It may also be used to solve the transportation system's problems [5]. WAVE and dedicated short-range communication (DSRC) can be used to collect and share data between vehicles. Cooperative ITS (C-ITS) refers to cooperative sensing or collective perception [6, 7].

Using cooperative V2X short-range communications, the CAR 2 CAR Communication Consortium deploys C-ITS everywhere, anytime. A cooperative V2X communication is carried out using the ETSI ITS G5 standard, derived from the IEEE 802.11p WLAN standard. WAVE/DSRC is not the only quality service provider available for V2X applications. LTE-A also offers quality of service [8, 9]. It is feasible to improve and evolve radio technology by using a successor standard, IEEE 802.11b, which leads to efficient spectrum utilization and continuous service provision [10]. A cooperative V2X system monitors and warns drivers about potential dangers based on the information they receive. The following guidelines can help avoid road safety situations as well as accidents that may result.

Our response was to adapt and summarize their requirements to meet the basic road safety requirement and the advanced V2X service requirement [11, 12].

ICVs (awareness, sensing, and cooperative driving) will require the introduction of 6G technology, as described in [13]. Several studies have suggested that existing 5G networks cannot provide the best connectivity for fully autonomous connected vehicles. Further, manual cars can benefit from V2X cooperation along with assistance and automation all the way up to fully autonomous, cooperative driving. Furthermore, services such as real-time traffic light predictions and barriers along the route play a crucial role in ensuring smooth and comfortable travel. It is also possible to reduce noise and emissions by avoiding fast accelerations and decelerations. Vehicles and infrastructure equipped with sophisticated sensors can detect other road users. In addition to reducing fuel/energy consumption, avoiding fast accelerations or decelerations lowers noise and emissions [14]. The purpose of this survey is to examine XAI applications for detecting and mitigating intrusions in ICV networks. Security in C-ITS relies on trust, as defined by ETSI (European Telecommunication Standards Institute).

An Arizona woman was killed by a self-driving vehicle on Mar. 18, 2018, when she crossed the street and was hit by the vehicle. An investigation into how the accident occurred was difficult. Many questions have been raised about the functionality and safety of AI's integration with ITS, which encompasses driverless or autonomous vehicles, intelligent controls, and intelligent vehicles [15]. It is not clear what causes ICV misbehavior, but intrusion and potential harm caused by compromised ICVs and IoVs have been suggested as possible contributors. Furthermore, traditional AI solutions are designed in a "black box" manner, making investigation challenging [16, 17].

As an innovative technology, artificial intelligence has been developed as a means of creating sophisticated systems that can learn and understand as humans do. Through the innovative use of artificial intelligence, we can now develop sophisticated systems that understand and learn just like humans [18]. A system using artificial neural networks (ANNs) can learn new things and make accurate decisions using machine learning algorithms [19]. In the transportation industry, the role of Intelligent Transportation Systems has been redefined by Artificial Intelligence. A mobility technology based on artificial intelligence, or ITS, serves end users, markets, and society as a whole by meeting their needs [20]. A number of emerging and disruptive artificial intelligence technologies supporting transportation are being recognized by the World Economic Forum.

These technologies include simulated annealing, fuzzy logic models, and optimization of ant colonies. In addition, these technologies improve the efficiency of transportation by reducing passengers' journey times [21]. As a result of AI systems being applied to transportation subsystems like traffic management, public transport, safety management, manufacturing, and logistics, transportation efficiency can also be improved [22].

With the growth of AI, massive models have been developed to meet humanity's day-to-day needs. Although AI models offer many benefits, their transparency, assimilability, fidelity, and compactness are challenges. Because the systems are black-boxed, they enable accurate decision-making but fail to justify the actions taken. As a result, explainable AI (XAI) was developed, a revolutionary technological advancement. As a result of this novel idea, AI-based systems become more reliable and open. With the concept of XAI, businesses can make AI more transparent and acceptable by fostering its adoption. XAI simplifies artificial intelligence, making it useful in a variety of fields, from security to healthcare to transportation [23]. XAI has, therefore, been the subject of several research papers. There are many publications on XAI, but it can be challenging to situate growth in XAI with the unique needs of each sector. The XAI field has published several review papers to address this issue. Although XAI has generally been reviewed, stakeholders still have to deal with problems specific to their sectors. Moreover, in this review, we have explained how each XAI model addresses ITS security uniquely.

7.2 Related Work

Over the past decade, artificial intelligence has advanced significantly and continuously, allowing algorithms to solve once unsolvable problems. Despite this achievement, the use of opaque black-box AI models and increased model complexity are important considerations. It is, therefore, necessary for AI systems to be able to address such a challenge in critical and sensitive areas (such as health care and security) where accuracy alone is insufficient.

It has been proposed that explainable artificial intelligence (XAI) can enable greater transparency in AI and, as a result, prevent AI from limiting the adoption of critical domains [24, 25]. XAI provides end-users with explainable techniques for better understanding, trusting, and managing the new age of artificial intelligence, according to [26]. It dates back to early attempts to explain expert systems and Bayesian networks that explanations were needed [27]. In recent years, deep learning has made

XAI an extremely thriving area of research. Many studies on XAI are published each year. There are also a number of reviews published on XAI that cover many topics, both general and specific. Numerous challenges and directions for research are addressed in these review studies. In terms of challenges and possible research directions, some have been identified, but they are scattered across a variety of fields. As far as we know, this is the first meta-survey covering XAI challenges and research directions. Researchers dealing with challenges and directions in XAI will find useful reference material in this meta-survey.

7.2.1 Why Explainable AI is Needed?

The use of black-box AI systems is everywhere in our world today, from autonomous vehicles to social networks to medical devices. There is a tendency to make decisions without understanding why they are made. Several factors have been identified as factors affecting black-box AI systems.

AI systems should not explain the reasons for every decision they make, including decreased efficiency and higher costs; according to [24], explainability and interpretability are not generally required in two situations [28]: In the case of an advertisement system and postal code sorting, the black-box system makes reliable decisions based on research and practice. Thus, it's important to consider the reasons and circumstances for explaining and interpreting [24]. Here are five reasons why XAI is needed.

- **Regulatory perspective:** Our lives are filled with applications of black-box artificial intelligence, but their decisions may be unacceptable, especially those with legal implications. As a result, the law will have to deal with this new challenge. From a regulatory perspective, GDPR demonstrates the importance of XAI. Algorithm use regulation establishes this right to explanation, allowing users to request for explanations of algorithm decisions that are critical to their lives [29, 30]. A right to explanation can be difficult to implement without technologies that provide explanations, and without explanations, it remains a theoretical concept.
- **Scientific perspective:** It is necessary to develop an approximate function when developing black-box AI models. Due to the black-box AI model, knowledge rather than data is represented [31]. By making use of XAI, we can uncover the scientific knowledge extracted from black-box AI models, allowing new concepts to be discovered in various fields.

- **Industrial perspective:** Hence, XAI may be useful in revealing scientific knowledge extracted by black-box AI models, which may lead to the discovery of novel concepts in a wide variety of fields. To comply with regulations, the industry may have to use less accurate and more interpretable models [32]. Performance and interpretability can both be improved with XAI by overcoming these challenges. The development costs and the deployment costs can be increased.
- **The development perspective of a black-box AI system:** In addition to lacking training data, biased training data, outliers, adversarial data, and overfitting, poor results can be obtained for many reasons. The benefits and risks of black-box AI systems are therefore important to understand, especially when those systems have an impact on humans. As part of XAI, Robustness, Safety, and User Trust will be enhanced, faulty behavior, bias, and unfairness will be reduced or eliminated, and black-box AI systems will be improved to meet these goals [33]. XAI can also reveal the factors that influenced a model's decision-making to select models with similar performance [34]. A proxy function can also be derived from XAI because it is optimized to cope with incomplete objectives. AI optimization without considering the risk factors associated with adherence can lead to ineffective or even harmful outcomes.
- **End-user and social perspectives:** It has been shown in the literature of deep learning [35] that altering an image in a way that is indiscernible to humans can lead to the model incorrectly classifying a class. By contrast, DL models can identify images that humans could never recognize with high confidence. The reliability of AI models built in black boxes could be questioned in light of these findings. AI systems with black boxes are also susceptible to making unfair decisions. With data displaying human biases and prejudices, black-box AI systems could be developed. We will increase trust by creating explanations and improving the interpretability of black-box AI systems by understanding the reasoning behind the decisions they make. We will know if the system is serving the purpose for which it was designed rather than the purpose for which it was trained [36]. Moreover, black-box AI systems often require interpretable models to ensure fairness, which cannot be assured by error measures [37].

As part of intrusion detection, systems and networks are monitored and analyzed for indications of potential events, such as lack of compliance with security regulations and breaches of acceptable standards [38]. There are several types of intrusion detection techniques, which can be categorized according to the type of activity that they detect and the method by which they are detected. A host-based intrusion detection system (IDS) and a network-based intrusion detection system (NIDS) are two types of IDSs. An intrusion detection approach is either based on specifications or signatures or based on anomalies. In addition to monitoring and assessing events, IDS normally collects activity information, issues warnings and alarms in the event of critical occurrences, and generates reports accordingly. For networks to be protected from non-adversarial attacks and malicious traffic, intrusion detection systems are essential. Recent developments in IDS research have been assisted by machine learning (ML).

As reported by IoT Analytics [39], three security holes (CVEs-2022-29831, CVEs-2022-29832, and CVEs-2022-29833) were discovered by Nozomi researchers in August 2022 that can compromise Mitsubishi GX Works3 (configuration and programming software for certain Mitsubishi PLCs) safety CPU modules. Several BMW series-specific infotainment systems were also found to have 14 vulnerabilities. ITSs today include interconnected vehicles, such as the V2V, the V2I or the V2X. In addition to their vulnerability to attacks, connected things pose a risk of falsifying data during transmission [40, 41]. As a result, ICVs must be protected from these attacks by detecting and mitigating them.

7.2.2 Evaluation of Explainable-AI (XAI) Frameworks and Results

To improve security on the ICV network, an integrated, adaptive, and automated IDS is required. It is necessary to develop explainable AI (XAI) for defence against vulnerabilities and intrusion detection, as well as to gain a better understanding of emerging threats and attack mechanisms [31, 42, 43]. Figure 7.1 depicts how XAI develops ML algorithms that produce more understandable techniques while retaining efficiency. A model's actions and decisions can be understood by humans as well. As an area that is rapidly expanding, it facilitates the extraction of information and the visualization of results [44].

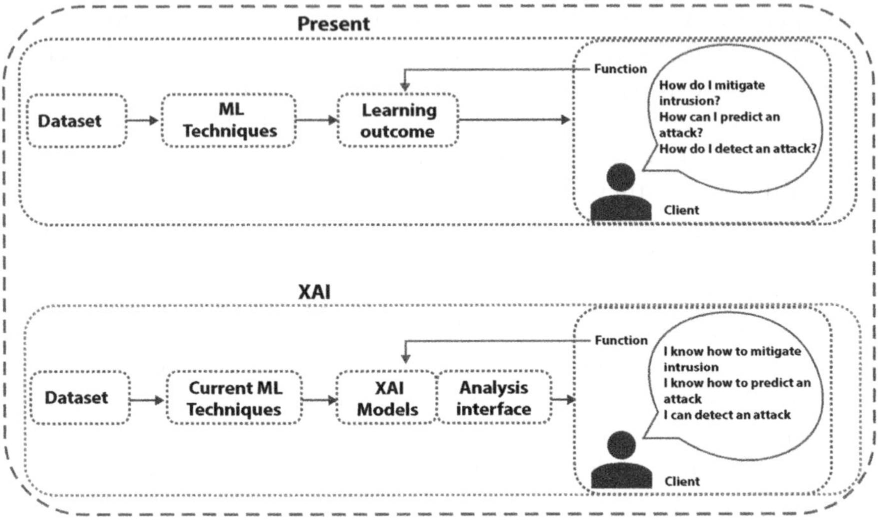

Figure 7.1 The concept of XAI.

The National Institute of Standards and Technology (NIIST) is a government agency within the U.S. Department that lists five kinds of artificial intelligence explanations. Following is a list of them:

- **Describe an algorithm to the subject:** To ensure system security against intruders, the AI model in the IDS for ITS needs to be explained in detail.
- **Keep up with compliance and regulatory requirements:** Industry regulators require AI algorithms to demonstrate compliance before they can be used. For AI algorithms to drive themselves, they should explain how they adhere to traffic laws.
- By focusing on outcomes rather than explanations, AI systems can build social trust. It is possible to describe algorithms based on their goals, development processes, data used, and sources, as well as their advantages and disadvantages.
- **Future system development assistance:** AI systems must be improved by understanding where and why they produce incorrect results.

- **In the owner's interest:** The adoption of artificial intelligence across all sectors is expected to yield significant benefits. An easy-to-follow recommendation system keeps viewers subscribed to a streaming service.

7.3 The Need for Explainable AI in Transportation

Transparency is increasingly demanded as AI is becoming more ingrained in transportation. Achieving a low carbon footprint and reducing resource consumption is especially important in green transportation, which is a shared responsibility between public and private entities. It is possible to gain a better understanding of the "why" of environmental decisions through explainable artificial intelligence [25].

The following are some reasons why explainable AI is important for greener transportation:

- **Enhanced Trust and Adoption:** A transportation company's adoption of AI will be impacted by whether the decisions it makes are in line with sustainability goals. In order for AI to foster trust, it needs to be able to explain its decision-making process.
- **Safety and Accountability:** Public safety must be taken into consideration when using AI for autonomous vehicles, routing, or emission control. It is easier to ensure the safe deployment of these systems when stakeholders are able to understand their limitations.
- **Regulatory Compliance:** A transparent decision-making process is often necessary to comply with environmental and safety regulations. Transportation providers can demonstrate their commitment to green policies using explainable AI.
- **Improved Efficiency and Resource Allocation:** Organizations can reduce emissions and energy consumption by identifying and addressing inefficiencies with transparent AI systems.

7.4 AI's Potential for Transforming Smart Cities and its Limitations

Urban crises, such as those caused by climate change, pandemics, natural disasters, or socioeconomic conditions, are commonly mitigated with smart and innovative digital technologies [45, 46]. Artificial intelligence has provided cities with invaluable opportunities to enhance infrastructure efficiency and predictive analytics capabilities, which has improved quality of life and sustainability. Due to artificial intelligence, our world has been revolutionized, impacting economies, societies, the environment, and governance.

According to [47], artificial intelligence plays an increasingly important role in smart cities by enabling better delivery of infrastructure, services, and amenities. Intelligent urban technologies, such as chatbots, virtual agents, cognitive robotics, and autonomous vehicles, can benefit from applications of artificial intelligence. Intelligent urban technologies, such as chatbots, virtual agents, cognitive robotics, and autonomous vehicles, can benefit from applications of artificial intelligence. There are several types of cognitive security analytics: expert systems, identity analytics, digital assistants, knowledge management systems, predictive analytics, data visualization, process automation, recommendation systems, speech analytics, threat intelligence, expert systems, identity analytics and more [48]. Several studies have been conducted on the potential of artificial intelligence in smart cities [49].

Its potential benefits, such as improved operational infrastructure or service efficiency, and AI technology can also damage cities and citizens through opaque decision-making processes and privacy violations. Among them are automating inequality, creating algorithmic bias due to poor or limited data and training, removing or limiting human responsibility, and omitting transparency and accountability, among others [50]. As compared to other technologies, AI faces a number of data-related challenges, such as acquiring data, streaming large amounts of data, heterogeneous data, complex dependencies, noisy and incomplete data, storing, processing, training, and protecting data privacy.

Mishaps involving AI solutions implemented as part of smart city projects have sparked public distrust in those solutions, including the following:

- In Pittsburgh, PA, USA, AI misdiagnoses child maltreatment and prescribes ineffective solutions.

- A recruiting tool developed by Amazon based on artificial intelligence, which favored men.
- Americans of color are disproportionately represented in hospital decisions made by AI algorithms.
- An image recognition database developed by Clear View AI with pictures from social media was breached in 2020, resulting in privacy concerns for citizens in democracies and an Orwellian nightmare for citizens in autocracies.
- It was revealed that Robodebt, an automated debt recovery program run by the Australian government, malfunctioned, ultimately leading to AUD 721 million being unlawfully charged to 400,000 Australians [50].

7.5 Explainable AI Supports Greener Transportation

There are numerous applications of explainable artificial intelligence in transportation, from fleet management to public transportation. To drive greener transportation, explainable AI can be used in the following ways:

7.5.1 Optimizing Traffic Flow and Reducing Emissions

In addition to air pollution, traffic congestion also contributes to fuel consumption and emissions. A large dataset of city sensors, GPS information, and traffic cameras can be analyzed to predict and alleviate congestion. AI that explains why certain routes or measures are recommended can help urban planners make informed decisions that reduce traffic congestion and emissions. AI-driven explainable can, for instance, demonstrate how adjusting traffic signal timing can ease bottlenecks in high-emission zones, which can lead to targeted interventions [51].

7.5.2 Managing and Reducing Fleet Emissions

In fleet management, artificial intelligence is used to monitor vehicle performance, plan routes, and optimize fuel consumption. An explainable AI model can break down these decisions into their parts so that fleet operators can see the impact of each factor (such as vehicle speed, route, or condition) on emissions. In the case of adjustments, explainability allows for actionable insights, such as altering speeds or maintenance schedules to reduce carbon emissions [46].

7.5.3 Enhancing Predictive Maintenance

Predictive maintenance powered by AI reduces downtime and emissions from inefficient vehicles by identifying potential failures before they occur. A number of factors contribute to vehicles deteriorating, and XAI helps fleet managers understand why they may require repairs sooner, allowing them to extend vehicle life accordingly [52].

7.5.4 Supporting Autonomous Vehicles and Green Routing

With AI, autonomous vehicles (AVs) and green routing systems are able to make real-time decisions, such as adjusting routes and managing energy consumption. AVS must understand AI decision-making, especially in areas that are sensitive to the environment. Explainable AI allows stakeholders to verify that autonomous vehicles adhere to eco-friendly practices, like avoiding idle time or choosing the best routes to minimize emissions [53].

7.5.5 Facilitating Transparent Data Sharing

The sharing of data among cities, organizations, and researchers is essential for creating greener transportation systems. It is possible to facilitate cross-sector collaboration with explainable AI by making shared insights understandable and actionable. Using data that stakeholders understand and trust will lead to the reduction of emissions and enhancement of sustainability [54].

7.6 Benefits of Explainable AI in Greener Transportation

A well-designed artificial intelligence system can be beneficial to society, the environment, and the economy in a variety of ways, including:

- **Reduced Environmental Impact:** By supporting data-driven decisions, explainable AI helps transportation networks reduce emissions and energy consumption.
- **Improved Decision-Making:** Transport planners and operators can reduce waste and optimize energy usage when AI decision-making processes are clearer.

- **Enhanced Public Trust and Engagement:** New transportation technologies will be more trustworthy if AI is transparent. Seeing how AI makes decisions may lead the public to support green transportation programs.
- **Regulatory Adherence and Policy Alignment:** In addition to demonstrating compliance with environmental regulations, explainable AI simplifies how organizations approach sustainable transportation.

7.7 Challenges of Implementing Explainable AI in Greener Transportation

In spite of explainable AI's advantages, implementing it in green transportation presents several challenges:

- **Complexity of AI Models:** Transportation AI models are often inherently complex, which makes them an excellent tool for predictive analytics and optimization. Transparency without oversimplification is a challenge when it comes to these models.
- **Balancing Accuracy with Explainability:** The accuracy or effectiveness of models can be reduced if they are simplified to make them more understandable. High-stakes applications, like autonomous vehicles, require the balance of these trade-offs.
- **Data Privacy and Security:** It is common for transportation data to contain sensitive information. In shared transportation systems, transparency must be balanced with privacy and security, but this is a challenge.
- **Resource and Infrastructure Limitations:** Some regions or organizations do not have the resources, skilled personnel, or advanced infrastructure required to build and deploy explainable AI models.
- **Opportunities for Advancing Explainable AI in Transportation:** Sustainable transportation demands more integration and improvement of explainable AI.
- **Interdisciplinary Collaboration:** It is possible to develop practical, explainable solutions tailored to real-world

problems by collaborating with AI researchers, urban planners, transportation experts, and policymakers.

- **Innovation in Explainability Techniques:** LIME (Local Interpretable Model-Agnostic Explanations) and Shapley Additive Explanations (SHAP) help understand complex transportation models more effectively.
- **Public-Private Partnerships:** To promote greener transportation, public-private partnerships are key to advancing XAI technology adoption and exchanging ideas and resources.
- **AI Policy and Regulation:** It can be beneficial for transportation companies to prioritize explainable AI solutions, thereby improving the safety and sustainability of the system.

7.8 Conclusion

Sustainable transportation could be made more efficient and environmentally friendly with explanatory AI. As a result of XAI, stakeholders, from regulatory bodies to end users, will be able to achieve greater transparency and reliability of advanced models. In this chapter, we highlight the ways in which XAI can enhance trust, facilitate compliance with regulatory requirements, and make transport networks safer and more environmentally friendly. It is, however, essential to ensure data privacy and balance model complexity with interpretability to fully leverage XAI in transportation. It will be critical to develop new XAI methods and collaborate with other disciplines to unlock new opportunities in sustainable transportation, ultimately helping cities reduce emissions and use resources more efficiently.

References

1. Wang, P., Ye, R., Zhang, J., Wang, T., An Eco-Driving Controller Based on Intelligent Connected Vehicles for Sustainable Transportation. *Appl. Sci.*, 12, 9, 4533, Apr. 2022, doi: 10.3390/app12094533.
2. Yu, M., Construction of Regional Intelligent Transportation System in Smart City Road Network via 5G Network. *IEEE Trans. Intell. Transp. Syst.*, 1–9, 2022, doi: 10.1109/TITS.2022.3141731.
3. Rani, P., Hussain, N., Khan, R.A.H., Sharma, Y., Shukla, P.K., Vehicular intelligence system: time-based vehicle next location prediction in software-defined internet of vehicles (SDN-IOV) for the smart cities, in: *Intelligence*

Things: AI-IoT Based Critical-Applications Innovations, vol. 23, 7, pp. 35–54, 2021.

4. Montoya-Torres, J.R., Moreno, S., Guerrero, W.J., Mejía, G., Big Data Analytics and Intelligent Transportation Systems. *IFAC-PapersOnLine*, 54, 2, 216–220, 2021, doi: 10.1016/j.ifacol.2021.06.025.
5. Garg, T., Kaur, G., Sant Longowal Institute of Engineering and Technology, India, A Systematic Review on Intelligent Transport Systems. *JCCE*, 2, 3, 175–188, Jun. 2022, doi: 10.47852/bonviewJCCE2202245. Department of Computer Science and Engineering, Sant Longowal Institute of Engineering and Technology, India.
6. Anyanwu, G.O., Nwakanma, C., II, Lee, J.-M., Kim, D.-S., Falsification Detection System for IoV Using Randomized Search Optimization Ensemble Algorithm. *IEEE Trans. Intell. Transp. Syst.*, 24, 4, 4158–4172, Apr. 2023, doi: 10.1109/TITS.2022.3233536.
7. Rani, P., Verma, S., Yadav, S.P., Rai, B.K., Naruka, M.S., Kumar, D., Simulation of the lightweight blockchain technique based on privacy and security for healthcare data for the cloud system. *Int. J. E-Health Med. Commun. (IJEHMC)*, 13, 4, 1–15, 2022.
8. Chekkouri, A.S., Ezzouhairi, A., Pierre, S., Connected vehicles in an intelligent transport system, in: *Vehicular Communications and Networks*, pp. 193–221, Woodhead Publishing, Cambridge, United Kingdom, 2015, doi: 10.1016/B978-1-78242-211-2.00010-6.
9. Rani, P. and Sharma, R., Intelligent Transportation System Performance Analysis of Indoor and Outdoor Internet of Vehicle (IoV) Applications Towards 5G. *Tsinghua Sci. Technol.*, 29, 6, 1785–1795, Dec. 2024, doi: 10.26599/TST.2023.9010119.
10. Kaffash, S., Nguyen, A.T., Zhu, J., Big data algorithms and applications in intelligent transportation system: A review and bibliometric analysis. *Int. J. Prod. Econ.*, 231, 107868, Jan. 2021, doi: 10.1016/j.ijpe.2020.107868.
11. Chen, S., Hu, J., Shi, Y., Zhao, L., Li, W., A Vision of C-V2X: Technologies, Field Testing, and Challenges With Chinese Development. *IEEE Internet Things J.*, 7, 5, 3872–3881, May 2020, doi: 10.1109/JIOT.2020.2974823.
12. Rani, P. and Sharma, R., IMFOCA-IOV: Intelligent Moth Flame Optimization based Clustering Algorithm for Internet of Vehicle, in: *2023 14th International Conference on Computing Communication and Networking Technologies (ICCCNT)*, IEEE, pp. 1–6, 2023.
13. Nguyen, V.-L., Hwang, R.-H., Lin, P.-C., Vyas, A., Nguyen, V.-T., Toward the Age of Intelligent Vehicular Networks for Connected and Autonomous Vehicles in 6G. *IEEE Netw.*, 37, 3, 44–51, May 2023, doi: 10.1109/MNET.010.2100509.
14. Tsolaki, K., Vafeiadis, T., Nizamis, A., Tzovaras, D., Utilizing machine learning on freight transportation and logistics applications: A review. *ICT Express*, 9, 3, 284–295, Jun. 2023, doi: 10.1016/j.icte.2022.02.001.

15. Van Brummelen, J., O'Brien, M., Gruyer, D., Najjaran, H., Autonomous vehicle perception: The technology of today and tomorrow. *Transp. Res. Part C: Emerging Technol.*, 89, 384–406, Apr. 2018, doi: 10.1016/j.trc.2018.02.012.
16. Jaigirdar, F.T., Rudolph, C., Oliver, G., Watts, D., Bain, C., What, Information is Required for Explainable AI? : A Provenance-based Research Agenda and Future Challenges. *2020 IEEE 6th International Conference on Collaboration and Internet Computing (CIC)*, IEEE, Atlanta, GA, USA, pp. 177–183, Dec. 2020, doi: 10.1109/CIC50333.2020.00030.
17. Singh, A., *et al.*, Smart Traffic Monitoring Through Real-Time Moving Vehicle Detection Using Deep Learning via Aerial Images for Consumer Application. *IEEE Trans. Consum. Electron.*, 1–1, 2024, doi: 10.1109/TCE.2024.3445728.
18. Boukerche, A., Tao, Y., Sun, P., Artificial intelligence-based vehicular traffic flow prediction methods for supporting intelligent transportation systems. *Comput. Netw.*, 182, 107484, Dec. 2020, doi: 10.1016/j.comnet.2020.107484.
19. Rani, P. and Sharma, R., Intelligent transportation system for internet of vehicles based vehicular networks for smart cities. *Comput. Electr. Eng.*, 105, 108543, 2023.
20. Nikitas, A., Michalakopoulou, K., Njoya, E.T., Karampatzakis, D., Artificial Intelligence, Transport and the Smart City: Definitions and Dimensions of a New Mobility Era. *Sustainability*, 12, 7, 2789, Apr. 2020, doi: 10.3390/su12072789.
21. Abduljabbar, R., Dia, H., Liyanage, S., Bagloee, S.A., Applications of Artificial Intelligence in Transport: An Overview. *Sustainability*, 11, 1, 189, Jan. 2019, doi: 10.3390/su11010189.
22. Iyer, L.S., AI enabled applications towards intelligent transportation. *Transp. Eng.*, 5, 100083, Sep. 2021, doi: 10.1016/j.treng.2021.100083.
23. Saeed, W. and Omlin, C., Explainable AI (XAI): A systematic meta-survey of current challenges and future opportunities. *Knowledge-Based Syst.*, 263, 110273, Mar. 2023, doi: 10.1016/j.knosys.2023.110273.
24. Adadi, A. and Berrada, M., Peeking Inside the Black-Box: A Survey on Explainable Artificial Intelligence (XAI). *IEEE Access*, 6, 52138–52160, 2018, doi: 10.1109/ACCESS.2018.2870052.
25. Barredo Arrieta, A., *et al.*, Explainable Artificial Intelligence (XAI): Concepts, taxonomies, opportunities and challenges toward responsible AI. *Inf. Fusion*, 58, 82–115, Jun. 2020, doi: 10.1016/j.inffus.2019.12.012.
26. Gunning, D., Broad agency announcement explainable artificial intelligence (XAI). *Defense Advanced Research Projects Agency (DARPA), Tech. Rep.*, 2016.
27. Biran, O. and Cotton, C., Explanation and justification in machine learning: A survey, in: *IJCAI-17 workshop on explainable AI (XAI)*, pp. 8–13, 2017, Accessed: Nov. 06, 2024. [Online]. Available: http://www.cs.columbia.edu/~orb/papers/xai_survey_paper_2017.pdf.

28. Doshi-Velez, F. and Kim, B., Towards A Rigorous Science of Interpretable Machine Learning. Mar. 02, 2017. arXiv: arXiv:1702.08608. Accessed: Nov. 06, 2024. [Online]. Available: http://arxiv.org/abs/1702.08608.
29. Goodman, B. and Flaxman, S., European Union Regulations on Algorithmic Decision Making and a 'Right to Explanation'. *AI Mag.*, 38, 3, 50–57, Sep. 2017, doi: 10.1609/aimag.v38i3.2741.
30. Samek, W. and Müller, K.-R., Towards Explainable Artificial Intelligence. in Explainable AI: Interpreting, Explaining and Visualizing Deep Learning, in: *Lecture Notes in Computer Science*, vol. 11700, W. Samek, G. Montavon, A. Vedaldi, L.K. Hansen, K.-R. Müller (Eds.), pp. 5–22, Springer International Publishing, Cham, 2019, doi: 10.1007/978-3-030-28954-6_1.
31. Das, S., Agarwal, N., Shiva, S., DDoS Explainer using Interpretable Machine Learning, in: *2021 IEEE 12th Annual Information Technology, Electronics and Mobile Communication Conference (IEMCON)*, IEEE, Vancouver, BC, Canada, pp. 0001–0007, Oct. 2021, doi: 10.1109/IEMCON53756.2021.9623251.
32. Veiber, L., Allix, K., Arslan, Y., Bissyandé, T.F., Klein, J., Challenges Towards {Production-Ready} Explainable Machine Learning, in: *2020 USENIX Conference on Operational Machine Learning (OpML 20)*, 2020, Accessed: Nov. 06, 2024. [Online]. Available: https://www.usenix.org/conference/opml20/presentation/veiber.
33. Confalonieri, R., Coba, L., Wagner, B., Besold, T.R., A historical perspective of explainable Artificial Intelligence. *WIREs Data Min. Knowl.*, 11, 1, e1391, Jan. 2021, doi: 10.1002/widm.1391.
34. Samek, W. and Müller, K.-R., *Explainable Artificial Intelligence: Understanding, Visualizing and Interpreting Deep Learning Models*, Aug. 28, 2017, arXiv: arXiv:1708.08296. Accessed: Nov. 06, 2024. [Online]. Available: http://arxiv.org/abs/1708.08296.
35. Szegedy, C., *et al.*, *Intriguing properties of neural networks*, Feb. 19, 2014, arXiv: arXiv:1312.6199. Accessed: Nov. 06, 2024. [Online]. Available: http://arxiv.org/abs/1312.6199.
36. Burkart, N. and Huber, M.F., A Survey on the Explainability of Supervised Machine Learning. *Jair*, 70, 245–317, Jan. 2021, doi: 10.1613/jair.1.12228.
37. Lipton, Z.C., The Mythos of Model Interpretability: In machine learning, the concept of interpretability is both important and slippery. *Queue*, 16, 3, 31–57, Jun. 2018, doi: 10.1145/3236386.3241340.
38. Ghosh, S. and Sampalli, S., A Survey of Security in SCADA Networks: Current Issues and Future Challenges. *IEEE Access*, 7, 135812–135831, 2019, doi: 10.1109/ACCESS.2019.2926441.
39. Guan, T., Han, Y., Kang, N., Tang, N., Chen, X., Wang, S., An Overview of Vehicular Cybersecurity for Intelligent Connected Vehicles. *Sustainability*, 14, 9, 5211, Apr. 2022, doi: 10.3390/su14095211.

40. Anyanwu, G.O., Nwakanma, C., II, Lee, J.M., Kim, D.-S., Novel hyper-tuned ensemble Random Forest algorithm for the detection of false basic safety messages in Internet of Vehicles. *ICT Express*, 9, 1, 122–129, Feb. 2023, doi: 10.1016/j.icte.2022.06.003.
41. Xiong, S. and Qi, X., The Research on Security Model Algorithm in Intelligent Connected Vehicles. *2022 IEEE International Conference on Sensing, Diagnostics, Prognostics, and Control (SDPC)*, IEEE, Chongqing, China, pp. 372–376, Aug. 2022, doi: 10.1109/SDPC55702.2022.9915868.
42. Zebin, T., Rezvy, S., Luo, Y., An Explainable AI-Based Intrusion Detection System for DNS Over HTTPS (DoH) Attacks. *IEEE Trans. Inform. Forensic Secur.*, 17, 2339–2349, 2022, doi: 10.1109/TIFS.2022.3183390.
43. Marino, D.L., Wickramasinghe, C.S., Manic, M., An Adversarial Approach for Explainable AI in Intrusion Detection Systems, in: *IECON 2018 - 44th Annual Conference of the IEEE Industrial Electronics Society*, IEEE, DC, USA, pp. 3237–3243, Oct. 2018, doi: 10.1109/IECON.2018.8591457.
44. Srivastava, G., *et al.*, *XAI for Cybersecurity: State of the Art, Challenges, Open Issues and Future Directions*, 2022, arXiv. doi: 10.48550/ARXIV.2206.03585.
45. Shin, D.-H., Ubiquitous city: Urban technologies, urban infrastructure and urban informatics. *J. Inf. Sci.*, 35, 5, 515–526, Oct. 2009, doi: 10.1177/0165551509100832.
46. Ortega-Fernández, A., Martín-Rojas, R., García-Morales, V.J., Artificial Intelligence in the Urban Environment: Smart Cities as Models for Developing Innovation and Sustainability. *Sustainability*, 12, 19, 7860, Sep. 2020, doi: 10.3390/su12197860.
47. Ullah, Z., Al-Turjman, F., Mostarda, L., Gagliardi, R., Applications of Artificial Intelligence and Machine learning in smart cities. *Comput. Commun.*, 154, 313–323, Mar. 2020, doi: 10.1016/j.comcom.2020.02.069.
48. Yigitcanlar, T., Butler, L., Windle, E., Desouza, K.C., Mehmood, R., Corchado, J.M., Can Building 'Artificially Intelligent Cities' Safeguard Humanity from Natural Disasters, Pandemics, and Other Catastrophes? An Urban Scholar's Perspective. *Sensors*, 20, 10, 2988, May 2020, doi: 10.3390/s20102988.
49. Zhang, Y., Geng, P., Sivaparthipan, C.B., Muthu, B.A., Big data and artificial intelligence based early risk warning system of fire hazard for smart cities. *Sustain. Energy Technol. Assess.*, 45, 100986, Jun. 2021, doi: 10.1016/j.seta.2020.100986.
50. Yigitcanlar, T., Desouza, K., Butler, L., Roozkhosh, F., Contributions and Risks of Artificial Intelligence (AI) in Building Smarter Cities: Insights from a Systematic Review of the Literature. *Energies*, 13, 6, 1473, Mar. 2020, doi: 10.3390/en13061473.
51. Shukla, A.K., Janmaijaya, M., Abraham, A., Muhuri, P.K., Engineering applications of artificial intelligence: A bibliometric analysis of 30 years (1988–2018). *Eng. Appl. Artif. Intell.*, 85, 517–532, Oct. 2019, doi: 10.1016/j.engappai.2019.06.010.

52. Dwivedi, Y.K., *et al.*, Artificial Intelligence (AI): Multidisciplinary perspectives on emerging challenges, opportunities, and agenda for research, practice and policy. *Int. J. Inf. Manage.*, 57, 101994, Apr. 2021, doi: 10.1016/j.ijinfomgt.2019.08.002.
53. Huntingford, C., Jeffers, E.S., Bonsall, M.B., Christensen, H.M., Lees, T., Yang, H., Machine learning and artificial intelligence to aid climate change research and preparedness. *Environ. Res. Lett.*, 14, 12, 124007, Dec. 2019, doi: 10.1088/1748-9326/ab4e55.
54. Tambe, P., Cappelli, P., Yakubovich, V., Artificial Intelligence in Human Resources Management: Challenges and a Path Forward. *Calif. Manage. Rev.*, 61, 4, 15–42, Aug. 2019, doi: 10.1177/0008125619867910.

8

Demystifying the Aspects of Edge Computing and Edge AI for Real-Time Insights

Abhilasha Jadhav[1], Heena Madan[2], Mohammed Y. Al-khuzaie[3], Ruby Pant[4], Nidhi Singh[5]* and Hassan M. Al-Jawahry[6]

[1]*Department of Computing Science and Artificial Intelligence, NIMS Institute of Engineering and Technology, NIMS University Rajasthan, Jaipur, India*
[2]*Department of Computer Science Engineering, Chandigarh Engineering College, Chandigarh Group of Colleges, Jhanjeri, Mohali, Punjab, India*
[3]*College of Technical Engineering, National University of Science and Technology, Dhi Qar, Iraq*
[4]*Department of Mechanical Engineering, Uttaranchal Institute of Technology, Uttaranchal University, Dehradun, Uttarakhand, India*
[5]*College of Computing Sciences & IT (CCSIT), Teerthanker Mahaveer University, Moradabad (U.P.), India*
[6]*Department of Computer Science Engineering, The Islamic University, Najaf, Iraq*

Abstract

We analysed how edge computing and Edge Artificial Intelligence are enabling real-time insights and decision-making across multiple industries in this chapter. Organisations confronting limitations in traditional cloud models, particularly latency and bandwidth, are turning to edge computing to solve these problems. Through local processing, it is possible to reduce delays, improve security, and increase bandwidth, making it ideal for applications like autonomous vehicles, healthcare, industrial automation, and smart cities. Edge AI incorporates local AI algorithms for millisecond-level intelligence. As a result of these technologies, real-time data analysis and privacy are improved, as is operational resilience. Throughout this chapter, we will demystify the underlying technology, key benefits, and application areas of edge computing and edge AI. This chapter

**Corresponding author*: nidhisingh.bittoo@gmail.com

Pethuru Raj, Sudesh Yadav, Manas Kumar Mishra, Satya Prakash Yadav and Victor Hugo C. de Albuquerque (eds.) Artificial Intelligence Technologies for Smart and Sustainable Urban Transportation: Integrated Platforms and Use Cases, (127–144) © 2026 Scrivener Publishing LLC

examines technical challenges and the potential of integrating emerging technologies such as 5G and TinyML to broaden its scope.

Keywords: Autonomous vehicles, edge artificial intelligence, edge computing, emerging technologies, smart cities, tinyML, 5G integration

8.1 Introduction

The development and integration of heterogeneous distributed computing paradigms, including cloud, fog, and edge, are occurring as artificial intelligence and computing undergo fundamental changes. Clouds orchestrate, fogs act as middle layers, and edges provide cloud services to end users. Three specific models must be developed to implement advanced AI and machine learning (ML) systems effectively. We will explore these three models in greater detail to gain a deeper understanding of their purpose, the roles they play in augmenting cloud-based AI and machine learning, and their interactions and complementarities [1]. With its massive storage capacity and powerful processing power, cloud computing has contributed significantly to the development of artificial intelligence. Now that engineers have access to large amounts of data, they are able to complete complex tasks more quickly and efficiently. Due to this, cloud-based systems have become increasingly important: AI and machine learning are evolving rapidly [2]. As IoT grows, there is an increased demand for real-time data processing, which has exposed some limitations of cloud computing, including latencies and bandwidth [3]. Because of the massive amounts of data, they process in real-time, fog and edge computing can be useful for solving this problem.

With real-time data processing becoming more and more complex, edge computing and edge AI are becoming essential technologies. IoT devices and other edge resources generate vast amounts of data, which traditional cloud models find difficult to analyse due to latency and bandwidth constraints. Using edge computing, data can be processed closer to the source, minimising delays, while edge artificial intelligence can make localised decisions at the network edge [4]. Data is processed by network edge processors near the data source. Unlike a traditional data centre model, which sends data to a central location, cloud computing uses distributed servers [5]. By leveraging edge processing, data collected at the edge can be analysed in real time, and actions can be taken at the moment it is collected. There are many applications for this method, including autonomous vehicles, industrial automation, and health monitoring systems, which require low latency [6]. Processing at the edge reduces bandwidth consumption and response times by minimising data transfers over the network [7].

A number of edge computing devices, including sensors, actuators, and embedded systems, possess computational capabilities that enable localised data processing. In addition to filtering, aggregating, and analysing data, these devices can also transmit only the most relevant data to the cloud. The method can be applied to autonomous vehicles, industrial automation, and healthcare monitoring systems, among many others. Despite disruptions in network connections to the cloud, distributed intelligence offers greater resilience because local processing continues [8].

8.1.1 Importance of Real-Time Processing in AI

Whenever there is an application requiring immediate feedback or action, real-time processing is essential. Real-time processing has become increasingly important in various fields:

- **Autonomous Vehicles:** As a result of real-time information collected by sensors such as LiDAR and cameras, self-driving cars are able to make split-second decisions. Accidents and fatalities can occur if this data is not processed in a timely manner.
- **Healthcare:** Continuous streams of data generated by wearables and remote monitoring systems require real-time analysis to detect anomalies, like irregular heartbeats and glucose levels, and alert medical professionals.
- **Industrial Automation:** Sensor data must be processed in real-time to improve operational efficiency and prevent equipment failures in manufacturing. Analysing machine data can help predict issues and prevent downtime before they occur.
- **Smart Cities:** The real-time processing of smart city applications enables them to adjust to changing conditions in real-time, such as traffic management, energy distribution, and public safety.

A powerful real-time analytics ecosystem is created by integrating edge computing and edge AI. Our goal is to demystify these technologies and uncover their core components, their impact on industries, how they enhance performance and security across industries, and user experiences.

The author is experiencing rapid development of the Internet of Things and communications technologies like WiFi and 5G, which will enable our future world to be interconnected to provide uninterrupted services for consumers and businesses. According to Cisco, 50 billion Internet-connected devices will exist by 2020 [9]. With the growing number of IoT devices, a great deal of valuable data will be generated, which can be processed to create ground-breaking applications. In traditional computing architectures, cloud computing provides sufficient computational capacity and energy sustainability. With the growing number of IoT devices, a great deal of valuable data will be generated, which can be processed to create ground-breaking applications.

The physical distance between two points can sometimes cause unacceptable latency, especially for applications such as autonomous driving or virtual reality gaming, where low latency is required. Data communication consumes a large amount of overhead and costs, which service providers bear. It is possible to resolve this problem through edge computing [10]. Fog computing [11], and cloudlets [12], are two forms of edge computing. As opposed to sending data to high-capacity places for processing, data is processed near the source. At the edge of the network, many servers can be deployed to offload tasks from IoT end devices. There are numerous advantages of edge computing over cloud computing. Calculations are closer to the data source, thereby reducing communication latency and making it possible for latency-dependent applications to operate more efficiently.

Furthermore, local computation can improve the security of data and applications. Furthermore, network edge data processing can effectively reduce traffic on the backbone network, resulting in a reduction in network stress. The amount of data generated in the cloud is growing rapidly, which poses challenges in the areas of latency, bandwidth, and security. In the future, Edge Computing and Edge AI will enable real-time insight and faster decision-making, resulting in a better experience for users. Demonstrations of how Edge Computing and Edge AI can be used in industries requiring low-latency, high-efficiency operations will be provided. A powerful perception capability has enabled deep learning to make notable advances in recent years [13].

Various applications have been developed using it, such as computer vision and natural language processing [14, 15]. Atari Games [16, 17] are two examples of computer games where their performance even exceeds that of humans' best players. By combining edge computing and deep learning, better applications will be able to address current challenges. Aside from edge computing, there are numerous complex scenarios, such as video analytics, that demand the powerful processing capabilities of deep learning [18], transportation control [19], etc. The Nvidia Jetson TX2 developing kit, for example, is a lightweight hardware platform that allows computing at the edge to be better supported by deep learning. Edge computing applications powered by deep learning are still in their infancy, but many pioneering efforts have been made towards this field.

Several studies have examined the convergence of deep learning and edge computing [20]. Edge computing and deep learning were explained from five perspectives, but an overview of applications was lacking. Moreover, [21], provided a general abstraction for deep learning and edge computing but analysed them from multiple perspectives. Training and inference of deep learning models using edge computing: Additionally, there have been surveys on mobile edge computing [22, 23] and deep learning [24, 25], all of which focused on one of them without addressing how they might work together. These factors necessitate a comprehensive review of deep learning applications and possible future directions for edge computing.

Since edge computing has been able to reduce latency and cost in many fields, it has seen success in recent years. In contrast to cloud computing, edge computing processes data at the network edge, whereas cloud computing sends the data to the cloud. Additionally, data computation is made closer to the data source, enabling applications with sensitivity to delays to be developed much more easily. Due to the local processing, much data is not transmitted, reducing network traffic and reducing costs considerably. Several edge computing paradigms are described in the following section, along with some benefits of edge computing.

8.1.2 A Paradigm for Edge Computing

Computing at the edge usually takes place on edge devices, such as edge servers, which handle processing, communication, caching, and other tasks. Edge computing is not the only paradigm or technology that shares these concepts. The next section of this paper discusses and differentiates some typical edge computing paradigms.

- **Cloudlet:** In an effort to process, offload, and cache information close to mobile devices, such as buildings and shopping centres, Carnegie Mellon University has developed Cloudlets. Mobile applications are usually better supported by Cloudlets, thanks to virtualisation technology. Cloudlet is aimed at bringing the advances of the cloud to mobile devices [26], resulting in improved latency and resourcefulness. The micro data centre (MDC) [27] is similar in size to cloudlets.

By connecting different MDCs, a backbone network enables more efficient, intelligent computation, caching, and management. In addition to managing numerous IoT devices, MDC serves as an essential part of the IoT ecosystem [28].

- **Fog Computing:** Through Cisco's fog computing concept, enterprises can extend cloud computing services to their edges. Usually, fog computing uses nodes at the network gateway to process data. A distributed fog network enables the interconnection of numerous Internet of Things devices to provide collaborative services using fog computing [29]. In addition to providing better support for IoT devices, fog nodes are also designed to improve network connectivity. It can be argued from this perspective that fog computing is aligned more with IoT than other similar edge computing paradigms and emphasises the end user more.

8.1.3 Mobile Edge Computing (MEC)

ETSI's MEC has been standardised as a means of providing sufficient computing power within radio access networks [30]. In cellular networks, wireless base stations contain computers. Mobile edge computing can simplify IoT data processing, as base stations serve as main access points to IoT devices.

8.1.3.1 Understanding Edge Computing

- **What is Edge Computing?**

Computing resources can be placed close to devices and sensors that are connected to the IoT or Edge Computing. Data is processed on demand with edge computing rather than in data centres, as with traditional cloud computing. It reduces latency, reduces bandwidth costs, and improves reliability.

8.1.3.2 The Architecture of Edge Computing

Three layers typically make up Edge Computing's architecture:

Device Layer: Various devices and devices that generate data, including mobile devices, IoT sensors, and industrial machines.
Edge Layer: Data processing and filtering are done on local servers and edge nodes or gateways.

Cloud Layer: In cloud data centres, data can be analysed and stored for long periods.

- It reduces latency, reduces bandwidth costs, and improves reliability.

Key Benefits of Edge Computing

Low Latency: Edge computing reduces latency so that it can be used for real-time monitoring and autonomous vehicles using edge computing.
Reduced bandwidth usage: Reduces bandwidth and cost by minimising data sent to cloud servers.
Enhanced Security and Privacy: By avoiding network transmission of sensitive data, local processing offers better privacy and security.
Reliability and Resilience: Since edge computing is not entirely dependent on cloud connectivity, it can continue to operate even when there is a connectivity issue.

8.1.4 Advantages of Edge Computing

Cloud computing offers many benefits over edge computing, including low latency, energy savings, context-aware services, and privacy protection. The following is a summary of them.

- **Low Latency:** When edge devices are placed near end devices, which are both data sources and recipients of results, transmission latency can be reduced in cloud computing environments. It is common for a cloud server's transmission latency to be several milliseconds or even microseconds, whereas an end user's is typically tens (or hundreds) of milliseconds or even microseconds. Edge computing advances due to 5G's low latency transmission, allowing autonomous driving, virtual reality, and augmented reality, as well as healthcare applications [31].
- **Energy Saving:** Due to their size and usage scenarios, IoT devices usually have limited power supply. However, they are also often capable of performing highly complex tasks that require a considerable amount of power. Given that batteries must be charged and discharged frequently, developing a cost-efficient powering solution for distributed IoT devices is challenging. As a result of IoT edge computing, billions

of IoT devices consume less energy and are more efficient because computation tasks are offloaded to edge servers.

- **Context-Aware Service:** Using context-aware computing is becoming increasingly popular in IoT and edge computing applications [32] since it is critical to understand the context of the data. By collecting more context information, edge servers can support the data processing process better because of their proximity. For example, Amazon Go supermarkets use video cameras that not only record products but also analyse customers' locations, durations, and behaviours to predict their interest in products.
- **A Secure and Privacy Concept:** It offers greater security and privacy protection for users' data and applications than cloud computing. An edge server is, in many cases, a cluster of computers that are geographically distributed. Users are generally responsible for managing and maintaining them. To protect sensitive information more strictly, we must monitor and manage it more closely. However, since small data centres are less visible than large ones, they are less likely to be attacked [33].

8.2 Edge AI

AI that is deployed directly on hardware at the edge of the network, rather than using cloud services, is referred to as edge AI. Data processing and decision-making are becoming more necessary in real-time, causing a paradigm shift. Data can be analysed at the edge of the network using smartphones or IoT devices to reduce latency, increase privacy, and reduce bandwidth consumption.

When immediate responses are needed, traditional AI methods require data to be transmitted from edge devices to central servers to avoid unacceptably long delays. To combat these challenges, edge AI utilises on-site data processing, enabling real-time decisions to be made. Data privacy and security concerns can also be addressed with this approach because there are no network interconnections. AI-based edge technology, or edge AI, allows innovative solutions to be delivered across a range of industries, including healthcare, manufacturing, and transportation.

8.2.1 Decision-Making in Real-Time: Why it's Important

The success of edge AI is based on its ability to make real-time decisions. Various applications require results to be analysed and processed instantly. For autonomous vehicles to navigate, detect obstacles, and avoid collisions, real-time processing is required. It is possible to have catastrophic consequences if you delay making a decision. In addition to real-time data analysis, industrial automation also uses predictive maintenance to minimise downtime, maximise efficiency, and maximise profits [34]. Health care is another field in which real-time decision-making is required. Medical personnel can be alerted to any anomalies in vital signs using wearable devices that integrate edge AI. As well as optimising traffic flow and reducing energy consumption, real-time sensor data can enhance public safety in smart cities. With 5G, edge AI will be able to transfer data faster and with less latency. As a result of this technological synergy, more sophisticated applications can be developed that require immediate processing and action. Real-time decision-making is increasingly important as edge devices generate more and more data that can be tapped into for actionable insights.

8.2.2 Purpose and Scope of the Paper

Edge AI as a transformative technology across various industries is presented in this paper, which examines its multifaceted dimensions, significance, and transformative impact. Traditionally, cloud-based AI models pose inherent challenges to real-time decision-making, which must be overcome. In this study, we try to understand how edge AI can assist real-time decision-making. This paper will examine specific case studies and applications in order to illustrate how edge AI is transforming modern technology landscapes practically and beneficially. In this chapter, underlying technologies, including hardware and software advances, are examined for edge AI. Furthermore, edge AI will be analysed in light of other emerging technologies, such as 5G, IoT, and blockchain, along with their interactions. A discussion of the impact of edge AI on ethics and privacy will also be discussed, emphasising the need for robust security and regulatory frameworks in regard to new technologies such as 5G, IoT, and blockchain.

The chapter also explores potential developments and challenges in edge AI going forward, in addition to offering insights into its future trajectory.

This paper provides researchers, practitioners, and policymakers with a comprehensive understanding of edge AI for effective application, leveraging, and implementation.

- **Edge Devices:** Edge AI relies heavily on edge devices. Peripheral devices generate data at the edges of the network. In addition to smartphones, IoT sensors, cameras, and industrial machines, edge devices also include IoT sensors. As early as possible, edge devices process data collected from their surroundings [20]. In today's edge devices, powerful processors, memory, and storage make it possible for them to run AI algorithms remotely. The Neural Engine or Pixel Visual Core, dedicated AI chips in smartphones, accelerate machine learning tasks, for instance. Aside from accelerometers, gyroscopes, and cameras, edge devices have a variety of sensors that provide data to AI algorithms. In addition to processing data locally, edge devices are often capable of communicating with other devices and cloud servers. Cooperative decision-making or complex tasks requiring extensive computing resources require this capability [35].
- **Edge Servers:** An edge server is an intermediary between an edge device and a cloud server. Consequently, these servers are usually closer to the data source, resulting in lower bandwidth usage and shorter latency times. As edge servers increase computational power and storage capacity, data processing at the edge can become more complex. The edge server also acts as an aggregator for data from different edge devices. Data aggregation can then be used to extract valuable information from the resulting data. A smart city's edge servers, for instance, can collect data from sensors to monitor traffic patterns, air quality, and energy consumption. The edge server is an important component of ensuring data reliability and redundancy as well. As a result, edge servers can handle the processing tasks when edge devices fail or go offline. Edge servers can also perform compression and data filtering prior to sending data to the cloud.
- **Real-time decision-making with Edge AI:** Using AI algorithms at the edge of devices allows for real-time decision-making. As a result, this section focuses on scalability, flexibility, and reliability, as well as reduced latency and security issues.

- **Reduced Latency:** Edge AI reduces latency when making real-time decisions. For data processing and algorithm execution, traditional AI systems typically use cloud servers. The time it takes to send data from edge devices to central servers makes this dependency a significant stumbling block. Data is processed locally by edge AI on the device, thus mitigating this issue. In autonomous vehicles, for example, safety depends on millisecond-level decisions. Due to edge AI, the sensors on these vehicles can be processed instantly without having to wait for a server to respond, enabling them to respond instantly to changing road conditions and potential hazards. It is crucial for applications such as emergency response systems, industrial automation, and medical diagnostics that require immediate data processing. The user experience in consumer applications is further enhanced by reduced latency. AR and VR systems must provide seamless and immersive experiences in order to succeed. In addition to reducing motion sickness and improving user satisfaction, edge AI helps these systems process data faster and in a more efficient manner. In addition, edge AI results in improved predictive maintenance due to reduced latency. Industrial and manufacturing sectors have real-time monitoring systems that can identify and correct anomalies in real-time. By using this proactive approach, downtime can be minimised, and operational efficiency can be enhanced.

8.3 Application of Edge AI in a Variety of Industries

With edge AI in place, a number of industries are experiencing a transformation as computation and data storage are brought closer to the user. As a result of this paradigm shift, systems become more responsive and intelligent, leading to an increase in efficiency and speed. In this paper, we examine edge AI's applications across a variety of sectors and assess its future impact and growth potential [35].

8.3.1 Manufacturing

Predictive Maintenance: Manufacturing companies rely on edge AI to perform predictive maintenance. Real-time equipment monitoring is possible by using edge devices with artificial intelligence. Machines have sensors that measure temperature, vibration, and pressure. As a result of this data, a machine failure prediction is made at the edge. This approach reduces downtime by completing maintenance before failures occur instead of on a fixed schedule or after breakdowns. In environments where latency can negatively affect production efficiency, edge at the source, AI enables rapid processing of large volumes of data. High-speed assembly lines can experience serious problems if a malfunction is not detected on time. AI at the edge ensures a continuous analysis and decision-making process, allowing equipment maintenance schedules to be optimised and its lifespan to be extended.

Quality Control: The quality control process in manufacturing plays a key role in ensuring that product specifications are met and customer expectations are exceeded. Edge AI makes real-time inspection and analysis possible, enhancing quality control processes. The production line uses high-resolution cameras and sensors to process images and data. Defects can be detected with high accuracy using these algorithms. As a result of implementing edge AI in quality control, inspections can be done more quickly and accurately, but they can also be corrected immediately. When a product defect is detected, an alert may be sent, or even a production line halted, preventing the production of further defective products. Proactively reducing waste, reducing costs, and maintaining high quality are the benefits of this approach.

8.4 Edge AI Challenges and Limitations

8.4.1 Challenges in Technology

Constraints Imposed by Hardware: There are usually limitations to the capabilities of edge AI devices due to their hardware. The physical size, power consumption, and heat dissipation limitations of edge devices are different from those of centralised cloud servers with massive computational resources. In contrast to cloud servers, edge devices typically have significantly lower processing power, memory, and storage capabilities. It imposes the following constraints:

- **Processing Power:** Due to the increase in power of edge devices, CPUs, including ARM-based CPUs and artificial intelligence accelerators, are becoming less powerful. However, they may not be able to handle high data volumes or complex artificial intelligence algorithms.

- **Memory and Storage:** Processors and storage devices have limited RAM and storage capacities. In AI models requiring large datasets or extensive features, this can have an adverse effect on performance.
- **Power Consumption:** The majority of edge devices, including IoT sensors and smartphones, are battery-powered or have a strict power budget. Often, these devices need to be recharged, or batteries need to be replaced because of their high-power consumption.
- **Thermal Management:** Edge devices can overheat if they do not have an adequate cooling system. During continuous processing, heat can damage performance and reliability.

To be able to operate within these hardware constraints, AI models need to be lightweight, efficient and scalable. For edge deployments, models are compressed, quantised, and pruned to reduce computational requirements.

- **Software Integration:** AI capabilities can only be incorporated into edge devices by overcoming several software challenges. The software environment of cloud-based systems is easily updated and maintainable, but the software environment of edge devices is often complex and fragmented.
- **Platforms with heterogeneity:** Various operating systems, architectures, and software stacks are available for edge devices. Different platforms can be challenging to integrate and ensure compatibility. Every device has its specific requirements, which must be met by custom solutions or adaptations of existing ones.
- **Real-Time Processing:** A variety of applications require real-time data processing, such as autonomous vehicles or industrial robotics. For real-time scenarios to be reliable and low latency, software solutions must meet rigorous timing constraints.
- **Management of Resources:** To successfully implement AI on edge devices, CPUs, memory, and power need to be managed efficiently. To balance the competing demands of multiple simultaneous requests, it is essential to employ advanced resource allocation and scheduling algorithms [35].
- **Deployment and Maintenance:** With a large number of distributed devices, it is hard to maintain edge devices.

The ability to update and patch remotely managed devices (without requiring physical access) allows updates and patches to be applied over-the-air (OTA) [36].

8.5 Future Directions and Trends

8.5.1 Federated Learning on the Edge

Data can be centralised or distributed without affecting what is being learned through federated learning. The federated learning approach preserves privacy and speeds up AI model improvement by training models locally and sharing only updates.

8.5.2 5G and Edge Synergy

Enhanced performance and reduced latency will be available with 5G connectivity for Edge Computing. As a result of edge-enabled applications, a variety of new advantages will arise, including immersive augmented reality, virtual reality, smart factories, and more.

8.5.3 TinyML for Edge AI

TinyML deploys machine learning models on edge devices that are highly efficient and small-scale. It allows even microcontrollers to run AI algorithms, expanding Edge AI to low-cost, low-power devices and enabling pervasive intelligence.

8.5.4 Integration with Blockchain for Security

In addition to providing tamper-proof data logs and device authentication, blockchain technology can help secure edge devices. With edge computing and blockchain, high-security applications will become possible, particularly in finance.

8.6 Conclusion

Applications requiring latency-sensitive processing are being transformed by edge computing and edge artificial intelligence. With localised data analysis and decision-making, these technologies are capable of reducing latency, enhancing security, and enhancing resilience, which makes them superior to traditional cloud-centric architectures. Industry adoption of Edge AI faces hardware limitations, software integration challenges, and resource management challenges. With edge computing, 5G, federated learning, and TinyML converged, new applications across industries are possible. A future where real-time insights are a vital part of every intelligent system will be made possible by the continued evolution of Edge AI, which will further empower industries with better data security and faster, context-aware data handling.

References

1. Rani, P. and Sharma, R., Intelligent transportation system for internet of vehicles based vehicular networks for smart cities. *Comput. Electr. Eng.*, 105, 108543, 2023.
2. Khan, T., Tian, W., Zhou, G., Ilager, S., Gong, M., Buyya, R., Machine learning (ML)-centric resource management in cloud computing: A review and future directions. *J. Netw. Comput. Appl.*, 204, 103405, Aug. 2022, doi: 10.1016/j.jnca.2022.103405.
3. Bhola, B., *et al.*, Quality-enabled decentralized dynamic IoT platform with scalable resources integration. *IET Commun.*, 2022.
4. Hussain, N. and Rani, P., Comparative studied based on attack resilient and efficient protocol with intrusion detection system based on deep neural network for vehicular system security, in: *Distributed Artificial Intelligence*, pp. 217–236, CRC Press, Boca Raton, 2020.
5. Ansari, G., Rani, P., Kumar, V., A novel technique of mixed gas identification based on the group method of data handling (GMDH) on time-dependent MOX gas sensor data, in: *Proceedings of International Conference on Recent Trends in Computing: ICRTC 2022*, Springer, pp. 641–654, 2023.
6. Yadav, S.P., Jindal, M., Rani, P., de Albuquerque, V.H.C., dos Santos Nascimento, C., Kumar, M., An improved deep learning-based optimal object detection system from images. *Multimed. Tools Appl.*, 83, 10, 30045–30072, 2024.
7. Shuvo, M.M.H., Islam, S.K., Cheng, J., Morshed, B., II, Efficient acceleration of deep learning inference on resource-constrained edge devices: A review. *Proc. IEEE*, 111, 1, 42–91, 2022.
8. Thakur, D., Saini, J.K., Srinivasan, S., DeepThink IoT: The Strength of Deep Learning in Internet of Things. *Artif. Intell. Rev.*, 56, 12, 14663–14730, Dec. 2023, doi: 10.1007/s10462-023-10513-4.
9. Kim, H.-S., Fog computing and the internet of things: Extend the cloud to where the things are. *Int. J. Cisco*, 2872, 5, 181–187, 2016.

10. Shi, W., Cao, J., Zhang, Q., Li, Y., Xu, L., Edge computing: Vision and challenges. *IEEE Internet Things J.*, 3, 5, 637–646, 2016.
11. Bonomi, F., Milito, R., Zhu, J., Addepalli, S., Fog computing and its role in the internet of things, in: *Proceedings of the first edition of the MCC workshop on Mobile cloud computing*, ACM, Helsinki Finland, pp. 13–16, Aug. 2012, doi: 10.1145/2342509.2342513.
12. Satyanarayanan, M., Bahl, P., Caceres, R., Davies, N., The case for vm-based cloudlets in mobile computing. *IEEE Pervasive Comput.*, 8, 4, 14–23, 2009.
13. Hussain, N., Rani, P., Kumar, N., Chaudhary, M.G., A deep comprehensive research architecture, characteristics, challenges, issues, and benefits of routing protocol for vehicular ad-hoc networks. *Int. J. Distrib. Syst. Technol. (IJDST)*, 13, 8, 1–23, 2022.
14. Krizhevsky, A., Sutskever, I., Hinton, G.E., Imagenet classification with deep convolutional neural networks. *Adv. Neural Inf. Process. Syst.*, 25, 1097–1105, 2012, Accessed: Nov. 08, 2024, [Online], Available: https://proceedings.neurips.cc/paper/2012/hash/c399862d3b9d6b76c8436e924a68c45b-Abstract.html.
15. Young, T., Hazarika, D., Poria, S., Cambria, E., Recent trends in deep learning based natural language processing. *IEEE Comput. Intell. Mag.*, 13, 3, 55–75, 2018.
16. Mnih, V., *et al.*, Human-level control through deep reinforcement learning. *Nature*, 518, 7540, 529–533, 2015.
17. Silver, D., *et al.*, Mastering the game of Go with deep neural networks and tree search. *Nature*, 529, 7587, 484–489, 2016.
18. Ananthanarayanan, G., *et al.*, Real-time video analytics: The killer app for edge computing. *Computer*, 50, 10, 58–67, 2017.
19. Abdoos, M., Mozayani, N., Bazzan, A.L., Traffic light control in non-stationary environments based on multi agent Q-learning, in: *2011 14th International IEEE conference on intelligent transportation systems (ITSC)*, IEEE, pp. 1580–1585, 2011, Accessed: Nov. 08, 2024, [Online], Available: https://ieeexplore.ieee.org/abstract/document/6083114/.
20. Wang, X., Han, Y., Leung, V.C., Niyato, D., Yan, X., Chen, X., Convergence of edge computing and deep learning: A comprehensive survey. *IEEE Commun. Surv. Tutor.*, 22, 2, 869–904, 2020.
21. Chen, J. and Ran, X., Deep learning with edge computing: A review. *Proc. IEEE*, 107, 8, 1655–1674, 2019.
22. Wang, S., Zhang, X., Zhang, Y., Wang, L., Yang, J., Wang, W., A survey on mobile edge networks: Convergence of computing, caching and communications. *IEEE Access*, 5, 6757–6779, 2017.
23. Mao, Y., You, C., Zhang, J., Huang, K., Letaief, K.B., A survey on mobile edge computing: The communication perspective. *IEEE Commun. Surv. Tutor.*, 19, 4, 2322–2358, 2017.

24. Mohammadi, M., Al-Fuqaha, A., Sorour, S., Guizani, M., Deep learning for IoT big data and streaming analytics: A survey. *IEEE Commun. Surv. Tutor.*, 20, 4, 2923–2960, 2018.
25. Ma, X., *et al.*, A survey on deep learning empowered IoT applications. *IEEE Access*, 7, 181721–181732, 2019.
26. Verbelen, T., Simoens, P., De Turck, F., Dhoedt, B., Cloudlets: bringing the cloud to the mobile user, in: *Proceedings of the third ACM workshop on Mobile cloud computing and services*, ACM, Low Wood Bay Lake District UK, pp. 29–36, Jun. 2012, doi: 10.1145/2307849.2307858.
27. Bilal, K., Khalid, O., Erbad, A., Khan, S.U., Potentials, trends, and prospects in edge technologies: Fog, cloudlet, mobile edge, and micro data centers. *Comput. Netw.*, 130, 94–120, 2018.
28. Aazam, M. and Huh, E.-N., Fog computing micro datacenter based dynamic resource estimation and pricing model for IoT, in: *2015 IEEE 29th international conference on advanced information networking and applications*, IEEE, pp. 687–694, 2015, Accessed: Nov. 08, 2024, [Online], Available: https://ieeexplore.ieee.org/abstract/document/7098039/.
29. Bonomi, F., Milito, R., Natarajan, P., Zhu, J., Fog Computing: A Platform for Internet of Things and Analytics. in Big Data and Internet of Things: A Roadmap for Smart Environments, in: *Studies in Computational Intelligence*, vol. 546, N. Bessis and C. Dobre (Eds.), pp. 169–186, Springer International Publishing, Cham, 2014, doi: 10.1007/978-3-319-05029-4_7.
30. Hu, Y.C., Patel, M., Sabella, D., Sprecher, N., Young, V., Mobile edge computing—A key technology towards 5G. *ETSI White Pap.*, 11, 11, 1–16, 2015.
31. Liu, S., Liu, L., Tang, J., Yu, B., Wang, Y., Shi, W., Edge computing for autonomous driving: Opportunities and challenges. *Proc. IEEE*, 107, 8, 1697–1716, 2019.
32. Perera, C., Zaslavsky, A., Christen, P., Georgakopoulos, D., Context aware computing for the internet of things: A survey. *IEEE Commun. Surveys Tutor.*, 16, 1, 414–454, 2013.
33. Wang, C., Liu, G., Huang, H., Feng, W., Peng, K., Wang, L., MIASec: Enabling Data Indistinguish ability Against Membership Inference Attacks in MLaaS. *IEEE Trans. Sustain. Comput.*, 5, 3, 365–376, Jul. 2020, doi: 10.1109/TSUSC.2019.2930526.
34. Al-Momani, T. and Al-Hussein, M., Real-Time Decision Making with Edge AI Technologies: Advanced Techniques for Optimizing Performance, Scalability, and Low-Latency Processing in Distributed Computing Environments. *J. Artif. Intell. Mach. Learn. Manage.*, 8, 2, 71–91, 2024.
35. Subramanya, T. and Riggio, R., Centralized and federated learning for predictive VNF autoscaling in multi-domain 5G networks and beyond. *IEEE Trans. Netw. Serv. Manage.*, 18, 1, 63–78, 2021.
36. Raith, P., Rausch, T., Furutanpey, A., Dustdar, S., *faas-sim*: A trace-driven simulation framework for serverless edge computing platforms. *Softw. Pract. Exp.*, 53, 12, 2327–2361, Dec. 2023, doi: 10.1002/spe.3277.

9

Elucidating the Strategic Significance of Smart Grids Towards Sustainable Cities

Abhilasha Jadhav[1], Heena Madan[2*], Mohammed Y. Al-khuzaie[3], Sanjeev Kumar Shah[4] and Mohammed I. Habelalmateen[5]

[1]*Department of Computing Science and Artificial Intelligence, NIMS Institute of Engineering and Technology, NIMS University Rajasthan, Jaipur, India*
[2]*Department of Computer Science Engineering, Chandigarh Engineering College, Chandigarh Group of Colleges, Jhanjeri, Mohali, Punjab, India*
[3]*College of Technical Engineering, National University of Science and Technology, Dhi Qar, Iraq*
[4]*Department of Electronics & Communication Engineering, Uttaranchal Institute of Technology, Uttaranchal University, Dehradun, Uttarakhand, India*
[5]*Department of Computer Science, The Islamic University, Najaf, Iraq*

Abstract

While urbanisation and energy needs continue to rise, smart grids have the potential to transform sustainable cities. With smart grids, electricity is managed efficiently, monitored in real-time, and integrated with digital communication, and automation. Besides promoting energy-conscious behaviour, smart grids also facilitate energy-conscious behaviour and integrate renewable energy sources. By employing energy-conscious behaviour and utilising solar and wind energy sources, smart grids minimise emissions, reduce energy waste, and enhance urban energy security. By improving energy efficiency, enhancing green technology opportunities, and providing equitable access to energy, smart grids support three essential pillars of sustainability–environmental, economic, and social. An examination of energy resilience, consumer empowerment, and the impact of innovative technologies is examined in the chapter, as well as the multifaceted benefits of smart grids for sustainable urban development.

**Corresponding author*: heena2111.research@cgcjhanjeri.in

Pethuru Raj, Sudesh Yadav, Manas Kumar Mishra, Satya Prakash Yadav and Victor Hugo C. de Albuquerque (eds.) Artificial Intelligence Technologies for Smart and Sustainable Urban Transportation: Integrated Platforms and Use Cases, (145–160)

Keywords: Energy efficiency, environmental sustainability, green technology, smart grids, sustainable cities, sustainable development

9.1 Introduction

A growing number of cities are undergoing rapid urbanisation, resulting in an increased demand for energy. This transition will be enabled by smart grid technology, a framework that integrates digital communication, automation, and intelligent energy management with traditional electricity networks. The smart grid of a sustainable city contains elements such as real-time monitoring, renewable energy integration, and grid resilience [1]. Energy distribution is only one strategic advantage of smart grids; they can also reduce carbon emissions, minimise energy waste, and improve energy security. Using data-driven insights, smart grids empower consumers to make energy-conscious choices and to manage demand for power, reducing climate change impacts. Furthermore, smart grids allow for the integration of electric vehicles (EVs) and distributed renewable energies like solar and wind, key components to creating a low-carbon urban landscape [2].

The paper illustrates the importance of smart grids in making cities sustainable. It is evident from this paper that smart grids play an important role in making cities sustainable. Using smart grids, renewable energy can be integrated and decentralised, grid reliability can be improved, and energy efficiency can be enhanced through demand-side management. Smart cities (SC) have received widespread attention from scholars and policymakers over the past two decades. Globalisation depends heavily on cities, so understanding and analysing their development paths is crucial. A key driver of global social and economic development is urbanisation, which significantly impacts environmental sustainability [3]. Using smart grids, renewable energy sources can be integrated and decentralised, grid reliability can be improved, and energy efficiency can be enhanced through demand-side management. By 2050, the number of megacities with populations exceeding 20 million will increase to 70% in Latin America, Africa, and Asia [4]. Cities are becoming more resource-intensive and have an increasingly negative environmental impact as they grow. According to UN-Habitat, more than 60% of global carbon emissions are produced in cities. A denser urban population will require more energy to meet its transportation and other needs [2, 4]. Waste and negative externalities are commonly generated in urban areas, which exacerbates social and economic problems [5]. To boost sustainability, natural resource reserves could be enhanced [6, 7]. In light of this approach, cities need to find sustainable ways to deal with socioeconomic challenges.

Urban services are being improved to respond to environmental pressures and stimulate economic growth in more and more cities throughout the world [8–10]. Public transportation is a key component of this undertaking, and it is adapted to the country's economic needs. For this to succeed, the country's economic needs must be taken into account when designing public transportation systems. As a result of technological advances, smart cities are becoming increasingly sustainable and intelligent. A number of SSCs provide sustainable and effective solutions on a variety of topics. A number of SSCs provide sustainable and effective solutions in various fields, including energy management, transportation, health care, and governance. Even though smart city initiatives have been incorporated into national plans and strategies, there seems to be some reluctance among urban policymakers.

There are many challenges facing cities today, such as the depletion of natural resources, the deterioration of air quality, the ineffective management of waste, traffic congestion, and inadequate infrastructure [11]. As a result of these challenges, SSCs do not develop sustainable but rather stagnant productivity levels and ineffective mitigation of climate change. Furthermore, SSC initiatives should improve their strategic planning knowledge base, including a detailed examination of their objectives, their theoretical foundations, and the challenges associated with their implementation. As urban populations grow at an unprecedented rate, energy demand is increasing, placing a strain on existing energy systems and raising environmental concerns. Creating sustainable cities requires innovative energy infrastructure and delivery systems that protect the environment, reduce resource consumption, and enhance quality of life. Smart grids are emerging as a critical enabler, providing a more intelligent, resilient, and efficient means to distribute and manage energy.

As part of the chapter's strategic analysis, smart grids are examined in terms of their role in promoting energy efficiency, integrating renewable energy, improving grid resilience, and encouraging communities to adopt sustainable energy behaviours. To understand the role smart grids play in sustainable urban development, we need to understand these contributions. To address these problems in urban areas, technology and computer software tools are becoming increasingly important. A multidimensional network involving the natural, the technical, the social, and the computer systems is a multimedia symbiotic network (ISN); according to [12], computer software tools (CS tools V2) play a crucial role in managing ISNs. ISNs can benefit from the sharing of utility services, wastes, and by-products between their stakeholders, which will enhance value, reduce costs, and enhance sustainability.

Additionally, the development of technologies is vital to the successful management of urban areas and the sustainability of SSCs. Furthermore, [13] argues that industrial symbiosis is essential for industrial ecology, emphasising geographical proximity and networks among companies. Industrial symbiosis can be hindered by environmental, geographic, and socioeconomic factors in the northeast of Italy. SSCs can use this information to address similar obstacles. As climate change, resource constraints, and urban population growth place tremendous strain on city infrastructures, resilient and sustainable energy systems are needed more than ever. Energy management is a challenge critical to sustainable urban development, and smart grids provide a transformative solution [14]. Electricity generation, distribution, and consumption have been revolutionised by smart grids in the modern city, leading to more efficiency, effectiveness, and sustainability. With this technology, information and power can flow both ways, enhancing grid responsiveness, security, and efficiency. Smart grids are more important than conventional improvements to the grid from a strategic perspective. Reduced greenhouse gas emissions and a transition to low-carbon economies can be achieved through renewable energy sources such as solar, wind, and hydropower. As electric vehicles (EVs) are widely adopted, smart grids help to manage the increased electricity demands and stabilise the grid during peak usage times while enabling vehicle-to-grid (V2G) energy flows that increase grid resilience even further. Solar, wind, and hydropower are essential for reducing greenhouse gas emissions, transitioning to a low-carbon economy, and supporting clean energy initiatives. Smart grids enable consumers to make informed energy choices based on real-time data, contributing to an economically, ecologically, and socially sustainable world. The benefits of smart grids for urban centres extend beyond their environmental benefits. It reduces energy losses, minimises power outages, and improves the reliability of electricity supplies. It is particularly important in densely populated cities, where power disruptions are plentiful. Moreover, smart grids generate green jobs and stimulate innovation in clean tech industries, which contribute to a robust and sustainable urban economy [15]. The implementation of smart grids facilitates energy-resilient, self-sufficient, and adaptive cities through demand response, decentralised generation, and microgrids. An examination of smart grids' strategic importance in the context of sustainable urbanisation is presented in this paper. In this chapter, we examine smart grids' multiple benefits to modern cities, including their role in promoting energy efficiency, environmental sustainability, economic resilience, and social equity [16] to fully appreciate the potential of smart grids for achieving ambitious environmental targets while improving residents' quality of life.

9.2 Related Work

In 2050, nearly 70% of people will live in urban areas, compared to 50% today. A city emits approximately 70% of the world's carbon dioxide (CO2) [17]. The COVID-19 pandemic has led to a rapid rebound in CO2 emissions, and the rise in 2021 is predicted to be the second-largest on record. Cities that generate 80% of the world's economy have the potential to accelerate progress towards ambitious climate goals [18]. Economic growth is primarily fuelled by cities, which serve as a pivotal hub for global economies. In urban areas, there are 8 billion people, which accounts for 56% of the world's population. They account for over 80% of the worldwide economy. Nearly 400 million urban dwellers are expected to be added to the global population between 2015 and 2020. In the EMDEs (Emerging Markets and Developing Economies), mainly Asia and Africa, urban expansion has accounted for approximately 90% of the expansion. India, China, and Nigeria have demonstrated the most rapid urban growth [19, 20].

The combined effects of urbanisation and the growing global population are rapidly reshaping societal structures like never before. It is estimated that by 2050, 70% of the world's population will live in urban areas, an increase of 1.8 billion people. Over the next three decades, Asia's population will increase by approximately one-third, while Africa's population will double [21]. In 1950, the total urban population was 1.1 million. In 2050, it is forecasted to be 2.8 million. A total of 1.8 billion people is expected to move to urban areas between 2024 and 2050, representing a 70% rise in urban population. Due to rapid population growth and urbanisation, urban areas produce a greater proportion of greenhouse gases today than they did in 2015, which was 62%. As a result of urbanisation, urban releases are projected to peak at almost 29 billion tonnes of carbon dioxide by 2023 [22–24]. In achieving sustainable urban development, smart grids play a key role in solving urban energy challenges as mentioned in Figure 9.1. Urban energy systems can be made more resilient and adaptable through smart grid technologies [25, 26], in addition to improving efficiency, reliability, and sustainability. Consequently, energy waste can be reduced, costs can

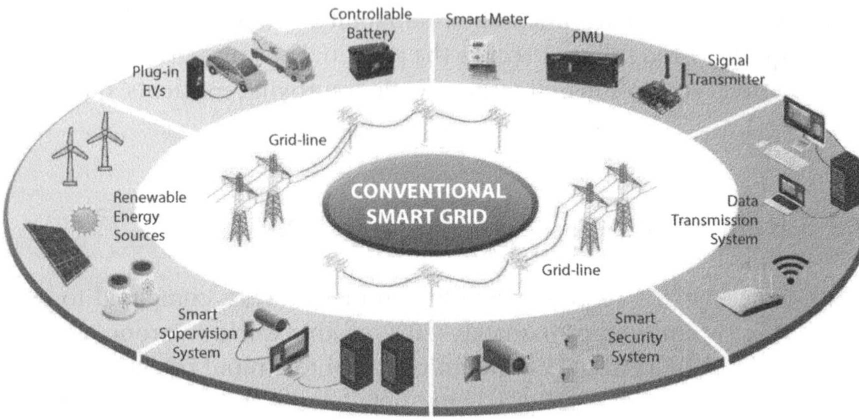

Figure 9.1 This diagram illustrates how a conventional smart grid works.

be reduced, and greenhouse gas emissions can be reduced, aligning with broader environmental and sustainability objectives. A high demand for electricity and complex infrastructure makes urban areas especially susceptible to power outages.

It is possible to detect, diagnose, and repair faults automatically with smart grids, which minimises power outages and their impact. Solar and wind power can be more easily integrated into urban energy supplies, thanks to smart grids. Smart grid technologies can support sustainable and climate-resilient urban policies. Their services include dynamic pricing, the management of electric vehicle charging networks, and the establishment of new regulatory frameworks for energy trading and carbon credits. They also assist cities in reducing their carbon footprints. In a number of studies, smart grid technologies are being integrated into sustainable urban planning. The majority of these studies focus on making cities more reliable, efficient, and sustainable through smart grids. The integration of solar power, demand response, distributed generation, and advanced metering infrastructure can reduce urban carbon footprints and improve urban energy management. In addition to escalating energy demands, climate change requires a paradigm shift regarding energy efficiency and environmental protection [27]. A crisis is likely to result from unchecked burgeoning energy demand caused by advances in urban and national infrastructures. Power flow systems with advanced digital power are an option for resolving this impending energy crisis. In addition to displaying self-healing capabilities and facilitating interoperability, this system also features robust cyber defence aimed at preventing malicious intrusions.

Renewable energies, such as solar, wind, and energy storage, are seamlessly integrated into smart grids. Therefore, for smart grids to be adopted and effective, research and policymakers need to conceptualise them properly and stress their importance.

A recent paper examines the potential impacts of emerging technologies on reducing carbon emissions, including smart grids and nanotechnology [28]. Energy efficiency is the primary objective of the discussion of smart grids because they must have certain abilities and characteristics to be effective. Furthermore, the technologies that enable these innovations, as well as their anticipated benefits, are discussed. The evidence in this review indicates that smart grids will become increasingly important in the development of clean energy strategies in the future. The smart grid will become even more mature and optimised as nanotechnology plays an increasingly important role. As a result, they play a key role in the transformation of energy systems and the reduction of pollution in the environment.

- **Smart Grid:** As opposed to a smart grid, it typically refers to a network that delivers electricity to consumers. "Smart" is sometimes used to describe intelligence, efficiency, cleanliness, orderliness, or automation. At the moment, there is no universal definition of smart grids, which leaves the concept open to both simplistic and complex interpretations. Smart grids are electrical systems that have a brain in layperson's terms. NIST offers the following technical definition of smart grids [29]. With a modernised grid, energy will be delivered bidirectional, along with communication and control, making it possible to develop a variety of new applications and functions. As defined by IEEE definitions [30], a smart grid is an innovative system with new communication, control, and generation capabilities.

Listed below are the IEC definitions of smart grids [31].

- With the smart grid, users and the power grid can integrate their actions intelligently to ensure that power is reliable, cost-effective, and sustainable. The US Department of Energy defines a smart grid as a system of electric power supply that uses digital technologies to improve reliability, security, and efficiency (both economically and environmentally), as well as large-scale generators and utility systems, there is an increase in distributed energy and storage resources [32, 33]. Smart grids facilitate energy and information exchange through advanced infrastructure. As a result, utility companies have the opportunity to manage the transmission and distribution of energy efficiently while also empowering consumers to make informed decisions about their energy consumption [34, 35]. Smart grids maintain energy supply and demand balance through real-time communications and data from information technology. Utility companies and consumers will be able to exchange energy bidirectional and knowledge reciprocally through a smart grid, unlike a traditional power grid [36, 37]. In addition to improved reliability and infrastructure efficiency, the smart grid can help increase economic growth, and defend against cyberattacks and power failures. Accordingly, a smart grid may be defined as a digital, self-healing network of power grids that are interoperable, self-repairing, and capable of prediction under varying uncertainties. The grid is also protected from malicious attacks by implementing security measures. A smart grid increases the efficiency of electricity generation, distribution, and consumption by leveraging digital technology, communication networks, and automation. A smart grid monitors, analyses, and controls energy demand and supply in real-time, allowing it to react dynamically to fluctuations in energy demand. The following components make up a smart grid:

Advanced Metering Infrastructure (AMI): AMI is responsible for measuring, analysing, and collecting energy usage, which is done on a regular basis or as needed. These include hardware, software, communication systems, energy displays, controllers, customer-related systems, meter data management (MDM), software and supplier business systems [38]. AMI is becoming a key component of smart grid initiatives all over the world. Most utilities don't currently have two-way communication in their automated meter reading system, so smart grid implementation can't be achieved. In AMI, software used for managing meter data, revenue, and assets is interconnected with a complex communication network and information technology infrastructure. With AMI technology, you can charge based on time of use, respond to demand, and connect/disconnect loads remotely, ensures transparency and demand-responsiveness by collecting energy usage data in real-time.

Distribution Automation Systems: Power distribution grids can be monitored, controlled, and optimised using distribution automation (DA) systems, which use sensors, processors, networks, and switches. Power grids can become more reliable and efficient with DA systems, which can reduce outages and costs for customers. During faults, automate power rerouting to improve efficiency and reliability. Automating distribution systems is one of the most efficient and economical methods of improving power supply reliability, quality, and capacity. Additionally, it is an important foundation for achieving smart grid goals. Almost forty years have passed since distribution network automation was first introduced in industrialised countries. Electric power companies have benefited greatly from distribution network automation over the past few decades due to improved customer service and increased economic efficiency. In modern grid management, it has become an essential component. Rural and urban power grids in China have been upgraded, enhancing their reliability. Only by automating distribution networks can power consumers continue to receive a quality power supply and service. Automation of distribution networks combines modern electronic technology, communication technology, computer networking technology, and power system technology, as well as monitoring, protecting, controlling, and measuring distribution networks under normal and emergency conditions with organic management of power supply departments. As a result, it can more effectively and efficiently provide power supply services, build deeper relationships with energy consumers, and improve enterprise management.

Renewable Integration Technologies: A renewable energy integration process is used to integrate renewable energy into the electric grid. Solar energy, wind energy, and water power will be self-renewing sources of energy in the future. Future energy sources could include wind, sunlight, and water. A low or zero carbon footprint usually refers to natural or self-renewing resources.

Consumer-Oriented Technologies: A consumer-oriented technology is a digital device, tool, or application designed with a focus on improving the end-user's experience. Technology enables consumers to communicate, entertain, shop, manage finances, get healthy, and interact with others. E-commerce platforms and smart home systems are a few examples. The use of consumer-oriented technology tends to emphasise ease of use, accessibility, personalisation, and real-time functionality, allowing individuals to make better choices in their lifestyles. Using real-time price signals to adjust consumers' consumption in energy management.

9.3 Smart Grids as a Catalyst for Sustainability in Urban Environments

The implementation of smart grids has the potential to reshape cities by addressing economic, environmental, and social sustainability. Enhanced energy efficiency, reduced emissions, and economic growth are all benefits of smart grid implementation.

- **Environmental Sustainability**

Reduced Greenhouse Gas Emissions: As cities use smart grids, their electricity consumption is optimised, resulting in a reduction in fossil fuel usage. Renewable energy sources can greatly reduce greenhouse gas emissions in cities, which aligns with global climate goals.
Energy Efficiency and Conservation: A smart grid can provide consumers with real-time feedback and monitor their energy consumption [39]. Power plants' energy load can be lowered, and emissions can be further reduced by shifting electricity consumption to off-peak hours through advanced metering and demand response technologies.

- **Economic Sustainability**

Cost Savings for Utilities and Consumers: Smart grids reduce the costs associated with electricity distribution and grid maintenance through predictive maintenance and demand forecasting. These cost savings translate into lower electricity prices for consumers, promoting long-term economic sustainability.
Stimulus for Green Job Creation: Smart grids require skilled workers for design, installation, and maintenance [40]. The growth of the local economy is driven by jobs created in engineering, information technology, and data science as a result of this demand.

- **Social Sustainability**

Enhanced Resilience to Power Outages: As a result of smart grids, power can be restored more quickly during outages when disturbances occur, and power can be detected and responded to. Modern cities require this level of reliability, as interruptions can cause a decline in quality of life, economic losses, and public safety concerns.
Increased Energy Access and Equity: Microgrids can provide power to remote or underserved communities when smart grids are installed [41]. Providing more affordable and consistent access to energy is one of the

main benefits of smart grid deployments for enhancing social equity and improving urban well-being.

Mixing Renewable Energy Sources with Conventional Energy: A smart grid can incorporate renewable energy sources such as solar and wind. Achieving sustainable cities requires this capability since reducing fossil fuel reliance is of utmost importance. Rooftop solar panels can be used as a source of energy for smart grids with the help of distributed sources of energy [42]. As a result of this technology, cities are able to establish distributed energy networks that provide cleaner, localised power, reducing the transmission losses and emissions associated with long-distance power transmission.

Demand Flexibility: Consumers can adjust their energy use during peak demand periods and periods of supply constraints with the help of a smart grid. Energy supply from renewables and consumer demand can be balanced smoothly with smart grids because they encourage flexible usage patterns. In smart grids, battery storage technologies can be incorporated to store excess renewable energy when production exceeds demand [43]. In periods of low renewable generation, stored energy can be used, improving grid stability and reliability.

9.4 Smart Grid Technologies: Enabling Real-Time Decision Making

Smart grids are powered by advanced technologies that collect, analyse, and utilise data to optimise energy distribution and usage [44]. Some of these critical technologies include:

Internet of Things (IoT): The IoT allows efficient monitoring and organisation of energy resources across the grid, thanks to IoT devices.

Machine Learning (ML) and Artificial Intelligence (AI): AI and machine learning algorithms allow utilities to predict demand and consumption patterns to allocate energy more efficiently.

Blockchain Technology: Using blockchains, consumers can directly trade excess energy with their community members and support peer-to-peer energy trading.

Utility companies and consumers can respond promptly to shifts in energy supply and demand, thanks to these technologies, thereby increasing the overall sustainability of smart grids.

9.5 Challenges in Implementing Smart Grids for Sustainable Cities

Although smart grids provide many benefits, they face several challenges before they can be widely adopted:

High Initial Costs: Investing in sensors, software, and consumer interfaces is an important element of the smart grid infrastructure upgrade. It can be difficult for some cities to secure the necessary capital, even though the long-term savings are substantial.

Data Privacy and Cybersecurity: It is crucial to protect the data that smart grids generate, including consumption patterns and grid activity, from unauthorised access [45]. The grid's security is of utmost importance, as threats to the grid may disrupt citywide power supplies and compromise sensitive data.

Interoperability of Technologies: Integrated smart grids require a variety of technologies and protocols that work seamlessly. The efficiencies and reliability of a smart grid depend on its compatibility with diverse devices and systems.

9.6 Case Studies: Smart Grid Implementation in Sustainable Cities

To achieve sustainability, cities around the world are implementing smart grids. Among the most notable examples are:

Amsterdam, Netherlands: An advanced energy management system and smart meters are being used as part of a pilot project in Amsterdam involving a smart grid. In addition to promoting demand-response programs, the project incorporates renewable energy sources to reduce energy consumption.

San Diego, California: The smart grid initiative in San Diego aims to integrate renewable energy, improve grid reliability, and support electric vehicle charging stations. A significant amount of money has been invested in smart grid infrastructure in the city, making it an industry leader in sustainable energy. Smart grids can improve energy efficiency and pollution reduction in cities.

9.7 Conclusion

Sustainable cities require smart grids, which are highly responsive, efficient, and resilient when it comes to addressing urban energy challenges. With smart grids, renewable energy can be integrated and decentralised, grid reliability can be improved, and energy efficiency can be improved through demand-side management. A sustainable urban ecosystem is cultivated when consumers are empowered to make informed decisions, energy costs are reduced, and sustainable energy behaviours are cultivated. There are still challenges to smart grid implementations, including high initial costs and data security, but successful implementations in cities worldwide demonstrate this technology's potential to drive urban sustainability. To harness the potential of smart grids for sustainable, resilient, and low-carbon cities, we must address these challenges and foster collaboration.

References

1. Heidari, A., Navimipour, N.J., Unal, M., Applications of ML/DL in the management of smart cities and societies based on new trends in information technologies: A systematic literature review. *Sustain. Cities Soc.*, 85, 104089, Oct. 2022, doi: 10.1016/j.scs.2022.104089.
2. Almuqrin, A., Mutambik, I., Alomran, A., Zhang, J.Z., Enforcing Information System Security: Policies and Procedures for Employee Compliance. *Int. J. Semantic Web Inf. Syst. (IJSWIS)*, 19, 1, 1–17, 2023.
3. Bibri, S.E., Alexandre, A., Sharifi, A., Krogstie, J., Environmentally sustainable smart cities and their converging AI, IoT, and big data technologies and solutions: an integrated approach to an extensive literature review. *Energy Inform*, 6, 1, 9, Apr. 2023, doi: 10.1186/s42162-023-00259-2.
4. Salamanca-Fonseca, M., *et al.*, Effects of urban, peri-urban and rural land covers on plant functional traits around Bogotá, Colombia. *Urban Ecosyst.*, 27, 1, 251–260, Feb. 2024, doi: 10.1007/s11252-023-01429-6.
5. Peráček, T., Majerčáková, D., Mittelman, A., Significance of the waste act in the context of the right to protection of the environment. *International Multidisciplinary Scientific GeoConference Surveying Geology and Mining Ecology Management, SGEM*, vol. 1, pp. 979–986, 2016.
6. Chan, C.S. and Tsun, W.Y., Unleashing the potential of local brand equity of Hong Kong as a green–creative–smart city. *JPMD*, 17, 1, 21–48, Jan. 2024, doi: 10.1108/JPMD-12-2022-0122.
7. Mutambik, I., Digital Transformation as a Driver of Sustainability Performance—A Study from Freight and Logistics Industry. *Sustainability*, 16, 10, 4310, May 2024, doi: 10.3390/su16104310.

8. Raihan, A., Nexus between greenhouse gas emissions and its determinants: The role of renewable energy and technological innovations towards green development in South Korea. *Innov. Green Dev.*, 2, 3, 100066, Sep. 2023, doi: 10.1016/j.igd.2023.100066.
9. Govindarajan, H.K. and Ganesh, L.S., Renewable energy for electricity use in India: Evidence from India's smart cities mission. *Renew. Energy Focus*, 38, 36–43, Sep. 2021, doi: 10.1016/j.ref.2021.05.005.
10. Mutambik, I., Almuqrin, A., Liu, Y., Alhossayin, M., Qintash, F.H., Gender Differentials on Information Sharing and Privacy Concerns on Social Networking Sites: Perspectives From Users. *J. Glob. Inf. Manage.*, 29, 3, 236–255, May 2021, doi: 10.4018/JGIM.2021050110.
11. Skalová, B., Jaššo, M., Husár, M., Ondrejička, V., Cooperation of the city management with the public and private sector in relation to the implementation of climate mitigation and adaptation measures. *Proceedings of the 10th Workshop on Metallization and Interconnection for Crystalline Silicon Solar Cells*, 160001, 2022, doi: 10.1063/5.0105175.
12. Leśniewski, M., *et al.*, Towards TransitOriented Development for Sustainable Urban Mobility: Insights from a Central European City. *GaEE*, 15, 3, 39–53, Apr. 2021, doi: 10.7494/geom.2021.15.3.39.
13. Mutambik, I., Almuqrin, A., Alharbi, F., Abusharhah, M., How to Encourage Public Engagement in Smart City Development—Learning from Saudi Arabia. *Land*, 12, 10, 1851, Sep. 2023, doi: 10.3390/land12101851.
14. Roggero, M., Fjornes, J., Eisenack, K., Ambitious climate targets and emission reductions in cities: a configurational analysis. *Clim. Policy*, 25, 1, 29–43, Nov. 2023, doi: 10.1080/14693062.2023.2282488.
15. Nassar, Y.F., *et al.*, Carbon footprint and energy life cycle assessment of wind energy industry in Libya. *Energy Convers. Manage.*, 300, 117846, 2024.
16. McCann, L., Hutchison, N., Adair, A., The role of UK universities as economic drivers in a localisation agenda: A case study of City Deals. *Land Use Policy*, 134, 106938, 2023.
17. Minas, A.M., *et al.*, Advancing Sustainable Development Goals through energy access: Lessons from the Global South. *Renew. Sustain. Energy Rev.*, 199, 114457, 2024.
18. Ben-Ishai, G., Dean, J., Manyika, J., Porat, R., Varian, H., Walker, K., AI and the Opportunity for Shared Prosperity: Lessons from the History of Technology and the Economy, Feb. 01, 2024, arXiv: arXiv:2401.09718. Accessed: Nov. 09, 2024, [Online], Available: http://arxiv.org/abs/2401.09718.
19. Yang, J., A study of the economic growth effects of market integration: An examination of 27 cities in the Yangtze River Delta city cluster. *PLoS One*, 18, 11, e0287970, 2023.
20. Siokas, G. and Tsakanikas, A., The role of economic and innovation initiatives in planning a smart city strategy in Greece. *Sustainability*, 15, 20, 14842, 2023.

21. Hoornweg, D., Sugar, L., Trejos Gómez, C.L., Cities and greenhouse gas emissions: moving forward. *Environ. Urbanization*, 23, 1, 207–227, Apr. 2011, doi: 10.1177/0956247810392270.
22. Cronin, J., Anandarajah, G., Dessens, O., Climate change impacts on the energy system: a review of trends and gaps. *Clim. Change*, 151, 2, 79–93, Nov. 2018, doi: 10.1007/s10584-018-2265-4.
23. Abulifa, S., Elbar, M., Mohamed, M., Khoudiri, A., Khoudiri, S., Performance evaluation of MG systems interfaced with wind turbines employing DFIG technology. *Int. J. Electr. Eng. Sustain.*, 2, 2, 22–35, 2024.
24. Pereira, P., *et al.*, Nature-based solutions for carbon sequestration in urban environments. *Curr. Opin. Environ. Sci. Health*, 37, 100536, 2024.
25. Geng, Y., Zhang, N., Zhu, R., Research progress analysis of sustainable smart grid based on CiteSpace. *Energy Strat. Rev.*, 48, 101111, 2023.
26. Bhattarai, T.N., Ghimire, S., Mainali, B., Gorjian, S., Treichel, H., Paudel, S.R., Applications of smart grid technology in Nepal: status, challenges, and opportunities. *Environ. Sci. Pollut. Res.*, 30, 10, 25452–25476, Feb. 2022, doi: 10.1007/s11356-022-19084-3.
27. Dorji, S., Stonier, A.A., Peter, G., Kuppusamy, R., Teekaraman, Y., An extensive critique on smart grid technologies: Recent advancements, key challenges, and future directions. *Technologies*, 11, 3, 81, 2023.
28. Markovic, D., Branovic, I., Popovic, R., Smart Grid and nanotechnologies: a solution for clean and sustainable energy. *EECT*, 1–13, doi: 10.2147/EECT.S48124.
29. Jamshidi, M., Yahya, S., II, Nouri, L., Hashemi-Dezaki, H., Rezaei, A., Chaudhary, M.A., A super-efficient GSM triplexer for 5G-enabled IoT in sustainable smart grid edge computing and the metaverse. *Sensors*, 23, 7, 3775, 2023.
30. Khaleel, M., Yusupov, Z., Alfalh, B., Guneser, M.T., Nassar, Y., El-Khozondar, H., Impact of Smart Grid Technologies on Sustainable Urban Development. *Int. J. Electr. Eng. Sustain.*, 2, 2, 62–82, 2024.
31. Khaleel, M., Intelligent control techniques for microgrid systems. *Brilliance: Res. Artif. Intell.*, 3, 1, 56–67, 2023.
32. Khalid, M., Smart grids and renewable energy systems: Perspectives and grid integration challenges. *Energy Strat. Rev.*, 51, 101299, 2024.
33. Khaleel, M.M., Ghandoori, T.M., Ahmed, A.A., Alsharif, A., Alnagrat, A.J.A., Abulifa, A.A., Impact of mechanical storage system technologies: A powerful combination to empowered the electrical grids application, in: *2022 IEEE 2nd International Maghreb Meeting of the Conference on Sciences and Techniques of Automatic Control and Computer Engineering (MI-STA)*, IEEE, pp. 628–636, 2022, Accessed: Nov. 09, 2024, [Online], Available: https://ieeexplore.ieee.org/abstract/document/9837670/.
34. Kataray, T., *et al.*, Integration of smart grid with renewable energy sources: Opportunities and challenges–A comprehensive review. *Sustain. Energy Technol. Assess.*, 58, 103363, 2023.

35. Khaleel, M., Abulifa, S.A., Abulifa, A.A., Artificial intelligent techniques for identifying the cause of disturbances in the power grid. *Brilliance: Res. Artif. Intell.*, 3, 1, 19–31, 2023.
36. Hasan, M.R., Revitalizing the electric grid: A machine learning paradigm for ensuring stability in the USA. *J. Comput. Sci. Technol. Stud.*, 6, 1, 141–154, 2024.
37. Khaleel, M., Yusupov, Z., Nassar, Y., El-khozondar, H.J., Ahmed, A., Alsharif, A., Technical challenges and optimization of superconducting magnetic energy storage in electrical power systems. *e-Prime Adv. Electr. Eng. Electron. Energy*, 5, 100223, 2023.
38. Kabeyi, M.J.B. and Olanrewaju, O.A., Smart grid technologies and application in the sustainable energy transition: a review. *Int. J. Sustain. Energy*, 42, 1, 685–758, Dec. 2023, doi: 10.1080/14786451.2023.2222298.
39. Rani, P. and Sharma, R., Intelligent transportation system for internet of vehicles based vehicular networks for smart cities. *Comput. Electr. Eng.*, 105, 108543, 2023.
40. Rani, P. and Sharma, R., IMFOCA-IOV: Intelligent Moth Flame Optimization based Clustering Algorithm for Internet of Vehicle, in: *2023 14th International Conference on Computing Communication and Networking Technologies (ICCCNT)*, IEEE, pp. 1–6, 2023, Accessed: Nov. 09, 2024, [Online], Available: https://ieeexplore.ieee.org/abstract/document/10307646/.
41. Rani, S., Talwar, R., Malhotra, J., Ahmed, S.H., Sarkar, M., Song, H., A novel scheme for an energy efficient Internet of Things based on wireless sensor networks. *Sensors*, 15, 11, 28603–28626, 2015.
42. Rani, P., Verma, S., Yadav, S.P., Rai, B.K., Naruka, M.S., Kumar, D., Simulation of the lightweight blockchain technique based on privacy and security for healthcare data for the cloud system. *Int. J. E-Health Med. Commun. (IJEHMC)*, 13, 4, 1–15, 2022.
43. Rani, P., Hussain, N., Khan, R.A.H., Sharma, Y., Shukla, P.K., Vehicular Intelligence System: Time-Based Vehicle Next Location Prediction in Software-Defined Internet of Vehicles (SDN-IOV) for the Smart Cities, in: *Intelligence of Things: AI-IoT Based Critical-Applications and Innovations*, F. Al-Turjman, A. Nayyar, A. Devi, P.K. Shukla (Eds.), pp. 35–54, Springer International Publishing, Cham, 2021, doi: 10.1007/978-3-030-82800-4_2.
44. Deku, G., *et al.*, Exploring rat meat consumption patterns, and perception of risks regarding urban rats; implications for rat-borne zoonoses outbreaks and drug resistant pathogens spread in urban areas of Ghana. *medRxiv*, 2024–04, 2024.
45. Aboulnaga, M.M., Badran, M.F., Barakat, M.M., Global Informal Settlements and Urban Slums in Cities and the Coverage, in: *Resilience of Informal Areas in Megacities – Magnitude, Challenges, and Policies*, pp. 1–51, Springer International Publishing, Cham, 2021, doi: 10.1007/978-3-030-87794-1_1.

10

Describing the Needs for Connected Electric Vehicles for Better Air Quality

Shivakrishna Dasi[1], Jasgurpreet Singh Chohan[2*], Saroj Kumar Gupta[3], Rajesh Singh[4] and Myasar Mundher Adnan[5]

[1]Department of Computing Science and Artificial Intelligence, NIMS Institute of Engineering and Technology, NIMS University Rajasthan, Jaipur, India
[2]School of Mechanical Engineering, Rayat Bahra University, Mohali, India
[3]Department of Computer Science and Engineering, Chandigarh Engineering College, Chandigarh Group of Colleges-Jhanjeri, Mohali, India
[4]Uttaranchal Institute of Technology, Uttaranchal University, Dehradun, Uttarakhand, India
[5]Department of Computer Science and Technology, The Islamic University, Najaf, Iraq

Abstract
As part of this chapter, we examine how connected and autonomous electric vehicles (CAEVs) can improve urban air quality, reduce traffic congestion, and increase road safety. Vehicle-to-vehicle communication and 5G connectivity enable safer and more efficient transportation through the use of autonomous vehicles (AV) and vehicular ad-hoc networks (VANETs). Electric vehicles (EVs) are proven to reduce harmful emissions and improve urban air quality when compared to traditional vehicles. In this paper, different CAEV technologies, charging infrastructure needs, and location optimization models are discussed to support the wide-scale adoption of EVs in city areas. Additionally, the paper proposes an assessment framework that involves performance-based criteria for assessing CAEV's safety, mobility, and environmental impacts. The following framework has been developed to enhance public health and sustainability in urban transportation systems using CAEV technology.

**Corresponding author*: jaskhera@gmail.com

Pethuru Raj, Sudesh Yadav, Manas Kumar Mishra, Satya Prakash Yadav and Victor Hugo C. de Albuquerque (eds.) Artificial Intelligence Technologies for Smart and Sustainable Urban Transportation: Integrated Platforms and Use Cases, (161–176)

Keywords: Autonomous vehicles, electric vehicles, mobility, performance assessment, sustainability, traffic congestion, urban air quality, urban transportation, 5G connectivity

10.1 Introduction

Metropolitan cities face several social and economic challenges, including traffic congestion and vehicle accidents. The number of deaths caused by vehicle crashes in 2013 worldwide was 1.25 million. Over the past decade, autonomous and connected vehicles (CAVs) have undergone significant advancements, becoming increasingly sophisticated [1, 2]. The vehicles were now able to communicate, and therefore, drivers were provided with timely information, which resulted in fewer accidents. Vehicular ad hoc networks (VANETs) are the most commonly used model for connected vehicles (CVs) [3]. With VANETs, vehicles can access the internet in real-time. Through VANETs, drivers and vehicles can be made more aware and safe [4]. Moreover, LTE-V2X solutions are being deployed on open roads [5]. Five-gigabit automotive networks are expected to offer improved performance and low latency, according to the 5GAA [6]. There are two types of connections vehicles can make on the road: V2V and V2I [7]. As such communication becomes more efficient, it will lead to an increase in safety, efficiency, and sustainability in transportation networks. Vehicles' high speeds, density of users, and bandwidth requirements for real-time traffic information communication impose several challenges on connected vehicles because of their high speeds. When ICT (Information and Communication Technologies) are used alone, performance will be suboptimal. V2X services should be developed using a combination of state-of-the-art technologies and LTE networks [8]. To maximize the usefulness of V2X services, CVs need to be optimized in terms of perception, planning, and communication. With the help of 5G technology, CV systems should become more efficient and superior [9, 10].

An autonomous vehicle is a combination of multiple technologies that achieves a certain level of autonomy. It is anticipated that such a combination would lead to an elimination of car accidents, a reduction in congestion, a reduction in driver and vehicle awareness, an increase in vehicle efficiency, a reduction in emissions, and an increase in passengers' comfort and safety. In recent years, AVs have progressed rapidly up the scale of autonomy. Today's vehicles don't have features such as adaptive cruise control or lane-keeping assistance at level one, which is what your car should have.

As self-control features are added, such as steering, acceleration, and deceleration, partial automation is achieved. It is imperative that a human driver monitors the environment and intervenes as needed at all levels. A partial automation level, a conditional automation level, and a high automation level are the types of automation offered. Automated vehicles require less human intervention as they become more automated, eventually driving themselves [11]. The health and environmental effects of urban air quality are causing global concern [12]. A rational urban layout and transportation system are important to reducing air pollution by reducing emissions and reducing their spread [13]. Mobile pollution caused by motor vehicle exhaust has become more evident over the years [14], and its adverse impact on urban environments is becoming more evident. Researchers have studied topics such as traffic flow management, vehicle exhaust emission standards, and public transportation system designs in the context of monitoring and controlling traffic emissions [15]. A number of pollutants can be released from vehicle tailpipes, including nitrogen dioxide, respirable particulates, carbon monoxide, and volatile organic compounds. Within the transportation sector, researchers are evaluating the potential for electricity to reduce air pollution and address climate change by examining its environmental advantages and economic feasibility [16]. For cities to reduce air pollution, it is crucial to continue decarbonizing the transport sector, especially the private car, which accounts for the majority of vehicles in cities [17]. Thus, electrical vehicles (EVs) should be promoted and charging facilities should be expanded to improve urban air quality. As EVs are used over a long period, long-term reductions in carbon emissions will gradually expand despite some emissions produced during production. Furthermore, this will improve air quality as well as improve urban heat islands, which will benefit public health. In a study conducted by the author [18], the relationship between EV penetration and environmental promotion in key Chinese urban clusters was identified as being more beneficial to health and the environment than to the climate.

In transportation, shifting vehicles to electric power is the most effective way to meet GHG reduction targets. In addition to reducing carbon emissions, electric vehicles do not use excessive amounts of fossil fuels. Fuel cells, plug-in hybrids, and pure electric vehicles can all be classified as EVs based on their type. Charging stations for electric vehicles (EVCS) and stations for hydrogen fuel cells are among the types of infrastructure needed to promote and support clean energy transportation [19, 20]. A hydrogen fuel cell vehicle has lower emissions than a conventional vehicle [19], which reduces air pollution and improves urban air quality. Electric vehicles are more efficient at converting energy into power, meaning they can travel further while using less gasoline, resulting in lower total energy consumption and emissions.

10.2 Related Work

A number of studies have been conducted to address the concerns raised by the location of charging stations for electric vehicles. Planning charging stations involves determining how many chargers are needed at each station and supplying them with those chargers. Station locations are based on two common approaches. In the first approach, charging stations are clustered according to demand and deployed in accordance with that demand [21]. A second approach considers demands as flows in a network [22] and places charging stations at major intersections to capture as much flow as possible. The model is modified to take into account the charging behavior of the vehicles to locate the locations that will serve the most vehicles [23–25]. There is a good understanding of EV users' intercity trips according to models based on demand as flows or clusters at nodes in the network, but their urban trips are less well understood due to a combination of trips. An urban charging demand model based on trips is presented in this study to simulate EV charging behavior in urban areas. Users who travel intercity start with similar charges under the current models. Due to EVs' varying states of charge at the beginning of their urban trips, it is imperative to consider the charging demand more closely. This study contributes greatly to the understanding of variations in charging demand associated with travel types and lands. Urban networks can estimate charging demand using data-driven models, which take into account endpoints, distances, purposes, departure times, and arrival times from travel surveys. Based on charging demand, candidate charging stations are selected, and unfulfilled trips are minimized [26, 27]. Micro-simulation approaches based on travel surveys can be used to create data sets, including vehicles' travel paths. As a result of this approach, complex problems under various hypothetical scenarios can be solved, which, owing to the small sample size of survey data, are often impossible to resolve directly. The proposed method simulates vehicle trajectory and captures temporal variations in travel times and shortest paths using detailed information about the departure time and battery status. Planning and designing infrastructure in urban networks (especially charging stations for EVs) should take into account these characteristics, as demonstrated in this study's micro-simulation model. Using simulation tools, real-time data regarding EV charging states along their trip trajectories were generated. Simulated optimization models are used to locate charging stations to minimize investment costs and/or trajectory failures [28, 29]. Moreover, based on real-time taxi trajectory data, we developed a time-series simulation model to analyze plug-in hybrid electric vehicles charging behavior, which can be employed to optimize charging station development [30]. An optimization framework is applied to the development of a state-of-charge simulator to address the unavailability of charging and trip trajectory data.

10.2.1 Battery Electric Vehicles

A battery-powered vehicle was first invented in 1832 by Robert Anderson, while others were created by inventors in Hungary, the Netherlands, and the United States [31]. Since 2008, electric vehicle sales have increased by approximately 43% a year, reaching approximately 3.2 million in 2022 [32, 33]. According to Figure 10.1, there are over 500,000 electric vehicles on the road worldwide. A hybrid car, a plug-in hybrid car, a fuel-cell electric car, and a battery-electric car are all types of electric vehicles. In Figure 10.2, you can see different types of EV drives. For more information, see [34].

There are two ways to charge electric vehicles: unidirectional or bidirectional. Control units communicate not only charge amount, flow type, and charging time but also how the charging process is going. An electric vehicle (EV) combines the power of an ICE with the propulsion of an electronic motorized. Depending on the configuration, hybrid electric vehicles can be classified as parallel, series, or combined electric vehicles [35, 36]. PHEV can decrease pollution and transport prices [37]. PHEVs have been calculated using a novel approach, focusing on efficiency and accuracy in calculations of reliability. An analytical model based on PHEVs has been developed to assess smart grid reliability. These experiments showed that increasing the number of discretized states for PHEV traits could result in significant speed improvements in smart grid reliability computations. Electric vehicles produce no emissions since they are powered by electricity. Researchers are exploring the development of a workable electric vehicle model.

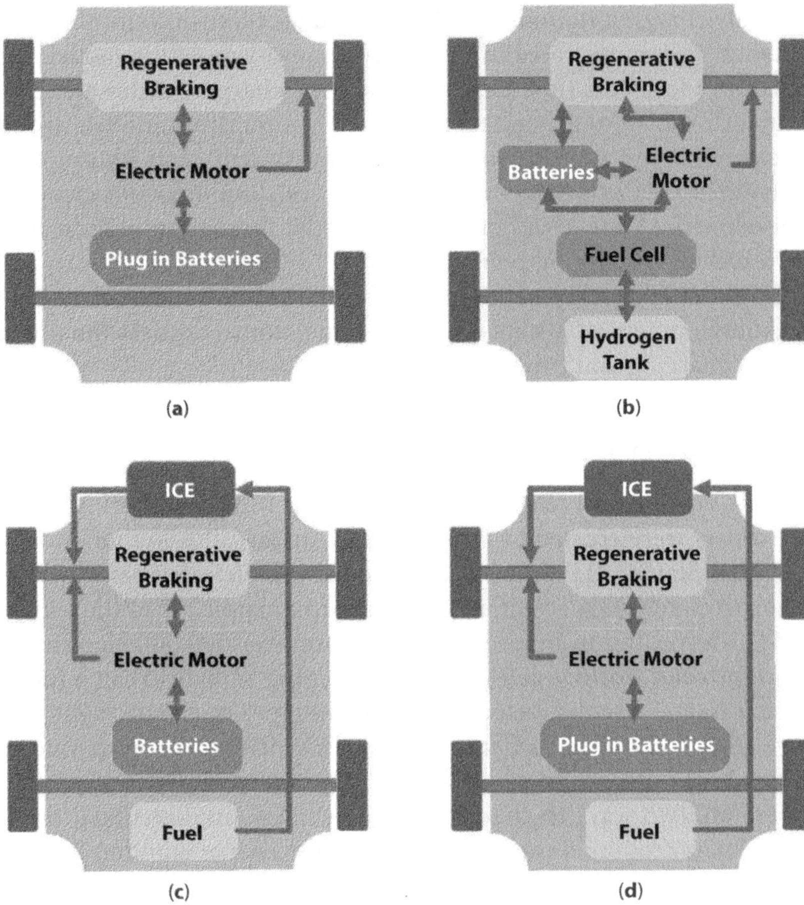

Figure 10.1 Energy knowledge in (a) BEVs, (b) FCEVs, (c) HEVs, and (d) PHEVs.

10.3 Performance Aspects of CAEVs

It has been decades since autonomous vehicles have been conceptualized [38], but high costs and technological infancy have prevented their large-scale production [39]. General Motors presented the first automated vehicle highway concept at the 1939 World's Fair, which led to the development of the first autonomous vehicle highway system. Early attempts at adapting both highway infrastructure and vehicles were most successful with the California State Department of Transportation's Automated Highway Systems (AHS). Several decades have passed since then. Before being connected, autonomous vehicle functions, such as safety systems, had been

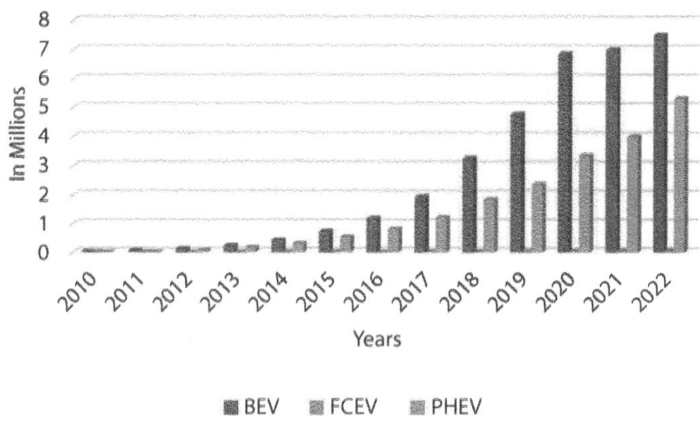

Figure 10.2 Around the world, how many electric vehicles are on the road.

developed, thanks to in-vehicle infotainment systems. Currently, research and development into advanced vehicle connectivity and autonomy are focused on reducing and eliminating road transportation's negative impacts, such as air pollution and automobile crashes. Several technologies were used in personal passenger vehicles to mitigate accidents, including seatbelts, airbags, crumple zones, and antilock brakes. Over the past few years, several automakers have introduced ADAS features, including adaptive cruise control and lane assistance. These systems are designed to reduce accidents to a minimum through the use of intelligent technologies. CAEV solutions based on these initial autonomous technologies are now becoming more advanced and integrated so that cars can fully operate without the involvement of a driver.

10.3.1 Autonomous Vehicles

Communications and robotics technology have greatly impacted our daily lives, including transportation. With these technologies, autonomous vehicles (AVs) are on the horizon, which will reduce crashes, energy consumption, pollution, congestion, and pollution while increasing accessibility to transportation. In spite of the fact that driverless cars have been discussed for decades, the cost has hindered their widespread use [39]. Research and development in the last decade have made significant progress toward AV development. AVs gained traction with the introduction of the Google car, for instance [40, 41]. Moreover, the auto industry invests more than 77 billion euros every year in research and development to remain competitive [42].

An ageing population and rapid technological developments in communication may make AV a necessity in developing countries and a key business paradigm. A number of scholars have pointed out that transportation is undergoing rapid change due to emerging technologies and ideas such as social networks, smartphones, and automated vehicles. A good example of this phenomenon is Uber, which has swept cities, so taxi companies are struggling to remain competitive and retain business. In addition to vehicle automation, the author lists ten disruptive technologies for the future. In addition to its positive societal impacts, AVs provide a safer, more affordable, and somewhat more convenient transport system for the non-ambulatory, people with disabilities, and those with lower incomes. Amounts between 0.2 and 1.9 trillion dollars are expected to be generated by 2025 in direct societal value. As a result of such positive impacts, AV technology will emerge as a viable economic model within the next few years. A multidisciplinary approach may be necessary to view AVs. As shown, the author provides two elements necessary for the success of AV: Connectedness and Big Data. A 'connected vehicle' is a vehicle that can communicate with pedestrians, authorities, and other vehicles and infrastructure. In this illustration, a connected system is conceptually represented. Tos for connected components to function properly, a number of sources of data will need to be accessed. Big Data refers to the necessity of dealing with such an unprecedented amount of information, which requires special hardware and software.

10.3.2 Connected Vehicles

By enabling road vehicles to communicate wirelessly between themselves, known as V2V, as well as the local infrastructure V2I, and the wider internet-of-things (IOT) (vehicle-to-everything – V2X), transportation systems can achieve significant safety improvements. V2X communications are fundamental to the advancement of ADAS and any CAEV ecosystem. However, despite CVs' V2X capability being an integral part of integrated and optimized CAEV and ITS services, neither autonomous nor electrical capabilities are required for CVs. It is one of the Wi-Fi standards defined by IEEE 802.11p. A total of seven channels are allocated to DSRC and operate between 5.850 and 5.925 GHz. The range is limited to 300m for DSRC, with data rates ranging from 3 to 27 Mb/s. As part of DSRC, vehicle locations, direction, speed, and braking are sent between vehicles, triggering safety warnings to the driver and automating actions such as emergency braking. The future of BSM is predicted to be fast, low-latency

5G, especially in large cities, since it can be updated and transmitted 10 times per second.

For CAEVs to be autonomous, satellite positioning technology and onboard sensor technology play the most important roles, whereas V2X technologies provide the connectivity that enhances positioning capabilities. By enabling V2X, neighboring road vehicles will be able to share information on position and velocity, as well as road and traffic conditions, within or outside their line-of-sight (LOS), to enhance CAEV situational awareness, which will prevent accidents. As a result of V2X technology, backup safety functions are provided should the primary positioning technologies malfunction or if the CAEV have difficulty functioning in extreme weather conditions. In addition to improving traffic management, they can also increase personal safety, enhance fuel efficiency, and reduce emissions. CV technologies still face challenges, especially in rural areas, due to their lack of network coverage and signal strength reliability. It is also difficult to integrate autonomous systems, meet the complex requirements of the wireless environment, be susceptible to interferences and cyber-attacks, have consistent regulations and standardization, and adapt to a fast-changing market.

10.3.3 Electric Vehicles

As oil prices have increased in the past decade, environmental concerns have contributed to the expansion of EVs. As an outcome, the EV's capacity for long-distance travel is limited by the battery's size and durability. EVs will only be able to travel a short distance, and finding charging stations will be constantly anxiety-inducing. Sharing AV fleets and EV technology have natural and organic synergies: Travel range anxiety can be resolved by AVs, as can charging infrastructure availability and charging time management. According to the author, fleet-managed AVs address these concerns by establishing charging stations that respond in real-time to travel demand. According to their estimates, charging infrastructure costs, vehicle capital and maintenance, electricity, insurance, and registration will take between $0.42 and $0.49 per occupied mile travelled. Low-mileage households can, therefore, access shared AV services at the equivalent cost per mile as if they owned their vehicles. In this way, shared AV services for households with low mileage can be offered at a cost equivalent to owning a private vehicle.

10.4 The Impact of Air Quality on Environmental Justice (EJ)

Our analysis allowed us to convert scenarios involving EV penetration and emission changes into concentrations of pollutants in the air. To separate two components of air pollutants, regional background concentrations were compared with projected on-road vehicle operations in the study area in 2040, and modeled concentrations were determined from the modeled data. The US Environmental Protection Agency's Air Quality System was used to calculate regional background concentrations of air pollutants. A detailed description of the method used to calculate the 2040 air quality projection based on monitored NOx levels from 2000 to 2020 can be found under the Methods section. In the simulations, the Research LINE (RLINE) source model was used to estimate on road vehicle concentration contributions based on scenario-specific emission estimates for on road vehicles.

10.4.1 Data Collection and Setup of Air Quality Modeling Systems

This study employed the WRF-SMOKE-CMAQ modeling system to simulate meteorological conditions, emissions, and chemical transport. USEPA CMAQ model version 5.0.2 was used for chemical transport, carbon bond 5 (CB05) and AERO6 models were used for gas and aerosol chemistry, and WRF version 3.7 was used for meteorology.

10.5 CAV Taxonomy Based on Performance

As modern vehicles become more sensor- and communication-enabled, CAV applications will significantly improve safety, increase mobility, and reduce the impact on the environment. This paper aims to develop a comprehensive performance measure evaluation framework by incorporating additional performance indicators previously employed in other research papers. In this section, we provide a brief overview of three major performance metrics, namely safety, mobility, and environmental impact. In CAV, a variety of applications focus on one or more of the three metrics for measuring performance. This list includes several applications published in 2015 and 2016. Several of these applications are examined in depth in Section IV.

- **Safety:** CAVs and ITSs primarily aim to ensure safety. CAV applications that focus on safety can help vehicles avoid road conflicts. Cooperative safety systems achieve collision prevention by notifying or warning. Examples include forward collision warnings and lane-keeping assistance.
- **Mobility:** CAV applications that focus on mobility enable transportation systems to become more efficient, which helps individuals travel more easily. Efficiencies in transportation resources, like roadway capacity and travel time, are needed to achieve acceptable transportation outputs, such as highway throughput and travel distance. A mobility-oriented CAV application is traffic signal coordination, along with platooning.
- **Ecological Impacts:** Emissions from the transport sector contribute significantly to air pollution and climate change. As far as transportation-related emissions are concerned, ITS and CAV technologies are widely accepted as contributing to a reduction of these emissions. CAV applications have been designed to reduce energy consumption and emissions associated with transportation over the last few years. Eco-route navigation is available alongside eco-driving assistance.

10.5.1 Connected and Autonomous Electric Vehicles (CAEVs)

Several methods are available for demonstrating quality of service to customers, including theoretical and practical ones. An electric vehicle's quality of service refers to improving an influencing factor and assessing user satisfaction through overall performance improvements. A number of aspects of quality of service are discussed in this review article, including the demands on energy, the costs, the scalability, the size, the scheduling of charging, and resource allocation. In a number of studies, conventional ICEs have been compared with electric vehicles (EVs) in terms of their technical performance. A vehicle powered by electricity is considered an advanced energy storage system since the excess energy can be transferred to the grid *via* converters. The optimal operation of an electric vehicle battery requires an energy management system (EMS). A number of energy storage systems have been assessed for EVs to increase battery life and improve the quality of service. In addition to battery swapping, battery management systems can be improved by improving service quality. The IEC installation standards were followed when installing several charging stations both indoors and outdoors. Energy management systems were studied in the early 2000s to enhance battery health and improve service quality since EV batteries were viewed as substandard in comparison to ICE batteries. Researchers are now studying how to improve the performance of charging infrastructures to enhance the quality of service for electric vehicles with the rise in popularity of electric vehicles.

10.5.2 The Quality of Experience Framework for CAEVs

To ensure that all important influencing factors are accounted for in sound models, the proposed CAEVs QoE taxonomy includes a wide variety of aspects. Models that focus solely on one factor may result in inaccurate assessment and evaluation because they neglect other factors. If you only look at service-related indicators, you may miss the underlying causes of user dissatisfaction, such as high costs, unwanted contexts, or unfavorable human circumstances. Application designers can, therefore, choose which factors to take into account and which indicators to incorporate when modeling the quality of experience. Using the proposed framework, we attempt to integrate the different factors affecting the quality of experience to enable ranking and classification. There is no statistical aggregation or mathematical formulation that attempts to form a QoE in the proposed framework. A mathematical formulation or statistical aggregation that attempts to formulate QoE are not included in the proposed framework. The paper presents sample formulations within analytic frameworks. This framework is intended to be embedded into CAEV applications.

In addition to the other influencing factors, a fourth entity represents a QoE cost model, which is incorporated with the taxonomy indicator measurements to form a central QoE Model. Hence, the QoE Model's outputs can be ranked by quality of experience and classifications of performance levels. Rankings and classifications of the outputs can be used as the basis for recommendations for adapting the embedded CAEV system and improving quality. Furthermore, the context entity represents a fixed element that cannot be modified based on recommendations, although options within the context can be revised. The QoS factor is the factor that feeds into the QoE model in terms of how well the current service is deployed and how high its quality level is. Here, the same information is provided to the QoS cost model. In addition, the QoS cost model provides the QoE model with details on the current adopted rate of services and other available service options. The QoS entity provides pricing alternatives based on the output recommendations. We combine alternate QoS options with human entity information to develop suitable cost options for the QoE model to rank and recommend new service quality options. A human entity is also able to adjust to changes in opinion and new social influences. Real-time implementation of the QoE model is possible, providing outputs that drive the whole system or enable optimization in real-time.

10.6 Conclusion

The CAEV technology offers a promising path to cleaner air and safer cities. The evolution of V2X communication, AV systems, and EV infrastructure has the potential to transform cities in a number of ways, including reducing traffic-related emissions, improving mobility, and minimizing accidents. As a means of accelerating CAEV adoption, policy, infrastructure development, and public awareness must be coordinated. Based on the proposed performance framework, users can assess how CAEVs contribute to safety, environmental impact, and satisfaction with the product. Providing environmental and societal benefits, CAEVs can play a pivotal role in the transformation of urban mobility as long as they remain technically advanced and supported by regulations.

References

1. Xu, P., Dherbomez, G., Hery, E., Abidli, A., Bonnifait, P., System Architecture of a Driverless Electric Car in the Grand Cooperative Driving Challenge. *IEEE Intell. Transp. Syst. Mag.*, 10, 1, 47–59, 2018, doi: 10.1109/MITS.2017.2776135.
2. Hussain, N., Rani, P., Chouhan, H., Gaur, U.S., Cyber security and privacy of connected and automated vehicles (CAVs)-based federated learning: challenges, opportunities, and open issues. *Fed. Learn. IoT Appl.*, 169–183, 2022.
3. Rani, P. and Sharma, R., Intelligent Transportation System Performance Analysis of Indoor and Outdoor Internet of Vehicle (IoV) Applications Towards 5G. *Tsinghua Sci. Technol.*, 29, 6, 1785–1795, Dec. 2024, doi: 10.26599/TST.2023.9010119.
4. Ahmad, I., *et al.*, VANET–LTE based heterogeneous vehicular clustering for driving assistance and route planning applications. *Comput. Netw.*, 145, 128–140, Nov. 2018, doi: 10.1016/j.comnet.2018.08.018.
5. Huawei, *WIoT 2018 Award for First LTE-V2X Commercial Solution launched by Huawei and industry partners. Technical Report*, 2018, URL. Accessed: Nov. 09, 2024, [Online], Available: https://www.huawei.com/en/news/2018/9/wiot-2018-award-lte-v2x-commercial-solution.

6. (Intel), D.S., (Intel), H.M., (Nokia), P.K., (Huawei), S.K., (Huawei), Z.Z., (Vodafone), A.L., (Bosch), C.T., Telecom), E.F.D., (Ford), I.V., (Ford), J.C., (Intel), V.Y., (Intel), S.J.T., (Qualcomm), V.P., (Qualcomm), M.V., (Ericsson), S.R., (Ericsson), S.S., *Toward fully connected vehicles: Edge computing for advanced automotive communications*, 2017, Accessed: Nov. 09, 2024, [Online], Available: https://5gaa.org/news/%20toward-fully-connected-vehicles-edge-computing-for-advanced-automotive-communications/.
7. Rani, P., Verma, S., Yadav, S.P., Rai, B.K., Naruka, M.S., Kumar, D., Simulation of the lightweight blockchain technique based on privacy and security for healthcare data for the cloud system. *Int. J. E-Health Med. Commun. IJEHMC*, 13, 4, 1–15, 2022.
8. Rani, P., *et al.*, Federated Learning-Based Misbehaviour Detection for the 5G-Enabled Internet of Vehicles. *IEEE Trans. Consum. Electron.*, 70, 2, 4656–4664, 2023.
9. Giust, F., *et al.*, Multi-access edge computing: The driver behind the wheel of 5G-connected cars. *IEEE Commun. Stand. Mag.*, 2, 3, 66–73, 2018.
10. kumar Agrawal, N., *et al.*, TFL-IHOA: Three-Layer Federated Learning Based Intelligent Hybrid Optimization Algorithm for Internet of Vehicle. *IEEE Trans. Consum. Electron.*, 70, 3, 5818–5828, 2023.
11. Margarita Martinez-Diaz, F.S. and Pérez, I., Autonomous driving: a birds eye view. *IET Intel. Transport Syst.*, 13, 563–579(16), 2019, [Online]. Available: https://digital-library.theiet.org/content/journals/10.1049/iet-its.2018.5061.
12. Zhao, Y., Tan, Y., Feng, S., Does reducing air pollution improve the progress of sustainable development in China? *J. Clean. Prod.*, 272, 122759, 2020.
13. Yuan, M., Song, Y., Huang, Y., Hong, S., Huang, L., Exploring the Association between Urban Form and Air Quality in China. *J. Plan. Educ. Res.*, 38, 4, 413–426, Dec. 2018, doi: 10.1177/0739456X17711516.
14. Miao, Z., Baležentis, T., Shao, S., Chang, D., Energy use, industrial soot and vehicle exhaust pollution—China's regional air pollution recognition, performance decomposition and governance. *Energy Econ.*, 83, 501–514, 2019.
15. Bajçinovci, B. and Bajçinovci, M., Prishtina: Urban Health Related to Air Quality. *Environ. Clim. Technol.*, 24, 1, 294–309, Jan. 2020, doi: 10.2478/rtuect-2020-0017.
16. Tong, Z., Chen, Y., Malkawi, A., Liu, Z., Freeman, R.B., Energy saving potential of natural ventilation in China: The impact of ambient air pollution. *Appl. Energy*, 179, 660–668, 2016.
17. Sanyal, S. and Wuebbles, D.J., The Potential Impact of a Clean Energy Society on Air Quality. *Earths Future*, 10, 6, e2021EF002558, Jun. 2022, doi: 10.1029/2021EF002558.
18. Liang, X., *et al.*, Air quality and health benefits from fleet electrification in China. *Nat. Sustain.*, 2, 10, 962–971, 2019.
19. Kavianipour, M., *et al.*, Electric vehicle fast charging infrastructure planning in urban networks considering daily travel and charging behavior. *Transp. Res. Part Transp. Environ.*, 93, 102769, 2021.

20. Rani, P., Singh, P.N., Verma, S., Ali, N., Shukla, P.K., Alhassan, M., An implementation of modified blowfish technique with honey bee behavior optimization for load balancing in cloud system environment. *Wirel. Commun. Mob. Comput.*, 2022, 1–14, 2022.
21. Ip, A., Fong, S., Liu, E., Optimization for allocating BEV recharging stations in urban areas by using hierarchical clustering, in: *2010 6th International conference on advanced information management and service (IMS)*, IEEE, pp. 460–465, 2010, Accessed: Nov. 09, 2024, [Online], Available: https://ieeexplore.ieee.org/abstract/document/5713494/.
22. Hodgson, M.J., A Flow-Capturing Location-Allocation Model. *Geogr. Anal.*, 22, 3, 270–279, Jul. 1990, doi: 10.1111/j.1538-4632.1990.tb00210.x.
23. Kuby, M. and Lim, S., Location of Alternative-Fuel Stations Using the Flow-Refueling Location Model and Dispersion of Candidate Sites on Arcs. *Netw. Spat. Econ.*, 7, 2, 129–152, Apr. 2007, doi: 10.1007/s11067-006-9003-6.
24. Kim, J.-G. and Kuby, M., The deviation-flow refueling location model for optimizing a network of refueling stations. *Int. J. Hydrog. Energy*, 37, 6, 5406–5420, Mar. 2012, doi: 10.1016/j.ijhydene.2011.08.108.
25. Upchurch, C., Kuby, M., Lim, S., A Model for Location of Capacitated Alternative-Fuel Stations. *Geogr. Anal.*, 41, 1, 85–106, Jan. 2009, doi: 10.1111/j.1538-4632.2009.00744.x.
26. Andrews, M., Dogru, M.K., Hobby, J.D., Jin, Y., Tucci, G.H., Modeling and optimization for electric vehicle charging infrastructure, in: *IEEE Innovative Smart Grid technologies Conference*, Citeseer, pp. 1–10, 2013, Accessed: Nov. 09, 2024, [Online], Available: https://citeseerx.ist.psu.edu/document?repid=rep1&type=pdf&doi=e436497888d356ddd77997c56b-3c150f57fe120b.
27. Chen, T.D., Kockelman, K.M., Khan, M., Locating Electric Vehicle Charging Stations: Parking-Based Assignment Method for Seattle, Washington. *Transp. Res. Rec. J. Transp. Res. Board*, 2385, 1, 28–36, Jan. 2013, doi: 10.3141/2385-04.
28. Dong, J., Liu, C., Lin, Z., Charging infrastructure planning for promoting battery electric vehicles: An activity-based approach using multiday travel data. *Transp. Res. Part C Emerg. Technol.*, 38, 44–55, Jan. 2014, doi: 10.1016/j.trc.2013.11.001.
29. Xi, X., Sioshansi, R., Marano, V., Simulation–optimization model for location of a public electric vehicle charging infrastructure. *Transp. Res. Part Transp. Environ.*, 22, 60–69, Jul. 2013, doi: 10.1016/j.trd.2013.02.014.
30. Li, M., Jia, Y., Shen, Z., He, F., Improving the electrification rate of the vehicle miles traveled in Beijing: A data-driven approach. *Transp. Res. Part Policy Pract.*, 97, 106–120, Mar. 2017, doi: 10.1016/j.tra.2017.01.005.
31. The History of Electric Car, Govt. Project Report, 2014. Available online: Accessed: Nov. 10, 2024. [Online]. Available: https://www.energy.gov/articles/history-electric-car.
32. IEA, *Global EV Outlook*, 2022, Available online: Accessed: Nov. 10, 2024. [Online]. Available: https://www.iea.org/reports/global-ev-outlook-2022.

33. IEA, *Global EV Outlook*, 2021, Available online: Accessed: Nov. 10, 2024. [Online]. Available: https://www.iea.org/reports/global-ev-outlook-2021.
34. Alternative Fuels Data Center, Project Report, Available online: Accessed: Nov. 10, 2024. [Online]. Available: https://afdc.energy.gov/vehicles/how-do-hybrid-electric-cars-work.
35. Suhail, M., Akhtar, I., Kirmani, S., Jameel, M., Development of Progressive Fuzzy Logic and ANFIS Control for Energy Management of Plug-In Hybrid Electric Vehicle. *IEEE Access*, 9, 62219–62231, 2021, doi: 10.1109/ACCESS.2021.3073862.
36. Wu, G., Zhang, X., Dong, Z., Powertrain architectures of electrified vehicles: Review, classification and comparison. *J. Frankl. Inst.*, 352, 2, 425–448, Feb. 2015, doi: 10.1016/j.jfranklin.2014.04.018.
37. Climent, H., Pla, B., Bares, P., Pandey, V., Exploiting driving history for optimising the Energy Management in plug-in Hybrid Electric Vehicles. *Energy Convers. Manag.*, 234, 113919, Apr. 2021, doi: 10.1016/j.enconman.2021.113919.
38. Bagloee, S.A., Tavana, M., Asadi, M., Oliver, T., Autonomous vehicles: challenges, opportunities, and future implications for transportation policies. *J. Mod. Transp.*, 24, 4, 284–303, Dec. 2016, doi: 10.1007/s40534-016-0117-3.
39. Fagnant, D.J. and Kockelman, K., Preparing a nation for autonomous vehicles: opportunities, barriers and policy recommendations. *Transp. Res. Part Policy Pract.*, 77, 167–181, Jul. 2015, doi: 10.1016/j.tra.2015.04.003.
40. Guizzo, E., How google's self-driving car works. *IEEE Spectr. Online*, 18, 7, 1132–1141, 2011.
41. Markoff, J., Google cars drive themseleves, in: *traffic*, vol. 9, N. Y. Times, 2010, Google test report, Accessed: Nov. 09, 2024, [Online], Available: https://cir.nii.ac.jp/crid/1572543026021081216.
42. ACEA, The automobile industry pocket guide. *Eur. Automob. Manufacturers Association.*, 2015, Blog Post, Accessed: Nov. 09, 2024, [Online], Available: http://www.acea.be/uploads/publications/POCKET_GUIDE_2015-2016.pdf.

11

Distilling the Convergence of AI and EVs Towards Self-Driving EVs

Shivakrishna Dasi[1], Heena Madan[2]*, Mohammed Y. Al-khuzaie[3], Anita Gehlot[4] and Ramy Riad Al-Fatlawy[5]

[1]Department of Computing Science and Artificial Intelligence, NIMS Institute of Engineering and Technology, NIMS University Rajasthan, Jaipur, India
[2]Department of Computer Science Engineering, Chandigarh Engineering College, Chandigarh Group of Colleges, Jhanjeri, Mohali, Punjab, India
[3]College of Technical Engineering, National University of Science and Technology, Dhi Qar, Iraq
[4]Uttaranchal Institute of Technology, Uttaranchal University, Dehradun, Uttarakhand, India
[5]Department of Computer Technology Engineering, The Islamic University, Najaf, Iraq

Abstract

Autonomous electric vehicles (EVs) will be the norm in the near future as artificial intelligence (AI) and electric vehicles (EVs) combine to revolutionize the industry. A number of artificial intelligence technologies, particularly deep learning (DL) algorithms, improve the efficiency, safety, and intelligence of autonomous vehicles. In addition to contributing to sustainable transportation, self-driving electric vehicles are an environmentally friendly way to commute. EV technology and artificial intelligence are brought together in this chapter, with an emphasis on the use of deep learning for behavior planning, motion planning, and trajectory prediction. EVs combined with AI can reduce urban congestion, reduce emissions, facilitate autonomous mobility, and transform smart cities and urban planning in the future. While advancements have been made, there are still several challenges remaining, such as complex driving scenarios, energy efficiency, and regulatory and cybersecurity issues. An in-depth analysis of self-driving electric vehicles is presented in this study, along with future research directions intended to support EV implementation safely and effectively.

**Corresponding author*: heena2111.research@cgcjhanjeri.in

Pethuru Raj, Sudesh Yadav, Manas Kumar Mishra, Satya Prakash Yadav and Victor Hugo C. de Albuquerque (eds.) Artificial Intelligence Technologies for Smart and Sustainable Urban Transportation: Integrated Platforms and Use Cases, (177–196) © 2026 Scrivener Publishing LLC

Keywords: Artificial intelligence, autonomous vehicles, electric vehicles, environmental sustainability, smart cities, sustainable transportation, urban congestion, urban planning

11.1 Introduction

EVs and AI are merging to reshape the automotive landscape, which is leading to an era of self-driving vehicles. Autonomous driving systems are being developed with AI technologies, which enhance vehicle intelligence, safety, and efficiency as they evolve. In conjunction with the growing number of electric cars, which are driving sustainable transportation, AI accelerates the development of eco-friendly self-driving cars. Our daily commutes, vehicle interactions, and vision of the future of mobility will be transformed by the convergence of these technologies. In this survey, we will consider how AI and electric vehicles are convergent, how self-driving technology plays a role, and how this impacts transportation.

Globally, research on autonomous vehicles (SDVs) has grown significantly over the last thirty years [1]. Due to recent improvements in sensor and processing technology, this manuscript has been approved for publication and has been reviewed and approved by an associate editor. These advancements are encouraged by the potential for altering vehicular mobility as well as the anticipated social benefit. Road accidents claim almost 1.3 million lives annually, according to WHO statistics. A non-fatal injury is estimated to result in impairment in 20–50 million people [2]. Driving under the influence of alcohol, distracted driving, or human error can all lead to accidents. It greatly reduces driver error and recklessness in vehicle collisions when SDVs are used. Individuals who are physically or visually handicapped and are unable to drive will also be able to travel independently. To reduce driving stress, SDVs can optimize the time spent in travelling. The government, industry, academia, and general public are all becoming increasingly interested in SDVs. The advancement of AI and computing hardware [3], makes SDVs a viable option for modernizing transportation. The adoption of SDVs in densely populated urban areas, in particular, has great potential for reducing traffic accidents and congestion [4]. In spite of experts' advances, it isn't easy to make SDVs reliable and safe. A wide range of sensors can improve performance, safety, efficiency, and situational awareness in small and medium-sized vehicles. The SDV, however, struggles to recognize and respond to complex situations even with the many sensors on board. To successfully implement self-driving technology, we need adaptable, safe, and resilient approaches [5].

In rule-based planning, the modular approach is well-established. Besides perception [6], localization, and control, SDV also relies on this method [7]. A simplified and modified version of the suggested strategy is provided here [8, 9]. A modular approach relies heavily on this method approach. In the event of unexpected or defective behavior, the modular approach framework can help identify faulty modules due to its interpretability. There is only one section on modular approach planning in Section II. Using a modular approach, you can plan from an initial point to a desired destination on a road level, as well as plan short-term behavior and trajectory. Although modular approach planning is widely used, it is time-consuming and relies heavily on manual heuristics. Researchers apply deep-learning-based behavior and trajectory planning algorithms to modules and E2EL. An in-depth review of the motion planning algorithms used in SDVs has been conducted. It is crucial to implement motion planning within a secure environment. Providing a path to move the ego vehicle from one point to another allows it to move from one endpoint to another. It is important to take into account many factors, including the road limits, the vehicle dynamics, and obstacles on the road [10]. As a result, SDV has gained substantial experience with motion planning algorithms, including graph search-based planners [11], sampling-based planners [12], and interpolating curve planners [13].

SDVs that use traditional regulations, however, are like vehicles that come pre-programmed with rules for every situation. Despite its effectiveness in simple situations, this technique fails when driving conditions are unpredictable, environmental conditions are complex, or decisions must be made in real-time. DL enhances the dynamic nature of learning. A model like this can predict drivers' behavior much better than pre-programmed restrictions because it is built upon a large set of driving data instead of being pre-programmed. Using sensor data, they can also make real-time adjustments to their plans to avoid unforeseen obstacles and driver behavior. As a result of DL, SDVs are able to adapt to real-world situations in a human-like manner, making them safer and more efficient. A distinction is made between traditional motion planning and motion planning based on DL.

A revolutionary evolution in transportation is underway, thanks to artificial intelligence (AI) and electric vehicles (EVs), which are self-driving EVs. A convergence of two transformative technologies can address a number of pressing challenges confronting society today, including urban congestion, energy sustainability, and road safety. Aside from advancing AI-powered systems, vehicles have become more capable of navigation, decision-making in real-time, and learning from dynamic environments than ever before. Meanwhile, EVs provide a sustainable alternative to traditional fossil fuel vehicles, improving energy efficiency and reducing emissions. Artificial intelligence-based technologies like machine learning, computer vision, and sensor fusion are increasingly being used to develop autonomous vehicles that are capable of navigating complicated road conditions without risking their safety. Our vision of autonomous, on-demand transport has become a reality. EVs have become more efficient and range-extended by optimizing their energy usage and adapting to changing driving conditions. In addition to reducing emissions of greenhouse gases and traffic congestion, self-driving electric vehicles will reduce the number of cars on the road and make transportation more accessible to all.

Self-driving electric vehicles and their long-term implications for transportation and urban planning are examined in this discussion. It explores the technological advancements and synergies that are driving the convergence of AI and EVs and explains how they can benefit us and pose challenges. In this chapter, DL-based behavior planning for SDVs is discussed in detail, with a focus on recent developments and future research directions [14]. Develop lane-changing and car-following prediction algorithms for self-driving vehicles based on deep neural networks. The authors utilized an attention mechanism and a temporal-convolution neural network (TCN) to develop a switch neural network (SNN). Using two-dimensional mathematical models, it was possible to predict vehicle trajectory. X and Y coordinates are used to describe a vehicle's trajectory, which indicates a target vehicle's position. The FIS_LSTM model developed by [15], tracks SDVs as they change lanes. With FIS_LSTM, the LSTMs and Fuzzy Inference Systems are integrated. Changing lanes is primarily determined by drivers and traffic conditions surrounding them. Integrated with the driver's behavior and the environment around them, a FIS monitors traffic conditions and driver behavior. To develop fuzzy rules, we take into account the drivers' cognitive abilities.

FFN and Autonomy hardware were used to develop a behavior selector for SDV [16]. This model allows us to adapt our behavior (e.g., following lanes, passing parked cars, braking, and parking) according to the traffic conditions surrounding us. We implemented the Yolov3 deep learning algorithm based on CNNs for environment perception. Using a residual MBI-LSTM network, [17] predicted lane change intentions using attention-enhanced residual MBI-LSTM. The high dataset is used to extract the trajectory of the ego-vehicle as well as the trajectory of the vehicles around it.

Throughout the 21st century, significant technological advancements have reshaped industries, economies, and lifestyles. Aside from artificial intelligence (AI), electric vehicles (EVs) have also transformative potential. As individuals, they each contribute to the advancement of sustainability, efficiency, and connectivity. The ultimate evolution in transportation will come from the convergence of AI and EVs, the self-driving electric vehicles of the future. Throughout this chapter, we examine how AI and EV technology can be combined for unique benefits, technical challenges, and environmental impacts of self-driving EVs.

11.1.1 The State of Electric Vehicles (EVs) Today

In the past century, electric vehicles have made great progress from their early prototypes. Over the last decade, the market for electric vehicles has grown exponentially due to advancements in battery technology, environmental awareness, and government incentives. In addition to reducing fossil fuel dependence, modern EVs produce zero emissions. In addition to battery life, charging infrastructure, and overall costs, there are still many challenges that need to be overcome. Manufacturers of EVs have addressed these challenges by developing efficient batteries, fast charging technology, and scalable production processes.

EV ecosystems are not limited to vehicles but also include charging stations, energy providers, and grid management systems. Nevertheless, a number of challenges remain. These include battery life, charging infrastructure, and overall costs. As renewable energy sources have become more prevalent, electric vehicles can now run on clean energy, enhancing their sustainability.

11.2 Related Work

SDV technology has been extensively researched in academic literature, and a wide range of dimensions have been explored. Among these investigations, DL algorithms have been extensively used to plan SDV motions [18, 19]. Existing surveys provide comprehensive overviews of this area, but they sometimes fail to differentiate between behavior planning and trajectory planning. An analysis of the existing surveys. Furthermore, we are incapable of understanding each component's nuances and challenges as a result of the lack of in-depth analysis. The literature lacks a comprehensive implementation analysis and metrics evaluation, which is another striking gap [20, 21]. To deploy a successful SDV system, theoretical frameworks and algorithmic advancements are essential, but so are practical implementations and performance metrics. Implementation strategies and performance metrics must be thoroughly examined and rigorously evaluated to ensure progress.

11.2.1 AI as the Backbone of Self-Driving Technology

Artificial intelligence has progressed significantly in recent years as a result of advances in machine learning, deep learning, and neural networks. Artificial intelligence is critical to developing self-driving technology. This technology analyzes vast amounts of data, recognizes patterns, makes decisions, and even learns from its experience. Autonomous vehicles rely on AI to process images from sensors, cameras, radars, and lidars, which provide an impression of the environment.

11.2.2 Machine Learning and Computer Vision

With the help of machine learning algorithms, autonomous vehicles are able to recognize and respond to objects and events in real time. Vehicles, pedestrians, signs, and obstacles can be detected by computer vision. Furthermore, autonomous vehicles use AI to make quick decisions to ensure their safety by anticipating what other road users will do.

11.2.3 Deep Reinforcement Learning

Deep reinforcement learning, for example, is an important subfield of artificial intelligence that is utilized by self-driving cars. As autonomous driving systems gain experience, their decision-making processes can improve. Taking this approach makes self-driving cars more capable of handling complex and unpredictable environments, such as city streets that are crowded or adverse weather.

For SDV technology to advance, these shortcomings must be addressed. SDV systems can be analyzed through separate surveys on behavior, trajectory, implementation, and metrics. In addition to increasing our understanding, holistic examinations will also support the development of SDVs that are capable of accurately navigating real-life scenarios [22, 23]. The main focus of this research is the DL-based motion planning (Trajectory and Behavior Planning) and learning end-to-end (IL and RL).

A number of technological trends are driving automotive innovation, including autonomous driving, connected vehicles, electric cars, and shared mobility. The low cost of driving and low emissions of electric vehicles make them a promising addition to intelligent transportation systems (ITS) [24, 25]. As EVs become more sophisticated, artificial intelligence (AI) will be a key component. The term "transport information systems" refers to a range of technologies designed to make transportation more secure, efficient, energy efficient, and environmentally friendly. As AV becomes more popular, intelligent decision-making poses challenges that humans often find incomprehensible. In the absence of transparency, AV technology has difficulty gaining widespread acceptance in society. AI systems that drive self-driving cars need to make quick, secure decisions, as well as explain those decisions in order to comply with local legal requirements.

Cloud computing plays an important role in artificial intelligence solutions for the storage and algorithmic processing of data. Thus, IoT platforms using the cloud are necessary for autonomous vehicles, but existing research points to a number of challenges associated with cloud computing [26]. In the age of interconnected devices, it is imperative to process data efficiently and to make robust decisions within strict latency limitations. The massive amounts of information generated by these devices cannot be transported to the cloud for analysis and storage despite the networks' efficiency and speed. The transfer of such vast data over cloud networks introduces overheads, which decrease throughput, increase energy consumption, and increase network traffic. In the cloud, AVs produce heterogeneous data from sensors and devices connected to the internet of things, making it more difficult to process the data [27].

Data streams from multiple AVs are sent to the cloud in real time, increasing its complexity. This challenge appears to be addressed by edge computing, which deploys distributed computation models near sources of data. A model for edge intelligence (EI) can make AV inference computation more accurate and faster. AVS is able to handle diverse, real-time data from interconnected devices by moving toward edge computing while addressing the limitations of centralized cloud processing. With autonomous driving becoming a reality, machine learning, specifically artificial intelligence, is coming together with MEC to create edge artificial intelligence (EI). As a result of this convergence of activities, various routine activities have been significantly enhanced [28–30]. Edge intelligence orchestrates the collaboration of various devices and servers to handle data generated near an edge device or server.

Additionally, AI seeks to mimic human intelligence by learning from data and replicating it in devices and machines. Considering the overlap between AI and edge intelligence, collectively called edge intelligence, the fusion of these two technologies is logical. An AV's edge intelligence enables it to recognize its surroundings from its point of view accurately. Data offloading is achieved by using additional powerful edge servers at the base station. As more and more information are generated and offloaded to the edges, edge intelligence is necessary. This requires healthy AI processes to be processed precisely. Therefore, edge intelligence models can significantly enhance the inference processing capabilities of AVs while reducing latency and enhancing precision. Despite this, academics and industries alike are still at the beginning of research on edge intelligence. As part of this field, there are a number of challenges, including broadcasting, calculating within thin bandwidths, data security, confidentiality, and energy use [31, 32]. Despite these challenges, edge intelligence continues to evolve and requires further investigation to integrate AI and edge computing for autonomous systems fully. To comprehend six kinds of edge intelligence, this study began with a thorough examination of those types of intelligence. Four key components were analyzed to enhance edge intellect efficiency: cache, train, interpret, and offload. The next step was to examine the methods devised to optimize these components. A description of how edge intelligence is applied to smart electric vehicles is provided in the document. In addition, it examines the challenges and prospects related to edge intelligence in electric vehicles, enhancements in performance, and emerging research directions.

11.3 The Convergence of AI and EVs: Key Enablers for Self-Driving EVs

AI's convergence with EVs represents a turning point in transport evolution. Digital connectivity, sensors, and sustainability all contribute to this fusion of technologies. Artificial intelligence and electric vehicles together are enabling the next frontier in mobility:

Autonomous Power Management: As a result of AI, EVs can manage their power consumption intelligently, optimizing battery usage according to driving patterns, route plans, and environmental conditions. By calculating energy-efficient routes, managing energy output, and balancing the load between power-hungry components, an AI-driven EV can extend battery life [33]. As a result, the vehicle is more efficient, and range anxiety is reduced, one of the main obstacles to the adoption of EVs.
Enhanced Safety Mechanisms: Autonomous electric vehicles must be as safe as possible, and artificial intelligence can play a critical role in

improving safety. Traffic conditions, weather, and driver behavior patterns can all be analyzed with machine learning models to predict how likely an accident will be [34]. As soon as AI detects a potential risk, it is able to trigger adaptive safety features, such as adjusting speed, steering, and braking to prevent a collision from occurring. Furthermore, the vehicle can react faster than a human driver due to real-time sensor analysis, potentially reducing accidents.

Continuous Learning and Over-the-Air Updates: Autonomous electric vehicles equipped with artificial intelligence continually learn from their own experiences and those of other vehicles. Machine learning allows them to improve their driving models [34], continuously. A further advantage of OTA updates is that EV drivers can receive real-time updates that improve their driving algorithms, battery management, and security protocols. In addition to adapting to evolving road regulations, traffic conditions, and cybersecurity threats, self-driving EVs also benefit from this capability.

Connected Infrastructure: EVs that drive themselves are equipped with V2X technology that allows them to interact with other vehicles, traffic signals, and smart city infrastructure. Developing a fully autonomous driving ecosystem is an integral part of the process [35]. Autonomous electric vehicles can reduce traffic congestion, reduce energy consumption, and improve traffic safety with AI-enabled communication.

11.3.1 Technical Challenges in the Path Towards Self-Driving EVs

Even though AI-powered EVs hold great potential, several technical challenges still need to be overcome before widespread adoption can take place. The following are some of the challenges:

Complex Driving Scenarios: People are unpredictable, cyclists are unpredictable, and weather conditions are unpredictable [36]. AI models must be trained using extensive data and sophisticated machine learning algorithms to handle the variety of scenarios described above. There are still some rare edge cases that prevent full autonomy, regardless of extensive testing [37].

Energy Efficiency and Processing Power: EVs that drive themselves will be able to analyze and make decisions at high speed because powerful processing units will power them. A vehicle's battery may be strained due to the high energy consumption of these processing units [38]. EVs that drive themselves must balance power efficiency with computational power to ensure that their range does not suffer from processing requirements [39].

Ethical and Regulatory Challenges: As self-driving EVs become more prevalent, safety and accountability concerns emerge [40]. Whenever there is an accident involving a self-driving vehicle, someone has to be held accountable, whether it's the OEM or the company developing the AI software. Autonomous electric vehicles must be deployed safely under regulatory frameworks that address these ethical concerns [1].
Data Privacy and Cybersecurity: The vulnerability of self-driving electric vehicles lies in their dependence on data, including personal information and driving patterns. The safety and privacy of passengers depend on preventing hackers from accessing these vehicles. The development of robust cybersecurity measures, such as encryption and secure communication channels, is essential for preventing malicious attacks.

11.4 The Impact of Self-Driving EVs on Society and the Environment

AI and EV technologies are poised to transform society, economics, and the environment. In the event that self-driving electric vehicles become widely adopted, the following impacts are expected:

Reduced Emissions and Pollution: As self-driving EVs eliminate emissions from gasoline and diesel engines, they contribute to a healthier urban environment by reducing air pollution. By optimizing EV efficiency using AI, energy waste can be minimized, ensuring that EVs are as sustainable as possible. EVs that drive themselves can reduce greenhouse gas emissions substantially as they become more mainstream.
Improved Mobility and Accessibility: EVs with self-driving capabilities could revolutionize mobility for a wide range of demographic groups, including seniors and people with disabilities. Driverless cars remove the need for manual operation, making them more independent for those with disabilities. As an additional benefit, autonomous EVs could be shared and used by multiple people at the same time, reducing the need for individuals to own their vehicles.
Economic Transformation: EVs with self-driving capabilities could have a significant effect on the job market, especially those in logistics, transportation, and the automotive industry. AI development, vehicle maintenance, and infrastructure management will create new opportunities, but certain driving-based jobs may also be eliminated. For affected individuals,

governments and industries will have to develop strategies to retrain and adapt their workforce.
Urban Planning and Traffic Optimization: Self-driving electric vehicles can reduce the need for extensive parking infrastructure in urban areas, allowing them to be used for green areas or other social needs. In addition, autonomous electric vehicles can reduce traffic jams and fuel consumption by streamlining traffic flow. As a result, urban planning would become more pedestrian-friendly and environmentally conscious.

Many researchers have been working on self-driving vehicles over the past decade in various laboratories and companies, testing them in diverse environments. The most common question in the minds of potential users is whether this technology can be relied upon or not, as well as what advantages it offers. We need to examine how many accidents and deaths are caused by non-autonomous vehicles every day to answer these questions. The World Health Organization reports that approximately 1.3 million people die in vehicle accidents every year, most of whom are children and young adults. Low-income and middle-income countries account for 93 per cent of these accidents. Several human behaviors contribute to crashes, such as driving while intoxicated or displaying a late reaction time. As such, self-driving technology is envisioned as replacing human-vehicle interaction. In a survey conducted by [41], 84.4% of respondents said they were supportive of self-driving vehicles in the future. A USD 54 billion value was also found to be associated with these vehicles in 2021, according to sales statistics [42]. The results of this study demonstrate that technology usage is becoming increasingly common. Aside from reducing pollution, energy consumption, and time consumption, this technology may also guarantee the safety and comfort of users [43].

A five-level approach to understanding the technological revolution that results from this technology has been developed by SAE. There is not much automatic service provided by the first two levels, so there is a low risk associated with this technology. From level 3, the system can avoid collisions, assist steering, brake, etc. Additionally, people should be practical when making decisions based on system alerts or hard decisions. In level four and level five, vehicles drive independently without interacting with anyone. The complexity of these levels increases when considering the dynamics and unpredictable nature of the surrounding environment, where collisions may occur from any direction [43]. Currently, no vehicle has demonstrated robust performance at these levels.

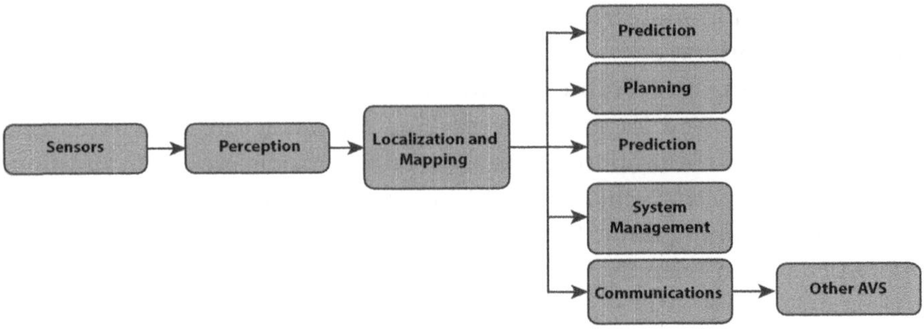

Figure 11.1 Steps for a self-driving car.

Self-driving vehicles are complex systems that require consistent performance of predetermined steps and robust adherence to the stated goals. The procedure of the self-driving car is mentioned in Figure 11.1. In order to use vehicles effectively without human interaction, it is essential to answer the following question: how can they do so? Vehicles need accurate sensors, such as GNSS, inertial measurement units (IMUs), wheel odometers, LiDAR, radar, cameras, etc., to continuously sense their environment. Further, in order to extract relevant information from the vast quantity of data collected, a high-performance computer is required. By utilizing this information, the vehicles are able to make better decisions and carry them out with the assistance of specific units designed for this purpose.

In addition to the Global Navigation Satellite System (GNSS), there are other classical methods. Utilizing three satellites, this technique utilizes trilateration for detecting positions anywhere in the world. The solution is also inexpensive. Signal interference can affect GNSS-based methods, and they are not appropriate for some environments, including those with obstacles blocking the line of sight. Additionally, there are more than three meters of positioning error.

As a consequence, driving can become unreliable and unsafe. In the literature, this issue has been addressed by several papers. RTK-based or D-GPS-based technologies, for instance, [44].

It is possible to create this communication using devices such as WiFi, cellular, and UWB radio technology [45]. I am curious to know what 'X' stands for and what exciting communications we can expect from it:

- **Vehicle-to-Vehicle (V2V):** Interconnect vehicles so they can share information, such as their position, speed, direction, etc., to avoid collisions and change lanes.

- **Vehicle-to-Infrastructure (V2I):** Creating a communication infrastructure so that any connected vehicle can share information between itself and the node.
- **Vehicle-to-people (V2P):** People use mobile devices, bicycles, and wheelchairs when travelling on the road, as well as vehicles.
- **Vehicle-to-Cloud (V2C):** Connected vehicles are enabled through cloud-based platforms and broadband cellular connections.
- **Vehicle-to-X (V2X):** Using this technology, vehicles can connect to everything, which is a great idea since it allows drivers to be notified of road users early enough to perform their tasks comfortably and check prioritizations so they can drive safely.

11.5 The Impact of Self-Driving Vehicles on the Environment

Our objective in this section is to explore how autonomous driving technologies can be used for localization and mapping. We conduct a wide range of research, including reducing energy consumption, reducing emissions, reducing noise, and reducing light pollution. We also optimize infrastructure to support autonomous driving technology, and we refine software algorithms.

Reducing emissions (air pollution) by consuming less energy: Sensors, computing systems, and data centers must be energy efficient so that self-driving vehicles can map. This may increase the overall energy demand for the vehicle [46]. The consumption of energy and the emission of CO2 exist in mutual causation (mutual cause). It seems that energy consumption affects CO2 emissions, so (in the near term) higher energy consumption can lead to higher CO2 emissions, and vice versa (over the longer term). According to the study, current autonomous electric vehicles have limited energy efficiency and power transmission capabilities due to their conventional propulsion systems. As a result, they may be difficult to adopt on a large scale in the future. An attempt was made to address this problem by investigating how to improve the energy consumption and efficiency of a medium-sized autonomous electric vehicle with wheel motors. In this study, we developed a numerical energy model, validated it against real driving data, and applied it to a case study. The energy analysis was conducted on flat ground, uphill terrain, and downhill terrain. The purpose of this study was to examine the energy consumption of the vehicle and to identify the potential for energy savings through the use of an in-wheel drive system.

Sound and light pollution: Besides sound pollution, autonomous vehicles can also contribute to luminous pollution.

Autonomous Vehicle Environment: "Safety and Emissions Impacts" is the topic of the paper [47]. Moreover, it presents an approach to evaluating driving behavior volatility in a connected vehicle setting as well as alerts that can help mitigate road conflicts and noise emissions. Although noise and light pollution impact ecosystems and communities in significant ways, autonomous driving literature tends to ignore them. There may be a reduction in noise pollution if drivers drive more quietly and reduce their vehicle volume. As an added benefit, AVs driving at night could reduce the need for artificial lighting. A greater focus should be placed on these considerations.

Smart cities: Smart cities involve the integration of several technologies, including immersive sensing, ubiquitous communication, robust computing, substantial storage, and highly intelligent systems (SCCSI). A host of applications use these sensors, including public safety, autonomous driving, health care, and smart homes. Based on the rapid rise of automated vehicles equipped with SCCSI in smart cities, a cost-efficient model for vehicle as a service (VaaS) is proposed [48]. The SCCSI-capable vehicles that enable smart cities are mobile servers and communicators that provide SCCSI functionality. SCCSI-capable vehicles provide SCCSI services and are mobile servers and communicators in smart cities.

Software optimization: Autonomous vehicles can also be mitigated by optimizing their software. It has been shown by [49], that edge processing reduces cars' mileage by 30%, making autonomous driving a challenging task for electric vehicles. Specifically designed to manage the computational power consumed by autonomous vehicles for tasks involving AI.

11.6 Conclusion

To achieve autonomous, sustainable transportation, AI and EVs need to be integrated. As autonomous electric vehicles become more prevalent, mobility will be improved, emissions will be reduced, and everyone will have better access to transportation. In this chapter, we have highlighted a number of key areas where deep learning and artificial intelligence can help autonomous EVs overcome technical challenges. For self-driving electric vehicles to reach full autonomy, further advances are needed to overcome issues such as energy management, regulatory compliance, and complex environments. To address technical, ethical, and environmental challenges associated with AI-powered electric vehicles, industry, government, and academia will need to cooperate and innovate together. As self-driving EVs become more prevalent, transportation will become safer, cleaner, and more efficient.

References

1. Rani, P. and Sharma, R., IMFOCA-IOV: Intelligent Moth Flame Optimization based Clustering Algorithm for Internet of Vehicle, in: *2023 14th International Conference on Computing Communication and Networking Technologies (ICCCNT)*, IEEE, pp. 1–6, 2023. Accessed: Nov. 09, 2024. [Online]. Available: https://ieeexplore.ieee.org/abstract/document/10307646/.
2. Bachani, A., Peden, M., Gururaj, G., Norton, R., Hyder, A., *Road traffic injuries*, 2017, Accessed: Nov. 09, 2024, [Online], Available: https://ora.ox.ac.uk/objects/uuid:4f270f40-cd2c-4c03-8c9b-69b061e05243/files/mbdd04c9df96bf72d6d3c038c4e7d3da6.
3. Chen, L., *et al.*, Milestones in autonomous driving and intelligent vehicles: Survey of surveys. *IEEE Trans. Intell. Veh.*, 8, 2, 1046–1056, 2022.
4. Wang, W., Wang, L., Zhang, C., Liu, C., Sun, L., Social Interactions for Autonomous Driving: A Review and Perspectives. *FNT Rob.*, 10, 3–4, 198–376, 2022, doi: 10.1561/2300000078.
5. Chen, L., Zhang, Y., Tian, B., Ai, Y., Cao, D., Wang, F.-Y., Parallel driving OS: A ubiquitous operating system for autonomous driving in CPSS. *IEEE Trans. Intell. Veh.*, 7, 4, 886–895, 2022.
6. Song, R., Ai, Y., Tian, B., Chen, L., Zhu, F., Yao, F., Msfanet: A light weight object detector based on context aggregation and attention mechanism for autonomous mining truck. *IEEE Trans. Intell. Veh.*, 8, 3, 2285–2295, 2022.
7. Gong, L., Wu, Y., Gao, B., Sun, Y., Le, X., Liu, C., Real-time dynamic planning and tracking control of auto-docking for efficient wireless charging. *IEEE Trans. Intell. Veh.*, 8, 3, 2123–2134, 2022.
8. Ziegler, J., *et al.*, Making bertha drive—an autonomous journey on a historic route. *IEEE Intell. Transp. Syst. Mag.*, 6, 2, 8–20, 2014.
9. González, D., Pérez, J., Milanés, V., Nashashibi, F., A review of motion planning techniques for automated vehicles. *IEEE Trans. Intell. Transp. Syst.*, 17, 4, 1135–1145, 2015.
10. Hu, X., Chen, L., Tang, B., Cao, D., He, H., Dynamic path planning for autonomous driving on various roads with avoidance of static and moving obstacles. *Mech. Syst. Signal Process.*, 100, 482–500, 2018.

11. Zhang, W., Wang, W., Zhu, J., Zhao, D., Multi-Vehicle Interaction Scenarios Generation with Interpretable Traffic Primitives and Gaussian Process Regression, in: *2020 IEEE Intelligent Vehicles Symposium (IV)*, IEEE, Las Vegas, NV, USA, pp. 1197–1204, Oct. 2020, doi: 10.1109/IV47402.2020.9304568.
12. Zhang, W. and Wang, W., Learning V2V interactive driving patterns at signalized intersections. *Transp. Res. Part C: Emerging Technol.*, 108, 151–166, 2019.
13. Chen, J., Liu, C., Tomizuka, M., FOAD: Fast Optimization-based Autonomous Driving Motion Planner, in: *2018 Annual American Control Conference (ACC)*, IEEE, Milwaukee, WI, pp. 4725–4732, Jun. 2018, doi: 10.23919/ACC.2018.8431104.
14. Shi, K., Wu, Y., Shi, H., Zhou, Y., Ran, B., An integrated car-following and lane changing vehicle trajectory prediction algorithm based on a deep neural network. *Physica A*, 599, 127303, 2022.
15. Wang, W., Qie, T., Yang, C., Liu, W., Xiang, C., Huang, K., An intelligent lane-changing behavior prediction and decision-making strategy for an autonomous vehicle. *IEEE Trans. Ind. Electron.*, 69, 3, 2927–2937, 2021.
16. González-Miranda, O. and Ibarra-Zannatha, J.M., Behavior selector for autonomous vehicles using neural networks, in: *2022 XXIV Robotics Mexican Congress (COMRob)*, IEEE, pp. 31–35, 2022, Accessed: Nov. 09, 2024, [Online], Available: https://ieeexplore.ieee.org/abstract/document/9962308/.
17. Wu, Z., Liang, K., Liu, D., Zhao, Z., Driver lane change intention recognition based on Attention Enhanced Residual-MBi-LSTM network. *IEEE Access*, 10, 58050–58061, 2022.
18. Zhou, H., *et al.*, Review of Learning-Based Longitudinal Motion Planning for Autonomous Vehicles: Research Gaps Between Self-Driving and Traffic Congestion. *Transp. Res. Record: J. Transp. Res. Board*, 2676, 1, 324–341, Jan. 2022, doi: 10.1177/03611981211035764.
19. Claussmann, L., Revilloud, M., Gruyer, D., Glaser, S., A Review of Motion Planning for Highway Autonomous Driving. *IEEE Trans. Intell. Transport. Syst.*, 21, 5, 1826–1848, May 2020, doi: 10.1109/TITS.2019.2913998.
20. Muhammad, K., Ullah, A., Lloret, J., Ser, J.D., De Albuquerque, V.H.C., Deep Learning for Safe Autonomous Driving: Current Challenges and Future Directions. *IEEE Trans. Intell. Transport. Syst.*, 22, 7, 4316–4336, Jul. 2021, doi: 10.1109/TITS.2020.3032227.
21. Aradi, S., Survey of Deep Reinforcement Learning for Motion Planning of Autonomous Vehicles. *IEEE Trans. Intell. Transport. Syst.*, 23, 2, 740–759, Feb. 2022, doi: 10.1109/TITS.2020.3024655.
22. Chib, P.S. and Singh, P., Recent Advancements in End-to-End Autonomous Driving Using Deep Learning: A Survey. *IEEE Trans. Intell. Veh.*, 9, 1, 103–118, Jan. 2024, doi: 10.1109/TIV.2023.3318070.
23. Wang, N., Li, X., Zhang, K., Wang, J., Xie, D., A Survey on Path Planning for Autonomous Ground Vehicles in Unstructured Environments. *Machines*, 12, 1, 31, Jan. 2024, doi: 10.3390/machines12010031.

24. Elvas, L.B. and Ferreira, J.C., Intelligent transportation systems for electric vehicles. *Energies*, 14, 17, 5550, 2021, MDPI Accessed: Nov. 09, 2024, [Online], Available: https://www.mdpi.com/1996-1073/14/17/5550.
25. Ahmad, K., Khujamatov, H., Lazarev, A., Usmanova, N., Alduailij, M., Alduailij, M., Internet of Things-Aided Intelligent Transport Systems in Smart Cities: Challenges, Opportunities, and Future. *Wireless Commun. Mobile Comput.*, 2023, 1–28, Apr. 2023, doi: 10.1155/2023/7989079.
26. Khayyam, H., Javadi, B., Jalili, M., Jazar, R.N., Jazar, Artificial Intelligence and Internet of Things for Autonomous Vehicles, in: *Nonlinear Approaches in Engineering Applications*, R.N. and L. Dai (Eds.), pp. 39–68, Springer International Publishing, Cham, 2020, doi: 10.1007/978-3-030-18963-1_2.
27. Pradhan, N.M., Chaudhari, B.S., Zennaro, M., 6TiSCH low latency autonomous scheduling for Industrial Internet of Things. *IEEE Access*, 10, 71566–71575, 2022.
28. Zhang, K., Zhu, Y., Leng, S., He, Y., Maharjan, S., Zhang, Y., Deep learning empowered task offloading for mobile edge computing in urban informatics. *IEEE Internet Things J.*, 6, 5, 7635–7647, 2019.
29. Chaudhari, B. and Borkar, S., Design considerations and network architectures for low-power wide-area networks, in: *LPWAN Technologies for IoT and M2M applications*, pp. 15–35, Elsevier, Academic Press, 2020, Accessed: Nov. 09, 2024, [Online], Available: https://www.sciencedirect.com/science/article/pii/B9780128188804000028.
30. Dai, Y., Xu, D., Maharjan, S., Qiao, G., Zhang, Y., Artificial intelligence empowered edge computing and caching for internet of vehicles. *IEEE Wireless Commun.*, 26, 3, 12–18, 2019.
31. Mendez, J., Bierzynski, K., Cuéllar, M.P., Morales, D.P., Edge Intelligence: Concepts, Architectures, Applications, and Future Directions. *ACM Trans. Embed. Comput. Syst.*, 21, 5, 1–41, Sep. 2022, doi: 10.1145/3486674.
32. Ghorpade, S.N., Zennaro, M., Chaudhari, B.S., Saeed, R.A., Alhumyani, H., Abdel-Khalek, S., Enhanced differential crossover and quantum particle swarm optimization for IoT applications. *IEEE Access*, 9, 93831–93846, 2021.
33. Olaverri-Monreal, C., Promoting trust in self-driving vehicles. *Nat. Electron.*, 3, 6, 292–294, 2020.
34. Gu, Z., Li, Z., Di, X., Shi, R., An LSTM-Based Autonomous Driving Model Using a Waymo Open Dataset. *Appl. Sci.*, 10, 6, 2046, Mar. 2020, doi: 10.3390/app10062046.
35. Pan, X., You, Y., Wang, Z., Lu, C., Virtual to Real Reinforcement Learning for Autonomous Driving, Sep. 26, 2017, *arXiv: arXiv:1704.03952*. Accessed: Nov. 09, 2024, [Online], Available: http://arxiv.org/abs/1704.03952.
36. Chen, J., Yuan, B., Tomizuka, M., Model-free Deep Reinforcement Learning for Urban Autonomous Driving, Oct. 21, 2019, *arXiv: arXiv:1904.09503*. Accessed: Nov. 09, 2024, [Online], Available: http://arxiv.org/abs/1904.09503.
37. Singh, A., Rahma, M.Z.U., Rani, P., Sharma, R., Kariri, E., Smart traffic monitoring through real-time moving vehicle detection using deep learning *via*

aerial images for consumer application. *IEEE Trans. Consum. Electron.*, 70, 4, 7302–7309, 2024. Accessed: Nov. 09, 2024, [Online], Available: https://ieeexplore.ieee.org/abstract/document/10638761/.

38. Rani, P. and Sharma, R., Intelligent transportation system for internet of vehicles based vehicular networks for smart cities. *Comput. Electr. Eng.*, 105, 108543, 2023.
39. Rani, P., Verma, S., Yadav, S.P., Rai, B.K., Naruka, M.S., Kumar, D., Simulation of the lightweight blockchain technique based on privacy and security for healthcare data for the cloud system. *Int. J. E-Health Med. Commun. (IJEHMC)*, 13, 4, 1–15, 2022.
40. Rani, P., Hussain, N., Khan, R.A.H., Sharma, Y., Shukla, P.K., Vehicular Intelligence System: Time-Based Vehicle Next Location Prediction in Software-Defined Internet of Vehicles (SDN-IOV) for the Smart Cities, in: *Intelligence of Things: AI-IoT Based Critical-Applications and Innovations*, F. Al-Turjman, A. Nayyar, A. Devi, P.K. Shukla (Eds.), pp. 35–54, Springer International Publishing, Cham, 2021, doi: 10.1007/978-3-030-82800-4_2.
41. Du, H., Zhu, G., Zheng, J., Why travelers trust and accept self-driving cars: An empirical study. *Travel Behav. Soc.*, 22, 1–9, 2021.
42. Nobles, C., Burrell, D.N., Burton, S.L., Waller, T., Driving into cybersecurity trouble with autonomous vehicles, in: *Handbook of Research on Cybersecurity Risk in Contemporary Business Systems*, pp. 255–273, IGI Global, 2023, Accessed: Nov. 09, 2024, [Online], Available: https://www.igi-global.com/chapter/driving-into-cybersecurity-trouble-with-autonomous-vehicles/321022.
43. Yurtsever, E., Lambert, J., Carballo, A., Takeda, K., A survey of autonomous driving: Common practices and emerging technologies. *IEEE Access*, 8, 58443–58469, 2020.
44. Sjafrie, H., *Introduction to self-driving vehicle technology*, Chapman and Hall/CRC, 2019, Accessed: Nov. 09, 2024, [Online], Available: https://www.taylorfrancis.com/books/mono/10.1201/9780429316777/introduction-self-driving-vehicle-technology-hanky-sjafrie.
45. Kuutti, S., Fallah, S., Katsaros, K., Dianati, M., Mccullough, F., Mouzakitis, A., A survey of the state-of-the-art localization techniques and their potentials for autonomous vehicle applications. *IEEE Internet Things J.*, 5, 2, 829–846, 2018.
46. Petit, J., Stottelaar, B., Feiri, M., Kargl, F., Remote attacks on automated vehicles sensors: Experiments on camera and lidar. *Black Hat Eur.*, 11, 2015, 995, 2015.
47. Coelho, M.C. and Guarnaccia, C., Driving information in a transition to a connected and autonomous vehicle environment: Impacts on pollutants, noise and safety. *Transp. Res. Procedia*, 45, 740–746, 2020.

48. Chen, X., *et al.*, Vehicle as a service (VaaS): Leverage vehicles to build service networks and capabilities for smart cities. *IEEE Commun. Surv. Tutorials*, 2024, Accessed: Nov. 09, 2024, [Online], Available: https://ieeexplore.ieee.org/abstract/document/10449899/.
49. Problem, A.C., *The Energy Consumption of Edge Processing Reduces a Car's Mileage with up to 30%*, 2019.

12

Explaining the Distinct Functionalities of Battery Management Systems (BMS)

Hawraa Ali Sabah[1], Shivakrishna Dasi[2], Jaspreet Kaur[3*], Devendra Singh[4] and Ahmad Radee Alawadi[5]

[1]*College of Technical Engineering, National University of Science and Technology, Dhi Qar, Iraq*
[2]*Department of Computing Science and Artificial Intelligence, NIMS Institute of Engineering and Technology, NIMS University Rajasthan, Jaipur, India*
[3]*Department of Computer Science Engineering, Chandigarh Engineering College, Chandigarh Group of Colleges, Jhanjeri, Mohali, Punjab, India*
[4]*Department of Computer Science & Engineering, Uttaranchal Institute of Technology, Uttaranchal University, Dehradun, Uttarakhand, India,*
[5]*Department of Computer Engineering, The Islamic University, Najaf, Iraq*

Abstract
Battery storage systems must be managed properly for them to be safe, efficient, and long-lasting, especially with the increased demand for reliable energy storage devices. The purpose of this chapter is to explore the functionality of BMS, especially as it relates to electric vehicles and smart grids, and to highlight the various roles that it plays in energy storage systems. Various topologies and components are discussed, ranging from centralized to modular to distributed, each tailored to specific applications. As part of its optimization of battery performance, this paper explores how BMSs estimate state of charge (SOC), state of health (SOH), and balance cells. A variety of BMS designs are explored in the chapter, including lithium-ion, nickel-metal hydride, and thermal and electrical modeling techniques. The chapter explores various BMS designs, including those for lithium-ion and nickel-metal hydride batteries, along with thermal and electrical modeling techniques.

**Corresponding author*: jaspreet2639.research@cgcjhanjeri.in

Pethuru Raj, Sudesh Yadav, Manas Kumar Mishra, Satya Prakash Yadav and Victor Hugo C. de Albuquerque (eds.) Artificial Intelligence Technologies for Smart and Sustainable Urban Transportation: Integrated Platforms and Use Cases, (197–214) © 2026 Scrivener Publishing LLC

Keywords: Battery management systems, battery performance, smart grids, electrical modelling, centralized topologies, reliable energy storage

12.1 Introduction

A power grid is one of the few supply systems where the generation and consumption of electricity occur simultaneously. The main reason is the lack of safe and reliable storage methods for electrical energy. In response to the changing global distribution and use of electricity, energy storage systems (ESSs) have become one of the fastest-growing electrical power systems products.

Depending on their maturity, electrical energy storage technologies can be classified as deployed, demonstrated, or early-stage [1]. In terms of energy storage, the most common are compressed air and hydroelectric power. Batteries and flywheels are also examples of energy storage systems. Superconducting magnetic energy storage, flow batteries, and electrochemical capacitors are among the technologies that have been tested for storing energy. There are several early-stage energy storage technologies, including adiabatic CAES, hydrogen, and synthetic natural gas. It is important to mention that batteries and capacitors, out of all the technologies mentioned above, are also subject to risks and safety concerns [2].

Energy can be stored for a long time in a battery, which is an energy storage device. Battery management systems (BMSs) are system control units that are designed to ensure the operational safety of battery packs in systems [3–5]. Battery management systems are primarily responsible for safeguarding batteries. Safety considerations, cell balance, and ageing issues make supervision of each cell imperative [6]. In addition, BMS ensures that corrective actions are taken when any abnormal conditions arise in the system infrastructure through pre-set corrective measures. A BMS also identifies the best way to control the temperature of the system since temperature has an impact on power consumption.

A discussion of electric and hybrid vehicles' battery management systems was provided in [7]. BMSs are currently facing a number of challenges and concerns. An evaluation of a battery's status, including its charge level, health, and life cycle, must be carried out by a battery management system. The latest methods for evaluating battery state are reviewed, and potential solutions are provided. Battery management systems are discussed in terms of their hardware concepts [8]. Battery management systems (BMS) are mainly used in electric vehicles and stationary batteries, and the paper examines the hardware of these systems. The paper [9] presented a novel bipolar-resonant LC battery equalizer that improved battery equalizing from multicell to multicell. The high speed and efficiency of this equalizer are demonstrated through an analysis of its mathematical properties and comparison with typical equalizers.

In [10], a battery management demonstration board using EMC simulations is presented. In this paper, the author describes how an EMC system simulation can be used to optimize a board's design in a fast-track manner. The direct power injection (DPI) method and radiated susceptibility measurements demonstrated that BMS integrated circuits can be tested for EMI susceptibility [11]. Results of experimental tests are discussed in terms of the different mechanisms that led to failures caused by EMI.

Using a battery pack under power demand, researchers examined the thermal behavior of the pack [12]. A thermal prediction model is proposed that accounts for Joules heating with equivalent resistance, reversible heat, and heat dissipation [13]. An equivalent resistance can be determined by determining the interval between successive states of charge using hybrid pulse power characterization. An in-depth study examining the effects and limitations of high-power charging on batteries was published in [14]. The power sources distribute the power optimally. Grid-connected storage systems can be managed to maximize their overall energy efficiency through a strategy that takes into account the battery's power requirements for charging and discharging [15]. This study explored the impact of pulse duration on the calculation of pulse power performance. A power generation system can use this method to store energy and electrify transportation [16].

12.2 Battery Management System (BMS)

Depending on the application, BMS may have different definitions. In general, BMSs consist of systems that monitor, control, and optimize a person's performance or a battery's performance. The BMS can disconnect modules from the system when there are abnormal conditions. System safety measures are used to enhance battery performance. A battery management system (BMS) integrates monitoring, control, and delivery of battery power (considering battery life as well). Energy management and protection from various hazards are provided by a BMS in automobiles. A BMS is composed of distinct functional components. Systems that manage battery storage and other structured resources, such as controllers, grids, and other distributed resources, are known as battery management systems (BMSs). Advanced circuitry, functional blocks, and smart architecture can extend battery life in a system. It is possible to purchase several commercial BMSs on the market. NUVATION Energy BMS provides flexible, modular, reliable, and modular energy storage applications with UL 1973 certification.

Smart grids and electric vehicles typically require BMSs to control and monitor batteries, as shown in Figure 12.1. To fulfil the requirements listed in "The Need for BMS in Smart Grids and Electric Vehicles," BMSs typically include the following features."

- **Cell Monitoring**

BMS tasks are accomplished by acquiring each cell's current, voltage, and temperature. Different battery technologies require different voltage and current measurements. As far as voltage accuracy is concerned, Li–FePO4 chemistry presents the greatest challenge. It is necessary to measure the cell voltage to an accuracy of 1–2 mV in order to estimate the SOC accurately [17]. A more accurate voltage measurement can be achieved with other Li-ion chemistry flavors, such as lithium-polymer, lithium-lithium titanate, and lithium-mangan (Li-Mn). Comparing Li-Po batteries with LiFePO4 batteries, the OCV-SOC curve is less flat. Many multicell battery monitoring ICs can usually achieve an accuracy of 5 mV.

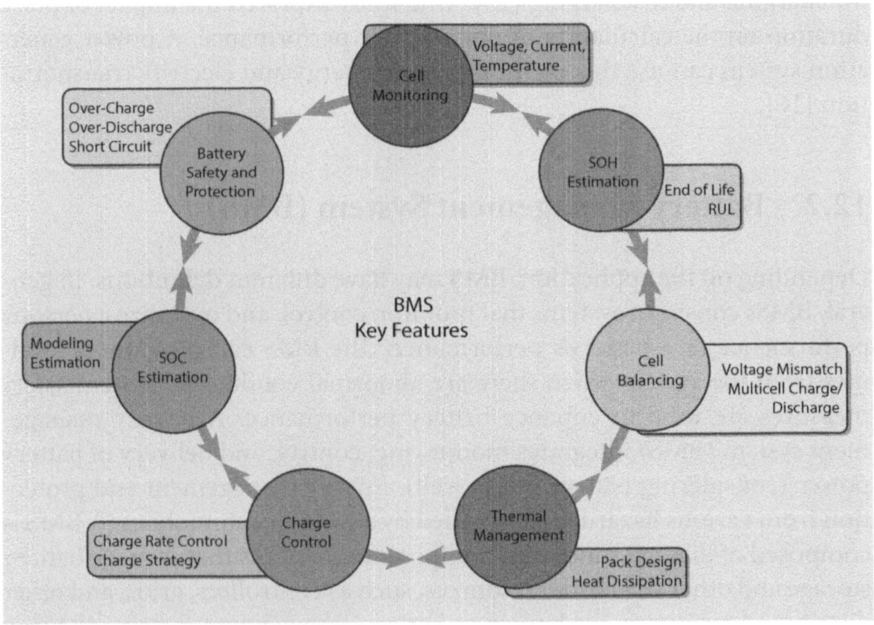

Figure 12.1 A BMS's features.

- **Battery Safety and Protection**

Furthermore, the BMS protects the battery from operating under hazardous conditions and ensures its safety [18]. Batteries can cause hazardous conditions due to their chemical properties. There is a possibility that the battery may experience dangerous situations if it is deeply charged, if its charge state is below a certain percentage when it is overcharged after full charge, when it is charged or discharged at charge rates (C-rates) that exceed the safe level for its chemistry, or when it is discharged at voltages exceeding its cutoff voltage. The operation temperature of Li-ion batteries is influenced both by the electrochemical reactions within the battery and the environment in which the batteries are used [19]. Batteries with Li-ion cells are protected from overcharging by a battery management system (BMS), which, for example, is 0-60°C for charging, and 20-60°C for storing and discharging.

- **SOC Estimation**

The SOC of a battery is a measure of the remaining charge in the battery in comparison to its full capacity. Along with optimizing energy management in electric vehicles and smart grids, it is important to estimate the SOC accurately to avoid deep discharges or overcharges of batteries, which may lead to dangerous situations and degrade battery life. In spite of its importance, it is not possible to measure the SOC directly from the battery terminals. The battery pack's state of charge and the individual cells' state of charge must be estimated using algorithms based on their measured data.

- **SOH Estimation**

Cells or packs can also measure SOH as an indicator of battery functionality. A battery's SOH indicates how many times it can charge and discharge before its life expires. To prolong battery life and arrange for the replacement of batteries simultaneously, EMS must have this information. A direct measurement of SOH from the battery terminals is not possible. A clear definition of SOH is also necessary. EVs and smart grids are now undergoing a lot of research into battery SOH, especially online estimation.

- **Cell Balancing**

Various battery cells are paralleled and linked in series to provide an appropriate amount of power and voltage. However, if the batteries are not connected in a way that matches their voltage and capacity, the entire battery pack will not work. A cell's discharge stops when it reaches below its cutoff voltage, preventing the charge from being utilized in other cells afterwards. Often, this type of mismatch is caused by cell capacity or SOC mismatches.

- **Charging Control**

It is the application that determines the battery's discharge rate, but the customer determines the battery's charging rate and normally charges the battery as fast as possible. Depending on their chemistry and structure, batteries have different limits on how fast they can accept a charge. To address those limitations, custom charging stations are made for specific batteries or general-use charging devices. Battery packs also need this built-in feature to optimize and manage their charging regimes when suitable chargers are not available.

12.3 An Overview of Components and Topologies

In a system infrastructure, a BMS cannot operate alone. Achieving the system objectives depends on its integration with other modules. BMMBIM manages batteries), battery units, and BSCs. It protects the battery pack, extends its lifespan, manages power demand, and interfaces with various networks.

As a BMS market player, you can choose from three different implementation models: centralized, distributed, and modular. There are multiple wires connected between a single control unit and the battery cells in a centralized topology. Dedicated control units are linked to individual

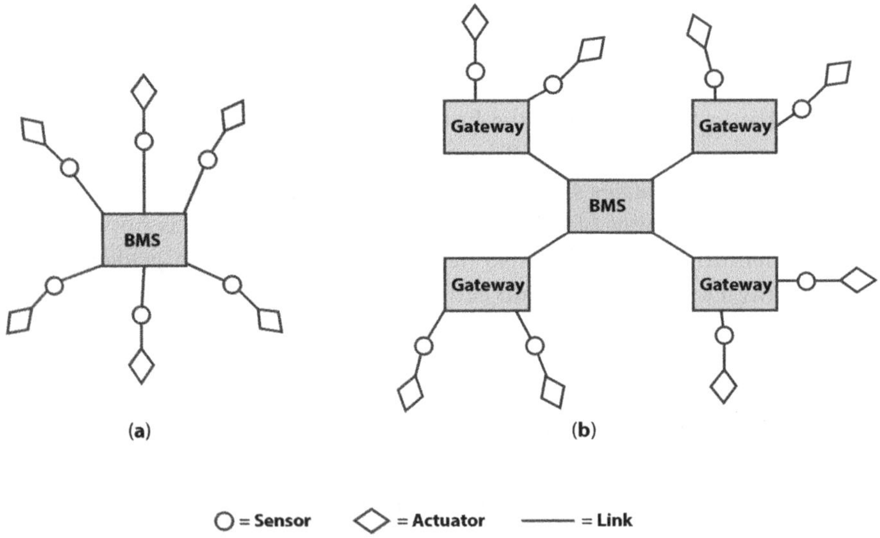

Figure 12.2 BMS implementation: (a) centralized and (b) distributed.

batteries by single communication cables in distributed topology. Last but not least, in a modular topology, multiple numbers of control units are interconnected, and each controls a specific battery cell. In terms of cost and expansion, the centralized system is the least expensive and most flexible. It is the most expensive BMS, but it offers the easiest installation and is the cleanest. There is more hardware and programming effort involved with the modular BMS, which is a conflict between each topology's advantages and disadvantages. Figure 12.2 above shows the topology of BMS implementation.

12.3.1 Software Architecture

The BMS software architecture supports multitasking. Different tasks could not be undertaken simultaneously in the past; one task was postponed for the sake of the other. BMS software architecture allows for the simultaneous execution of multiple tasks. As a BMS software architect, it is essential to complete the initial task of measuring voltage/current, protecting against overcurrent/voltage, measuring temperature, and activating protective relays promptly. RTOS has been introduced to implement real-time functionality in BMS software architecture. Figure 12.3 shows the software architecture of a business organization scheme [16].

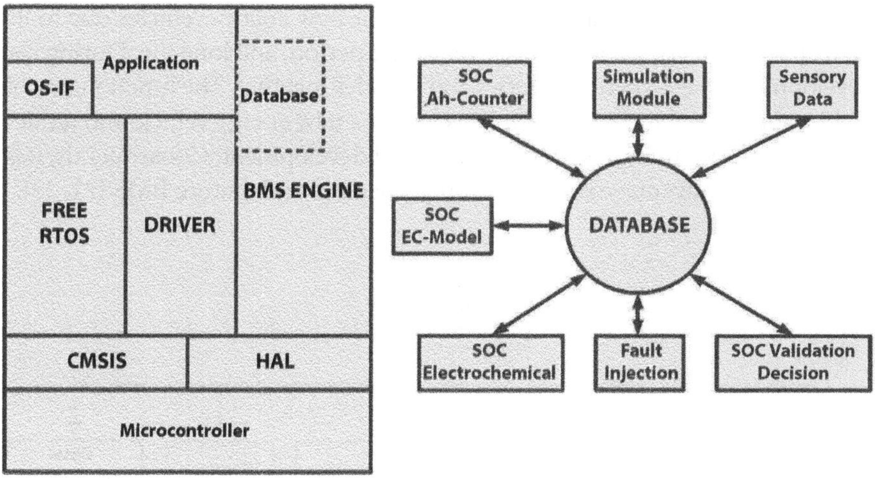

Figure 12.3 Architecting BMS software.

12.3.2 Functionalities

There are two types of battery management systems: internal and external. It is calculated based on a number of measurements, including the voltage of the cells, the current of the packs, the voltage of the packs, and the temperature of the packs. BMS determines SOC, SOH, DOD, and operating key parameters. As a result, battery life is increased, and the original power network's demands are met.

Design techniques and functional units are used in the construction of a BMS. BMS and charging schemes must be designed by identifying the correct battery architectures, functional units, and electronic circuits. Several factors can be taken into account when optimizing battery life [20]:

- This is a software application that manages and analyzes the performance of battery systems in different blocks of a computer system.
- Battery pack safety and performance.
- System resilience in the event of different types of accidents.
- Integrated battery/energy technology that integrates conventional and non-conventional sources.
- Energy management system controlled and monitored by internet of things (IoT).

Environmental pollution and the oil crisis have resulted in a rise in electricity consumption, creating opportunities and challenges for energy storage. A lithium-ion battery is an ideal candidate for new energy vehicles due to its high energy density, performance over a long period, and long life. During car development, the BMS ties batteries and vehicles together. The functionality of BMS is mentioned in Figure 12.4. Developing a new energy vehicle and reducing emissions requires in-depth research and development. Consequently, it is urgent to take advantage of a more comprehensive and mature BMS [21, 22].

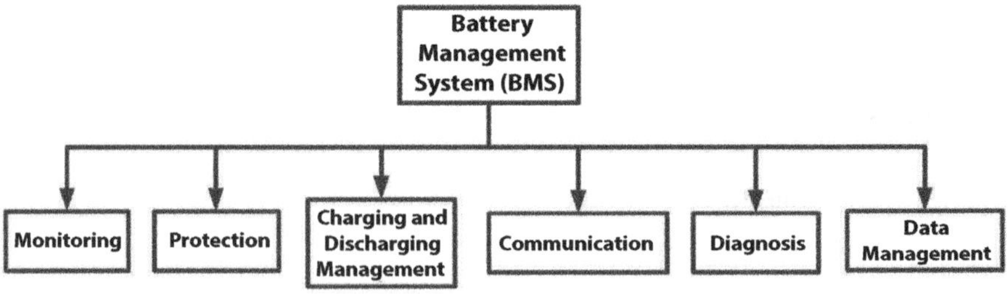

Figure 12.4 Functionality of BMS.

A wide range, a high capacity, and a series-parallel architecture characterize EVs. A BMS is required to control and achieve the battery pack effectively, thereby ensuring reliability (safety) and maximizing energy utilization (power), thereby reducing vehicle operating costs (durability) [23]. Different kinds of sensors, actuators, and controllers make up the BMS. Key technologies include state estimation for cells and battery packs (SOC, SOH, SOF, SOE), documentation and equalization of battery variation, current security, and fault diagnosis.

Battery packs are controlled by an integrated organic brain that performs many functions [24]. BADICHEQ, a battery management system, started running tests in 1991. As a result, BMS has evolved from a monitoring device to an intelligent and integrated device that can select the best battery pack operation for each specific operation. In addition to managing the battery pack, GM's "EVI" systems can also calculate power mileage and detect ground insulation. To optimize battery performance, American AC Propulsion's BatOpt battery management system combines the monitoring system on each battery with a central control unit.

12.4 Battery Models

Modeling a battery accurately is essential to understanding its behavior, as well as to estimating and managing its state. There is a strong nonlinearity in lithium-ion batteries, as well as a time-varying parameter set, which makes modeling them more difficult. This section provides an overview of battery mechanisms and existing battery models, along with a discussion on how to model batteries based on multiphysics, heat, electricity, ageing, and machine field abilities.

12.4.1 Thermal Modeling

When an LIB is operating under different conditions, its temperature will depend on the amount of heat it produces and the cooling method that it uses. To combine heat generation, heat transfer, and heat dissipation, a thermal model analyzes how temperatures are distributed between and within battery cells, modules, and stacks over time and space to design and develop a safe battery thermal management system.

Battery heat transfer is a typical example of an unsteady, time-varying subject with an internal heating source. In addition to defining the temperature generation rate q, the thermal conductivity and boundary conditions are key components of the battery model. According to [25], heat generation in batteries is calculated based on electrochemical reaction and phase transition heat, as well as heat mixing heat. A study by [26], found that heat is usually mixed with phase change and that the measure of entropy heat is equal to that of Joule heat in most cases. When the battery-to-external heat exchange is ignored, the basic form using average heat capacity is obtained.

BMS' thermal management block shows a boundary diagram that estimates battery temperature, as shown in Figure 12.5. Using a 1-D method for estimating temperature, the temperature is estimated [27]. A battery's temperature is monitored and controlled to prevent it from being damaged by very high or very low temperatures. This block controls several fans and electric heaters to maintain a stable battery temperature.

Despite the accuracy of prediction, the above methods add complexity to calculations. In vehicles, it cannot be applied [28]. This paper presents a thermal model that captures both the surface and core temperatures of a battery. Due to the simultaneous acquisition of both battery temperatures, this model is more accurate than lumped parameter models (single state). This model can also be referred to as a dual-state lumped parameter model. Models that account for differences in internal and surface temperature can be used to control batteries in real time. In a way that resembles the original intent of this method [29], applied equivalent circuit thermal models to battery temperature estimation for computer simulation of the electrical power system's thermal simulation. The heat source converts current into heat flow, increasing temperature as a result. The current, temperature by the voltage, heat transfer resistance by the resistance, and heat capacity by the capacitance represents a heat flow. Using this hypothesis, it is possible to convert the thermodynamic properties of the battery into equivalent electrical parameters. An LIB model based on analogy is presented in this paper. It was discovered that by using the heat balance equation, a battery's

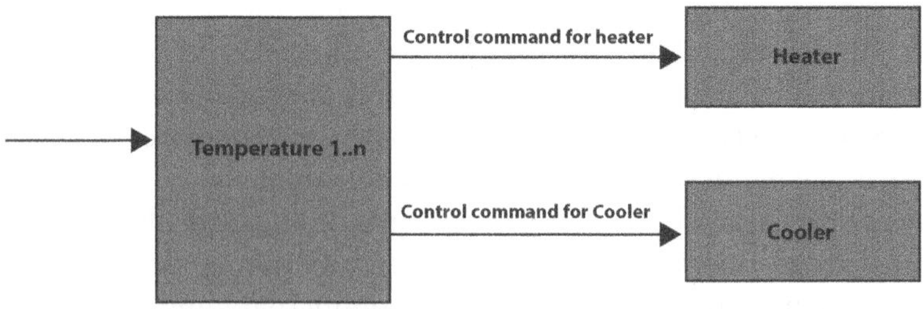

Figure 12.5 Managing the thermal environment.

core and surface temperatures could be determined according to the law of electricity.

Furthermore, convection phenomena between the battery surface and surrounding environments were taken into account. A thermal model for dual-state equivalent circuits was further simplified by [30]. According to current pulse experiments as well as full-charged and full-discharged experiments, the internal temperature of the model is within 1.5°C of its accurate estimation. It's easy to implement this approach in an electric vehicle's BMS. Those models are represented mathematically by equations and diagrams.

12.4.2 Electrical Modeling

As a result, a model of electrical properties can be used to test batteries for macroscopic electrical parameters, saving time and money, or the particle domain parameters within the microstructure can be analyzed to determine battery performance to meet the demands of a variety of management systems. An EV model can be either electrochemical or equivalent circuit-based.

Based on the internal reaction mechanisms of battery cells, kinetic partial differential equations (PDEs) are derived for electrodes and electrolytes. It is able to predict the macroscopic properties of batteries (for example, voltage and current). Still, it is also capable of stimulating the distribution of microscopic features (for instance, electrode current density or electrolyte overpotential). In spite of their accuracy, these models are costly to calculate. Due to this, electrochemical models can be used to optimize the design of batteries and analyze battery safety. As well as simplifying battery management technology, they can be used for a variety of applications.

Batteries are usually stacked or wound with multilayer structures (micron-sized electrode sheets and centimeter-sized bodies), making it difficult to create models of high quality. A three-dimensional model is typically developed by taking the average of a certain volume to calculate overall characteristics. Modeling frameworks for LIBs were presented at NREL30 and 32 by the US National Renewable Energy Laboratory. It inputs parameters from the high dimension to the low dimension as fields and parameters from the low dimension as volumes (current i, heat generation q, etc.). It was easier to calculate and update the mode since the particle mechanism and electrode models were packaged separately.

To meet the electricity consumption demands, the electric grid must be able to generate enough power. As a result of the wide spread between peak and base demand for electricity, matching generators to loads with broad peak-to-base spreads can be challenging [31]. A sufficient amount of installed power capacity is required for electricity providers to meet peak demand on a day-to-day basis. The capacity is generally operated at 20% above estimated demand to meet these obligations, and only 65% of the total installed generation capacity is utilized each year as a result [32].

12.5 Monitoring the Stack

A battery management system monitors battery current, voltage, temperature, concentration, and any commands or monitoring data provided by the SSC during operation. A battery's State of Charge (SOC) and State of Health (SOH) are calculated, system history is compiled, and the battery's service life and performance are determined using system commands. For the battery management systems to enhance battery performance and extend battery life, they need to have access to reliable information about the battery state of charge and discharge, particularly when it comes to SOC and SOH. Alternatively, the BMS must come up with algorithms that will be able to predict these states accurately in the absence of this information. A battery's state of charge (SOC) represents how much charge the battery has over its life cycle (SOC), and its state of health (SOH) refers to how much capacity remains from the battery's original state of charge (SOH).

Coulomb counting is impossible due to the inaccuracy of current meters, slow and variable losses in batteries, and side reactions in the system. The physical characteristics of the battery and measurement system must be taken into consideration to estimate SOC accurately, such as natural cell self-discharge and series string charge balancing.

Some BMSs use other methods of determining SOC in addition to tracking charges entering and leaving the terminal. A terminal voltage measuring open-circuit SOC is usually used. The terminal voltage is a measure of the battery state of charge, which decreases as SOC decreases. Suppose the chemistry differs from the terminal voltage. In that case, it is necessary to create complex nonlinear models between the terminal voltage and the saturation oxidation coefficient. Still, whichever method is used, this method can be more or less accurate [33]. The battery will degrade and lose capacity as a result of this approach, which can lead to overcharging or over-discharging.

12.5.1 Batteries for Grid Storage

While grid-scale energy storage has many benefits, it is imperative to consider its cost. Additional storage costs will be incurred by including a BESS. During storage, kilowatt-hours of energy will be more expensive. As an example, let's assume that the power generation source costs 10 cents/kWh, and the battery costs $400 per kilowatt-hour to store energy. For a truly accurate economic impact estimate, we would also have to analyze the energy storage costs (i.e., the dollar generated by the improvement).

Due to the high cost of energy storage, it is more economical to build additional generation instead of using energy storage in the past. It has become more viable to store energy since the cost of energy storage has continued to plummet, cycle life has increased, and environmental concerns have intensified. Battery technology has also improved in recent years. The main advantage of RFBs is that they can be decoupled from the power and energy systems; this enables them to provide large energy capacities and lower costs per kilowatt-hour than conventional chemistries. Additionally, as comprehension of fundamentals improves and predictive modeling approaches develop, redox-flow BESSs will become more viable. A cost-conscious approach is required to enhance battery lifespans and, thus, reduce storage cycle costs. It has been noted previously that BESS use a range of battery chemistries. In general, lithium-ion battery systems can be simulated using electrochemical engineering models incorporating proper system parameters that reflect the different chemical compositions and solid phase diffusions. Because the electrochemical systems are largely the same, models can be developed for different chemistries based on the same underlying physics. RFBs, which use an exterior electrolyte to pump through a structure that is similar to cathode-separator-anode batteries, have recently gained popularity within the storage sector.

12.5.2 A Modeling Approach to Lithium-Ion Batteries

Li-ion batteries use Li-ions to transport charge, and Li-based storage materials make up the anode and cathode. According to the consensus, a lithium-ion battery has a lithium-titanate cathode and anode made of lithiated graphite. An electrochemically active material, usually mixed with a binder, is coated onto a metal foil that functions as a current collector. Typically, copper is the metal of choice for the anode of the collector foil and aluminum for the cathode of the collector foil. A polymer separator containing electrolytes is used to assemble the anode and cathode electrodes. Intercalating lithium ions into a graphite layer during charging releases them from the cathode. When the reaction is discharged, it reverses.

12.5.3 Advanced Model-Based BMSs

The best way to ensure safety, long-term performance, and optimum battery life is to manage batteries in real-time. Modeling physics depends on the functionality needed (e.g., SOC and SOH estimations). Our understanding of the physics behind it, the scale involved, the time constants, etc., will allow us to develop faster numerical algorithms and better control schemes that will improve our effectiveness. Even though grid-connected battery systems are often expensive and large, even a little improvement in their utilization can save substantial sums of money. In light of this, optimal models, algorithms, and control schemes cannot be overstated. Therefore, if a system is to support dynamic power demands, its transient flow dynamics must be reconsidered rather than assuming steady-state flow development. An example of a Li-ion battery and a rechargeable battery is shown below. Advanced models are employed based on efficient simulation techniques and allow the system to react in real time when required.

12.6 State of Charge Estimation

The state space method is used in this paper to describe OCV using polynomial equations to predict SOC (status of the system). This paper presents a model based on the state space method. SOC is the state of the system. OCV is determined by solving polynomial equations. Based on the calculated terminal voltage of the battery and the predicted OCV, the battery's terminal voltage is calculated with the predicted OCV. A Kalman filter update is performed by calculating its gain using the differences between its calculated and real battery voltages. Kalman filter gain is determined by the difference between the calculated and actual battery voltage to update the state of charge. This method is used to estimate a battery's SOC. Modeling the battery management strategy for an electric vehicle is presented here. Using this model, you can simulate the state of charge of a battery. Based on data collected from the experiment, parameters were derived for the selected LiFePO4 battery.

Based on nominal capacity, SOH is calculated. Several factors contribute to a battery's SOH reduction, including age and charge-discharge cycles. A lithium-ion battery's SOH decreases by 80% after 1000 charge-discharge cycles.

12.6.1 The Need for BMS in Smart Grids and EVs

In spite of rapid advancements in battery technology, it continues to be impossible to guarantee a solution that will solve all the problems associated with electric vehicles and smart grids. The integration of battery storage into smart grids raises several concerns:

- **Cost:** Costs related to manufacturing, labor, maintenance, and operation.
- **Lifetime:** Calculated by charging and discharging the battery and calculating its calendar life.
- **Power delivery:** Charge-discharge rates, energy storage, ramp rates, and efficiency are measured.
- **Environmental impact and safety:** Battery safety/risk factors are dependent on chemical composition, temperature, and operating conditions

12.6.2 Challenges of BMS and Possible Solutions

In addition to demonstrating methods for assessing battery performance, their characteristics were also illustrated. Many BMS applications today are unfeasible with estimation and prediction methods because they require expensive and impractical hardware such as impedance measurement. A central processor, for example, is a costly item that is required for high computational complexity. As can be seen, a BMS must balance performance with feasibility. It is also important to note that most studies are carried out in laboratories and use full charge-discharge cycles.

12.7 Conclusion

Electric vehicles, grid storage, and many other applications require battery management systems to maintain battery health, ensure safety, and optimize performance. With an effective BMS design, battery life can be extended significantly, power delivery is enhanced, and overall system efficiency is improved. BMS strategies must evolve with battery technologies, including the development of more sophisticated thermal and electrical models to adapt to new chemistries and applications. Energy storage is a complex process that requires modern power networks to be equipped with advanced BMS architectures that are capable of handling its dynamic requirements in real-time. By advancing research and technology, BMS will play an increasingly integral role in providing sustainable energy solutions, facilitating the transition to renewable energy sources, and enhancing the feasibility of electric mobility on a large scale.

References

1. Rani, P., Verma, S., Yadav, S.P., Rai, B.K., Naruka, M.S., Kumar, D., Simulation of the lightweight blockchain technique based on privacy and security for healthcare data for the cloud system. *Int. J. E-Health Med. Commun. (IJEHMC)*, 13, 4, 1–15, 2022.
2. Gabbar, H.A., Othman, A.M., Abdussami, M.R., Review of battery management systems (BMS) development and industrial standards. *Technologies*, 9, 2, 28, 2021.
3. Jiang, J. and Zhang, C., *Fundamentals and applications of lithium-ion batteries in electric drive vehicles*, John Wiley & Sons, 2015, Accessed: Nov. 10, 2024, [Online], Available: https://books.google.com/books?hl=en&lr=&id=nVe-BgAAQBAJ&oi=fnd&pg=PR11&dq=Jiang,+J.%3B+Zhang,+C.+Fundamentals+and+Applications+of+Lithium-Ion+Batteries+in+Electric+Drive+Vehicles%3B+Wiley+Online+Library:+Hoboken,+NJ,+USA,+2015&ots=pzAqNoBMTg&sig=oR2fjX8xf80tRrOh3EU0PQS0Ad8.
4. Pop, V., Bergveld, H.J., Danilov, D., Regtien, P.P., Notten, P.H., *Battery management systems: Accurate state-of-charge indication for battery-powered applications*, vol. 9, Springer Science & Business Media, Springer Dordrecht, 2008, Accessed: Nov. 10, 2024, [Online], Available: https://books.google.com/books?hl=en&lr=&id=Qwxrq8tQRSAC&oi=fnd&pg=PR11&dq=Pop,+V.+(Ed.)+Battery+Management+Systems:+Accurate+State-of-Charge+Indication+for+Battery+Powered+Applications%3B+Springer:+Dordrecht,+The+Netherlands,+2008.&ots=07vgfA-lva&sig=TJNZyr-fAirQ6g0Y6nT8PCMkHfA.
5. Sung, W. and Shin, C.B., Electrochemical model of a lithium-ion battery implemented into an automotive battery management system. *Comput. Chem. Eng.*, 76, 87–97, 2015.
6. Bhola, B., *et al.*, Quality-enabled decentralized dynamic IoT platform with scalable resources integration. *IET Commun.*, 2022.
7. Xing, Y., Ma, E.W., Tsui, K.L., Pecht, M., Battery management systems in electric and hybrid vehicles. *Energies*, 4, 11, 1840–1857, 2011.

8. Lelie, M., *et al.*, Battery management system hardware concepts: An overview. *Appl. Sci.*, 8, 4, 534, 2018.
9. Luo, X., Kang, L., Lu, C., Linghu, J., Lin, H., Hu, B., An enhanced multicell-to-multicell battery equalizer based on bipolar-resonant LC converter. *Electronics*, 10, 3, 293, 2021.
10. Doridant, A., Abouda, K., Givelin, P., Thibaud, B., Battery management system demonstrator board design using EMC system simulation, in: *2019 International Symposium on Electromagnetic Compatibility-EMC EUROPE*, IEEE, pp. 427–432, 2019, Accessed: Nov. 10, 2024, [Online], Available: https://ieeexplore.ieee.org/abstract/document/8872061/.
11. Aiello, O., Crovetti, P.S., Fiori, F., Susceptibility to EMI of a Battery Management System IC for electric vehicles, in: *2015 IEEE International Symposium on Electromagnetic Compatibility (EMC)*, IEEE, pp. 749–754, 2015, Accessed: Nov. 10, 2024, [Online], Available: https://ieeexplore.ieee.org/abstract/document/7256257/.
12. Kang, T., Park, S., Lee, P.-Y., Cho, I.-H., Yoo, K., Kim, J., Thermal analysis of a parallel-configured battery pack (1S18P) using 21700 cells for a battery-powered train. *Electronics*, 9, 3, 447, 2020.
13. Singh, A., *et al.*, Smart Traffic Monitoring Through Real-Time Moving Vehicle Detection Using Deep Learning *via* Aerial Images for Consumer Application. *IEEE Trans. Consum. Electron.*, 70, 4, 7302–7309, 2024, doi: 10.1109/TCE.2024.3445728.
14. Abbas, M., Cho, I., Kim, J., Analysis of high-power charging limitations of a battery in a hybrid railway system. *Electronics*, 9, 2, 212, 2020.
15. Arnieri, E., Boccia, L., Amoroso, F., Amendola, G., Cappuccino, G., Improved efficiency management strategy for battery-based energy storage systems. *Electronics*, 8, 12, 1459, 2019.
16. Lee, S. and Kim, J., Power capability analysis of lithium battery and supercapacitor by pulse duration. *Electronics*, 8, 12, 1395, 2019.
17. Eichi, H.R. and Chow, M.-Y., Modeling and analysis of battery hysteresis effects, in: *2012 IEEE Energy Conversion Congress and Exposition (ECCE)*, pp. 4479–4486, 2012, Accessed: Nov. 10, 2024, [Online], Available: https://ieeexplore.ieee.org/abstract/document/6342212/, doi: IEEE.
18. Singh, A., *et al.*, Blockchain-Based Lightweight Authentication Protocol for Next-Generation Trustworthy Internet of Vehicles Communication. *IEEE Trans. Consum. Electron.*, 70, 2, 4898–4907, 2024.
19. Yadav, S.P., Jindal, M., Rani, P., de Albuquerque, V.H.C., dos Santos Nascimento, C., Kumar, M., An improved deep learning-based optimal object detection system from images. *Multimed. Tools Appl.*, 83, 10, 30045–30072, 2024.
20. Uno, M., Ueno, T., Yoshino, K., Cell voltage equalizer using a selective voltage multiplier with a reduced selection switch count for series-connected energy storage cells. *Electronics*, 8, 11, 1303, 2019.

21. Hu, X., Zou, C., Zhang, C., Li, Y., Technological developments in batteries: a survey of principal roles, types, and management needs. *IEEE Power Energy Mag.*, 15, 5, 20–31, 2017.
22. Wang, Q., Jiang, B., Li, B., Yan, Y., A critical review of thermal management models and solutions of lithium-ion batteries for the development of pure electric vehicles. *Renew. Sustain. Energy Rev.*, 64, 106–128, 2016.
23. Lu, L., Han, X., Li, J., Hua, J., Ouyang, M., A review on the key issues for lithium-ion battery management in electric vehicles. *J. Power Sources*, 226, 272–288, 2013.
24. Li, J., Wang, S., Fernandez, C., Wang, N., Xie, H., The battery management system construction method study for the power lithium-ion battery pack, in: *2017 2nd International Conference on Robotics and Automation Engineering (ICRAE)*, IEEE, pp. 285–289, 2017, Accessed: Nov. 10, 2024, [Online], Available: https://ieeexplore.ieee.org/abstract/document/8291396/.
25. Bernardi, D., Pawlikowski, E., Newman, J., A general energy balance for battery systems. *J. Electrochem. Soc.*, 132, 1, 5, 1985.
26. Thomas, K.E. and Newman, J., Heats of mixing and of entropy in porous insertion electrodes. *J. Power Sources*, 119, 844–849, 2003.
27. Gold, S., A PSPICE macromodel for lithium-ion batteries, in: *The Twelfth Annual Battery Conference on Applications and Advances*, IEEE, pp. 215–222, 1997, Accessed: Nov. 10, 2024, [Online], Available: https://ieeexplore.ieee.org/abstract/document/574106/.
28. Lin, X., *et al.*, A lumped-parameter electro-thermal model for cylindrical batteries. *J. Power Sources*, 257, 1–11, 2014.
29. Xiao, Y., Model-based virtual thermal sensors for lithium-ion battery in EV applications. *IEEE Trans. Ind. Electron.*, 62, 5, 3112–3122, 2014.
30. Forgez, C., Do, D.V., Friedrich, G., Morcrette, M., Delacourt, C., Thermal modeling of a cylindrical LiFePO4/graphite lithium-ion battery. *J. Power Sources*, 195, 9, 2961–2968, 2010.
31. Afridi, K., *et al.*, *The Future of the Electric Grid*, MIT Press, United Kingdom, 2011.
32. Strbac, G., Demand side management: Benefits and challenges. *Energy Policy*, 36, 12, 4419–4426, 2008.
33. Batteries, M.-F. and Berndt, D., Research Studies Press Ltd, Taunton, Somerset, UK, 1993.

13

Detailing How AI Empowers Battery Management Systems

Umesh Chandra Garjola[1]*, Ashish Singh[2], Hawraa Ali Sabah[3], Jaspreet Kaur[4] and Zaid Alsalami[5]

[1]*College of Computing Sciences & IT (CCSIT), Teerthanker Mahaveer University, Moradabad (U.P.), India*
[2]*Department of Electrical, Electronics & Electric Vehicle Engineering, NIMS Institute of Engineering & Technology, NIMS University Rajasthan, Jaipur, India*
[3]*College of Technical Engineering, National University of Science and Technology, Dhi Qar, Iraq*
[4]*Department of Computer Science Engineering, Chandigarh Engineering College, Chandigarh Group of Colleges, Jhanjeri, Mohali, Punjab, India*
[5]*Department of Cloud Computing and Technology, The Islamic University, Najaf, Iraq*

Abstract

To make sure batteries perform well and remain safe, Battery Management Systems (BMS) are essential in electric vehicles and renewable energy storage systems. BMS has evolved significantly as a result of the rise of artificial intelligence (AI), which has become more sophisticated in terms of monitoring, maintenance, and energy optimization. A BMS powered by AI can detect faults, estimate the state of charge, and predict the state of health in real time, addressing the limitations of traditional systems. In this chapter, we explore AI's role in the BMS, emphasizing its applications to performance monitoring, fault prediction, and energy efficiency. In line with global efforts to increase the use of renewable energies, BMS technology is integrating more AI-driven features to offer safer, more efficient, and sustainable solutions.

**Corresponding author*: umeshgarjola6@gmail.com

Pethuru Raj, Sudesh Yadav, Manas Kumar Mishra, Satya Prakash Yadav and Victor Hugo C. de Albuquerque (eds.) Artificial Intelligence Technologies for Smart and Sustainable Urban Transportation: Integrated Platforms and Use Cases, (215–232) © 2026 Scrivener Publishing LLC

Keywords: Artificial intelligence, battery management systems, electric vehicles, performance monitoring, sustainable solutions

13.1 Introduction

An electric vehicle ("EV") or renewable energy storage solution requires a battery management system (BMS) that can manage batteries efficiently, reliably, and safely. With the rapid development of artificial intelligence (AI), BMS has become more effective and safer while tackling limitations in traditional designs. AI has improved the performance of batteries, extended their lifespans, and ensured the safety of BMS. In this chapter, we discuss how AI is empowering BMS by enabling advancements in performance monitoring, predictive maintenance, and energy efficiency [1].

Carbon emissions can be minimized through the use of renewable energy, according to the United Nations (UN). Many countries are adopting renewable energy as part of this process. There has been considerable attention paid to ESS implementation in the past [2]. A monitoring system for these ESS, on the other hand, has made it possible to implement battery-management systems for alarm detection and fault detection [3–5]. Figure 13.1 shows that the global market for battery-management systems will grow by 54.8% over the next five years, owing to wireless connections bifurcating in various ways [6]. With the internet of things and the reduction of wires, wireless battery management systems are becoming increasingly popular.

Batteries-management systems also take care of special circumstances associated with device architecture. Lithium-ion batteries are faster to charge, have a greater energy density, and provide a greater power density, allowing them to last longer [7]. A lithium-ion battery has a smaller

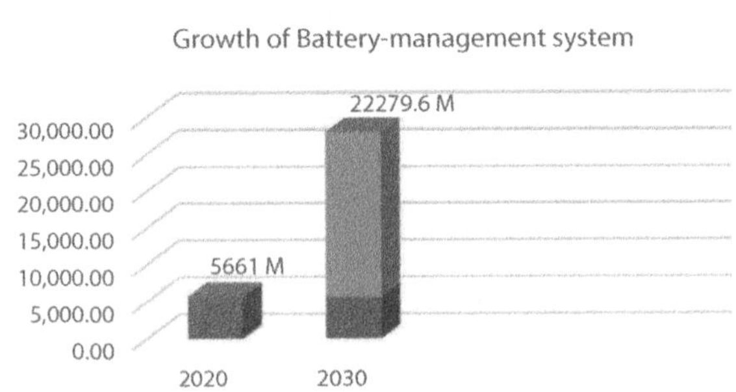

Figure 13.1 The rise of wireless battery management systems with IoT expansion [6].

self-discharge than a nickel-based battery, and they don't require prolonged priming (priming improves battery performance during use and after prolonged storage) [8]. Lithium-ion batteries are becoming more popular among electric vehicle manufacturers as they become more affordable [9].

In traditional charging methods, constant current (CC) and constant voltage (CV) are the most common. Constant current (CC) charging is used to recharge batteries. Despite CV charging's eco-friendliness, it damages the battery's capability because it is so dependent on the battery's technology. A hybrid charging strategy is used instead of CC or CV charging in CV [10]. During MCC charging, phases are separated from CC, and the current slowly decreases as the terminal voltage approaches a predetermined level. Once the battery meets the terminal's requirements, it is charged until it is fully charged [11]. In conventional charging methods, batteries are at risk of overheating, overvoltage, deep discharging, overcurrent, pressure, and mechanical stresses. Battery failure can be prevented, and potentially dangerous circumstances can be reduced by ensuring batteries work correctly for their intended purpose. A supervisory system must ensure batteries are working correctly to prevent failures and avoid potentially dangerous situations. Such monitors are known as battery-management systems [12]. Today's batteries come with a wide range of features to improve their efficiency and safety. Mobile devices can be monitored, protected from damage, assessed for health, assessed for charge, balanced, and charged.

Battery management systems are crucial for stand-alone systems due to safety, reliability, and performance issues [13]. A number of digital technologies, such as wireless networks, internet of things, cloud computing, artificial intelligence, machine learning, neural networks, deep learning, blockchains, big data, cyber security, etc., have gained recognition as tools for detecting faults, diagnosing faults, generating alerts, and predicting failures in real-time. Currently, electricity storage is expensive, and charging batteries takes a long time. A battery's price is also determined by its components. Public vehicles cannot be charged at the moment. Many technologies have shaped the management of batteries. Still, some restrictions remain, such as cell balancing, temperature control, charge control, environmental influences, and accurate readings of state of health (SoH), state of charge (SoC), and logbook function [14].

To achieve global energy transformation, smart grids, or smart grid technology, are essential for integrating renewable energies, as well as supporting and enabling them. Smart grids will enable renewable energy to be used in an efficient, sustainable way, lowering energy consumption and consumer costs at the same time. The energy system will benefit from a decentralized, open design and architecture [15]. Despite the complexity of today's energy systems, energy scholars acknowledge the challenges. In this sense, the electric grid is an enormous infrastructure with a number of limitations, including physical infrastructure [16]. As a rapidly emerging field of technology, artificial intelligence (AI) has transformed numerous industries and economies across the world. Artificial intelligence is regarded as a technology that brings innovation and enables progress [17, 18]. With artificial intelligence (AI), insight is extracted from multiple data sources, including "small data" generated by algorithms themselves, to make better decisions [15], and provide valuable information [19]. General-purpose technologies like artificial intelligence have the potential to influence technology, society, economy, and politics in significant ways [20]. Technological advancements, such as steam engines and electricity, have transformed many aspects of our lives [21].

In the report of the European Commission says tomorrow's control networks will consist of a network of interconnected, diverse, and continuously generating data devices. As recommended by EU regulations [23], smart meters have packets weighing 200 bytes on average, with a reading interval of 15 minutes. By 2020, 200 million smart meters will be installed in Europe [24], which means that the total amount of memory available in Europe is approximately 5,606 TB. It is estimated that 5 exabytes of data will be collected from smart meters alone in a year if sampling is done every second, resulting in almost real-time measurements. For the electric industry, big data capabilities and novel architectures are crucial to enhancing the control of complex, decentralized grids. Energy forecasting and optimization have been the subject of decades of research and analysis around the world [25].

13.2 Systems for Managing Batteries

In this section, battery management systems are outlined, and their environmental and technological impacts are compared. Systems for managing batteries are widely used, and they use a variety of sophisticated and innovative techniques, as shown in Figure 13.2. By monitoring and managing the battery modules of a garage device, battery management systems can optimize the overall performance of the device. An abnormal circumstance may result in the detaching of modules from a machine.

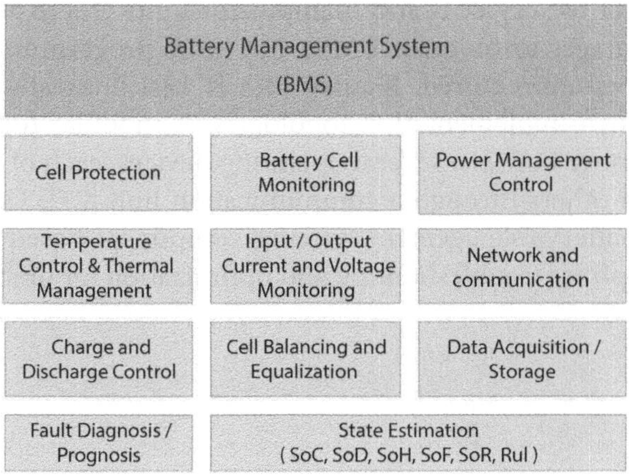

Figure 13.2 System applications for battery management.

It ensures that a battery pack performs optimally, is safe, and lasts as long as possible by monitoring and managing its operation. Among the key functions of BMS are:

Monitoring: Real-time monitoring of battery temperatures, voltages, and currents.
Balancing: Making sure cells are charged and discharged at an even rate to maximize performance and lifespan.
Diagnostics and Fault Detection: To prevent failures, early detection of issues like overheating or degradation of capacity is crucial.
State Estimation: In energy storage applications, measuring metrics such as SoC, SoH, and SoP is crucial for decision-making. Using AI-enabled business management solutions, companies can move from reactive to predictive management, enabling more intelligent, adaptive, and efficient processes.

13.2.1 Structure of Elements and Arrangements

In a machine infrastructure, a BMS cannot stand alone. Modules for managing batteries, power grid connections, energy storage packs, and energy consumption monitoring systems are included in smart electrical automation machines [26]. There are several styles of battery-management systems, such as centralized, distributed, and modular systems. Several cables connect battery cells and manipulator units. There are advantages and disadvantages to modular BMSs, but more programming and hardware resources are required. Figures 13.3 & 13.4 illustrate battery management system implementation topologies. A modular topology allows batteries to move between several operating devices, each of which can be connected to others through a communication link [27]. Unlike component-based battery management systems, component-based battery management simplifies troubleshooting and optimization for different network configurations.

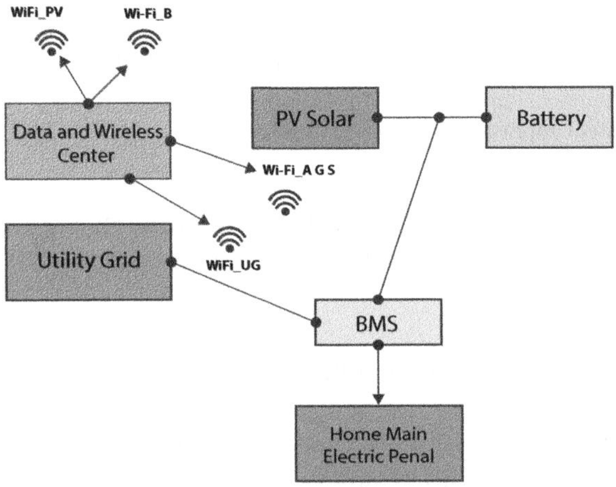

Figure 13.3 BMS integrates with other systems [28].

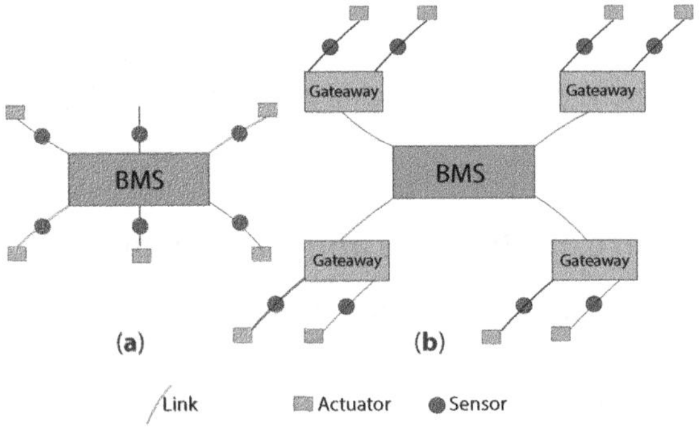

Figure 13.4 Developing a topology for implementing BMS. (a) Central (b) Dispersed.

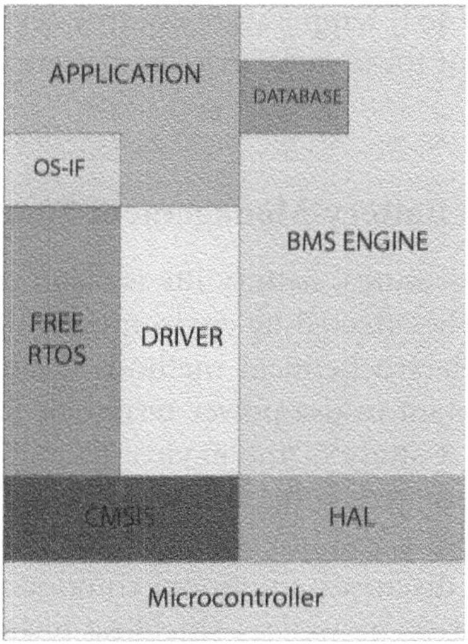

Figure 13.5 Software architecture for BMS.

13.2.2 Structure of Battery-Management System

Figure 13.5 illustrates how BMS recognizes activities quickly, simplifying multitasking [29]. In the past, both extraordinary commitments could not be sustained simultaneously; one mission required postponement to sustain the other. For one mission to be sustained, another mission had to be postponed. Previously, two extraordinary commitments could not be sustained simultaneously. The new architecture of battery-management systems is capable of performing multiple tasks simultaneously. A battery management system's architecture determines the first tasks, such as reading and calculating voltage and current, protecting against over currents and voltages, and actuating protection relays. Performing this task promptly will ensure battery management's safety. Microcontrollers are interfaced with microcontroller software interface standards (CMSIS) and hardware access layers (HAL). The BMS software architecture introduces RTOS for real-time functionalities [30].

13.2.3 System Functions to Manage Batteries

Batteries-management systems have a wide range of functions. Current, temperature, and voltage percentages are calculated to estimate battery capacity. Through the IoT, temperature control devices can be controlled remotely. As a result of these measures, the battery's life can be extended as well [31].

13.2.4 Impacts of Battery-Management Systems

As a result of these measures, battery life is also extended. Battery management systems can be affected by environmental and technical factors. EES structures use electrochemical methods. Batteries are included in this study. It is important to use power terminals and batteries correctly to avoid any danger. A study of the environmental effects of small-scale power garages was presented in this article. As a result of this investigation, wildlife extinctions and cadmium contamination of soil and water were identified. The paper discusses additional compounds in state-of-the-art electrochemical batteries, as well as shielding techniques for improving their security and sustainability [32]. With these movements, software networks and electric flows could be reduced, air pollution could be reduced, and other policies and societal effects could be improved. Electricity is an electrical energy formed by converting chemical energy into electrical energy, and it powers all the gadgets and technology we use every day. Batteries have an anode and cathode that form an electrical circuit that can supply power to digital devices. It's important to properly dispose of batteries once the electric circuit has been exhausted. Every year, this oversight results in the waste of tens of thousands of batteries [33]. Although it might seem harmless to dispose of batteries, doing so can have disastrous consequences. Mercury, lead, lithium, cadmium, and lithium are among the damaging, dangerous, and corrosive elements in batteries. When it comes to batteries, there are five facts you should know to help you understand their environmental impact. Listed below is a summary of concerns about battery management systems regarding efficiency, the environment, and other aspects of operation as mention in Table 13.1.

13.2.5 A Study of How AI Can Be Applied to Smart Grids and Renewable Energy

AI applications in energy research are primarily focused on optimizing, managing, and designing different operations, according to their definition [34]. Studies of artificial intelligence have been conducted in the fields of solar energy, wind energy, geothermal energy, and hydropower [35]. Several studies have been conducted to predict energy consumption [36, 37].

Table 13.1 Environmental and technical impacts of BMS.

Environmental impact	Tech-effectiveness impact
Emission reductions of CO2:	Analysis of real-time SoCs:
Additionally, storing off-height electricity would reduce CO2 emissions by 40%.	The system will also store electricity generated off-height, and a battery management system will be implemented.
GHGs are beneficial to the environment:	How to charge your device optimally:
It is possible to reduce greenhouse gas emissions from batteries by doubling their benefits by adding battery management systems to batteries, as well as smoothing off-top power instead of surges.	In the target layout, layout characteristics are the main focus, and the layout should provide exceptional time efficiency, security, and efficiency.
Deflationary effects of metals:	The ability to characterize quickly:
It is an excellent idea to use a battery management system to manage cycles and operating frequencies when it comes to charging and discharging batteries. With compounds that are highly environmentally and power invasive, this substance can have a significant effect.	BSI offers accurate characterization of SOCs and SOHs. In contrast to SOH's characterization, which uses a single cycle of data, SOC relies on a single cycle of data to model its conclusions.
Temperature regulation's impact:	Evaluation of oneself:
In addition to the electrochemical response temperature, the battery's ambient temperature can also be controlled, depending on the BMS.	In a BMS, complex battery functions, such as capacity, power, hysteresis effects, and temperature effects, are represented mathematically.

Forecast models can be classified into four categories [38]: A physical system, a data-oriented system, a hybrid system, or a large-scale system. Artificial neural networks have been used to predict energy consumption with high accuracy in energy prediction. It is possible to forecast loads accurately at a larger scale, but when scale is reduced, accuracy becomes a challenge. A neural network may consume as much as 40% of the energy consumed by a computer system, according to [39, 40].

In the field of solar energy, AI is being used: In this paper, artificial intelligence is reviewed in relation to solar energy [41, 42]. A variety of artificial neural network methods are usually used in solar modelling, including both single and hybrid methods [43]. Machine learning, for instance, is capable of improving solar forecast accuracy by 30% to 50% compared with conventional forecasting techniques [44, 45].
AI and wind energy: With the help of internet of things sensors, data networks, and advanced analytics, industry incumbents like GE are able to maximize the efficiency of wind turbines using artificial neural networks by 20%. In reviewing the academic literature [35], identify several studies related to physical models, statistical models, and artificial neural networks. As a general rule, AI techniques used in wind energy fall into three categories: neural, statistical, and evolutionary.

13.3 Traditionally, BMS Has Faced Many Challenges

BMS designs have limitations, especially for complex and dynamic applications like electric vehicles and grid storage. There are several challenges to overcome:

Inaccurate State Estimation: As a result of deterministic methods for calculating surface area, surface heat, and surface pressure, the results are inaccurate due to variations in temperature, load, and age.
Inefficient Cell Balancing: Without an intelligent approach, charging and discharging large battery packs uniformly is challenging, impacting effectiveness and life expectancy.
Slow Response to Anomalies: In traditional BMS, there is a rule-based response to anomalies, which can lead to a delay in correcting problems or failure to prevent them in real-time.
Limited Adaptability: In traditional BMS, there is a rule-based response to anomalies, which can lead to a delay in correcting problems or failure to prevent them in real-time.

BMS operations can be made more accurate, real-time, and adaptive with AI [46].

13.4 AI in Business Management Systems

13.4.1 Calculation of State of Charge (SoC) and State of Health (SoH)

The optimal operation of a BMS requires an accurate estimation of its state, but this is a complex process that depends on a variety of factors. Machine learning and deep learning can be used to improve the accuracy of estimations of SoHs and SoCs owing to their ability to handle complex, nonlinear relationships between variables [47].

Estimation of SoCs using AI models: AI models equipped with historical data can provide high-precision predictions of SoCs, taking temperature variations, cell ageing, and usage patterns into account.
Predicting SoH with AI: Tracking battery degradation involves several factors, including cycle count, temperature, and discharge rate, to estimate SoH. The use of AI-based models can detect degradation trends early, enabling maintenance to be performed pre-emptively and preventing sudden failures from occurring.

13.4.2 Balancing and Controlling the Temperature of Cells

A BMS powered by AI can optimize the balance of cells by dynamically adapting to varying cell conditions. In this way, BMSs are able to detect and adjust to minor variations in cellular performance, thereby promoting even cell utilization and preventing hotspots that would accelerate degradation [48].

Smart Cell Balancing: AI-based BMS improves energy efficiency by predicting which cells require balance and by how much [49].
Thermal Management: Heat generation can cause batteries to deteriorate. As a result of analyzing historical usage patterns, AI is capable of predicting the rise in temperature and proactively adjusting cooling systems to reduce overheating.

13.4.3 Predicting and Diagnosing Faults

BMS diagnostics are enhanced by artificial intelligence by identifying potential issues before they become a problem. Through the use of historical and real-time data, machine learning algorithms are capable of classifying and predicting faults.

Anomaly Detection: When trained on normal operating data, artificial intelligence models can detect anomalous behavior and flag deviations so that the BMS can alert operators or intervene appropriately.
Predictive Maintenance: In AI-driven BMS, data analysis ensures prompt maintenance and minimizes downtime by predicting when components will fail.

13.4.4 Optimizing Energy Efficiency and Extending the Range

Energy optimization is crucial for applications such as electric vehicles to extend their range and make their batteries last longer. With artificial intelligence algorithms, real-time data is used to optimize energy consumption.

Dynamic Power Distribution: With AI, battery cells can be dynamically allocated power, which enhances performance and extends battery life.
Usage Pattern Recognition: As a result of machine learning models, energy management can be adjusted based on typical usage patterns, resulting in improved efficiency and extending battery life.

13.5 BMS Powered by Artificial Intelligence

Using AI techniques to enhance BMS capabilities is crucial:

13.5.1 Machine Learning (ML) and Deep Learning (DL)

For BMS, ML and DL algorithms are central to predictive analysis, fault detection, and state estimation. Large datasets can be processed using methods like SVM RNNs to learn complex battery behaviors and predict future performance [50].

13.5.2 Reinforcement Learning (RL)

The Reinforcement Learning (RL) method trains models to make optimal decisions through trial and error, which is extremely useful for energy management and cell balancing. RL algorithms can learn which balancing and power distribution strategy works best under different conditions by exploring various options [51].

13.5.3 An Algorithm for Detecting Anomalies

The BMS can detect anomalous behavior in real time using algorithms such as isolation forests and autoencoders. As a result, early fault detection is possible and catastrophic failures can be prevented [52].

13.5.4 Digital Twins

Battery cells or packs can be modeled as digital twins by creating virtual representations of their physical behavior. Battery behavior can be simulated and analyzed using digital twins powered by AI, enabling predictive maintenance and real-time monitoring.

13.6 BMS with AI Enhancements: Benefits

13.6.1 BMS Integration with AI Offers Numerous Benefits

A more accurate estimate of state size: A BMS based on artificial intelligence provides more accurate estimates of the SoC and SoH of the battery, which are crucial for extending battery life and preventing overcharging and over-discharging [53].
Enhanced Safety and Reliability: The use of artificial intelligence enhances fault detection and anomaly detection, reducing the likelihood of critical failures such as thermal runaway.
Longer Battery Life: Artificial intelligence-enabled BMS extends the life of batteries by optimizing cell usage and reducing unnecessary stress on individual cells.
Increased Energy Efficiency: Using AI models, power distribution can be optimized, energy efficiency can be improved, and energy footprints can be reduced.
Predictive Maintenance: The use of AI-based predictive capabilities reduces downtime and extends the life of batteries by preventing problems from becoming critical [54].

13.6.2 Future Trends and Challenges

Although AI-enhanced BMS holds the potential to transform healthcare, there are challenges to overcome:

Data Quality and Availability: Data of high quality and quantity is essential for AI's effectiveness. Often, it isn't easy to achieve this level of data scale and diversity.
Computational Demand: Computational power is required for real-time data processing, which can result in higher costs and energy consumption.
Model Interpretability: The inexact nature of AI models, in particular deep learning, makes it difficult to interpret the decisions taken by a BMS, which is critical in safety-critical applications.

13.6.3 Future Prospects

Hybrid Models: A hybrid BMS model that combines AI and traditional models could improve accuracy and reduce computation time.
Edge AI: Through the use of AI on edge devices, the BMS can process data in real-time, reducing latency and energy consumption.
Advanced Predictive Models: In the future, advances in AI, including federated learning, could enable battery systems and applications to build even more robust and secure predictors.

13.7 Conclusion

The use of AI-enhanced battery management systems (BMS) represents a breakthrough in energy storage technology, allowing batteries to operate more intelligently, adaptively, and efficiently. In addition to real-time monitoring, estimation of SoC and SoH, and fault prediction, AI is allowing BMS to improve safety and battery life. While AI-driven BMS have challenges, including data quality and computational demands, they hold promise for extending battery lifespan and efficiency while supporting the transition to renewable energy sources. BMSs will continue to perform better in the future, thanks to developments such as hybrid AI models and edge AI, which will ensure strength and reliability in modern power systems.

References

1. Singh, A., *et al.*, Smart Traffic Monitoring Through Real-Time Moving Vehicle Detection Using Deep Learning *via* Aerial Images for Consumer Application. *IEEE Trans. Consum. Electron.*, 70, 4, 7302–7309, 2024, doi: 10.1109/TCE.2024.3445728.

2. Owusu, P.A. and Asumadu-Sarkodie, S., A review of renewable energy sources, sustainability issues and climate change mitigation. *Cogent Eng.*, 3, 1, 1167990, Dec. 2016, doi: 10.1080/23311916.2016.1167990.
3. Dai, H., Jiang, B., Hu, X., Lin, X., Wei, X., Pecht, M., Advanced battery management strategies for a sustainable energy future: Multilayer design concepts and research trends. *Renew. Sustain. Energy Rev.*, 138, 110480, 2021.
4. Rani, P., *et al.*, Federated Learning-Based Misbehaviour Detection for the 5G-Enabled Internet of Vehicles. *IEEE Trans. Consum. Electron.*, 70, 2, 4656–4664, 2023.
5. Rani, P. and Sharma, R., Intelligent transportation system for internet of vehicles based vehicular networks for smart cities. *Comput. Electr. Eng.*, 105, 108543, 2023.
6. Mohammadi, F. and Saif, M., A comprehensive overview of electric vehicle batteries market. *e-Prime Adv. Electr. Eng. Electron. Energy*, 3, 100127, 2023.
7. Balasingam, B., Ahmed, M., Pattipati, K., Battery management systems—Challenges and some solutions. *Energies*, 13, 11, 2825, 2020.
8. Meng, J., Luo, G., Ricco, M., Swierczynski, M., Stroe, D.-I., Teodorescu, R., Overview of lithium-ion battery modeling methods for state-of-charge estimation in electrical vehicles. *Appl. Sci.*, 8, 5, 659, 2018.
9. Bilgin, B., *et al.*, Making the case for electrified transportation. *IEEE Trans. Transp. Electrif.*, 1, 1, 4–17, 2015.
10. Chu, Z., Feng, X., Lu, L., Li, J., Han, X., Ouyang, M., Non-destructive fast charging algorithm of lithium-ion batteries based on the control-oriented electrochemical model. *Appl. Energy*, 204, 1240–1250, 2017.
11. Lin, Q., Wang, J., Xiong, R., Shen, W., He, H., Towards a smarter battery management system: A critical review on optimal charging methods of lithium ion batteries. *Energy*, 183, 220–234, 2019.
12. Zheng, T., Fault diagnosis of overcharge and overdischarge of lithium ion batteries. *Chem. Eng. Trans.*, 71, 1453–1458, 2018.
13. Banguero, E., Correcher, A., Pérez-Navarro, A., Morant, F., Aristizabal, A., A review on battery charging and discharging control strategies: Application to renewable energy systems. *Energies*, 11, 4, 1021, 2018.
14. Yadav, S.P., Jindal, M., Rani, P., de Albuquerque, V.H.C., dos Santos Nascimento, C., Kumar, M., An improved deep learning-based optimal object detection system from images. *Multimed. Tools Appl.*, 83, 10, 30045–30072, 2024.
15. Nagy, K. and Hajrizi, E., Beyond the Age of Oil and Gas–How artificial intelligence is transforming the energy portfolio of the societies. *IFAC-PapersOnLine*, 51, 30, 308–310, 2018.
16. Ramos, C. and Liu, C.-C., AI in power systems and energy markets. *IEEE Intell. Syst.*, 26, 2, 5–8, 2011.
17. van der Vegt, A.H., *et al.*, Systematic review and longitudinal analysis of implementing Artificial Intelligence to predict clinical deterioration in adult

hospitals: what is known and what remains uncertain. *J. Am. Med. Inf. Assoc.*, 31, 2, 509–524, 2024.
18. Hernández, J.L., Sanz, R., Corredera, A., Palomar, R., Lacave, I., A fuzzy-based building energy management system for energy efficiency. *Buildings*, 8, 2, 14, 2018.
19. Xu, Y., Ahokangas, P., Louis, J.-N., Pongrácz, E., Electricity market empowered by artificial intelligence: A platform approach. *Energies*, 12, 21, 4128, 2019.
20. Teece, D.J., Profiting from innovation in the digital economy: Enabling technologies, standards, and licensing models in the wireless world. *Res. Policy*, 47, 8, 1367–1387, 2018.
21. Rani, P., Verma, S., Yadav, S.P., Rai, B.K., Naruka, M.S., Kumar, D., Simulation of the lightweight blockchain technique based on privacy and security for healthcare data for the cloud system. *Int. J. E-Health Med. Commun. (IJEHMC)*, 13, 4, 1–15, 2022.
22. Grawitz, C., *Horizon 2020-Work Programme for Research & Innovation 2018-2020*, 2018, Accessed: Nov. 12, 2024, [Online], Available: https://bibliotecadigital.fgv.br/dspace/bitstream/handle/10438/24901/H2020%20BR%20(FGV).pdf?sequence=1.
23. Andreadou, N., Kotsakis, E., Masera, M., Smart meter traffic in a real LV distribution network. *Energies*, 11, 5, 1156, 2018.
24. Chren, S., Rossi, B., Pitner, T., Smart grids deployments within EU projects: The role of smart meters, in: *2016 Smart Cities Symposium Prague (SCSP)*, IEEE, pp. 1–5, 2016, Accessed: Nov. 12, 2024, [Online], Available: https://ieeexplore.ieee.org/abstract/document/7501033/.
25. Rani, P. and Sharma, R., An experimental study of IEEE 802.11 n devices for vehicular networks with various propagation loss models, in: *International Conference on Signal Processing and Integrated Networks*, Springer, pp. 125–135, 2022.
26. Abbas, M., Cho, I., Kim, J., Analysis of high-power charging limitations of a battery in a hybrid railway system. *Electronics*, 9, 2, 212, 2020.
27. Arnieri, E., Boccia, L., Amoroso, F., Amendola, G., Cappuccino, G., Improved efficiency management strategy for battery-based energy storage systems. *Electronics*, 8, 12, 1459, 2019.
28. Xing, Y., Ma, E.W., Tsui, K.L., Pecht, M., Battery management systems in electric and hybrid vehicles. *Energies*, 4, 11, 1840–1857, 2011.
29. Lee, S. and Kim, J., Power capability analysis of lithium battery and supercapacitor by pulse duration. *Electronics*, 8, 12, 1395, 2019.
30. Gabbar, H.A., Othman, A.M., Abdussami, M.R., Review of battery management systems (BMS) development and industrial standards. *Technologies*, 9, 2, 28, 2021.
31. Uno, M., Ueno, T., Yoshino, K., Cell voltage equalizer using a selective voltage multiplier with a reduced selection switch count for series-connected energy storage cells. *Electronics*, 8, 11, 1303, 2019.

32. Kokkotis, P., II, Psomopoulos, C.S., Ioannidis, G.C., Kaminaris, S.D., Environmental Aspects of Small Scale Energy Storage Systems, in: *Proceedings of the Fifth International Conference on Environmental Management, Engineering, Planning & Economics*, pp. 14–18, 2015, Accessed: Nov. 12, 2024, [Online], Available: https://www.academia.edu/download/38060821/_205__Environmental_Aspects_of_Small_Scale_Energy_Storage_Systems.pdf.
33. Pedersen, E.L., Navigating the Inflation Reduction Act: impacts on the battery industry, transatlantic trade and green transitions, Norwegian University of Life Sciences, 2023, Master's Thesis, Accessed: Nov. 12, 2024. [Online]. Available: https://nmbu.brage.unit.no/nmbu-xmlui/handle/11250/3076810.
34. Quan, X., II and Sanderson, J., Understanding the artificial intelligence business ecosystem. *IEEE Eng. Manage. Rev.*, 46, 4, 22–25, 2018.
35. Jha, S.K., Bilalovic, J., Jha, A., Patel, N., Zhang, H., Renewable energy: Present research and future scope of Artificial Intelligence. *Renew. Sustain. Energy Rev.*, 77, 297–317, 2017.
36. Ahmad, T. and Chen, H., Utility companies strategy for short-term energy demand forecasting using machine learning based models. *Sustain. Cities Soc.*, 39, 401–417, 2018.
37. Ahmad, T., Chen, H., Shah, W.A., Effective bulk energy consumption control and management for power utilities using artificial intelligence techniques under conventional and renewable energy resources. *Int. J. Electr. Power Energy Syst.*, 109, 242–258, 2019.
38. Ahmad, T., Chen, H., Guo, Y., Wang, J., A comprehensive overview on the data driven and large scale based approaches for forecasting of building energy demand: A review. *Energy Build.*, 165, 301–320, 2018.
39. Son, L.H., Pritam, N., Khari, M., Kumar, R., Phuong, P.T.M., Thong, P.H., Empirical Study of Software Defect Prediction: A Systematic Mapping. *Symmetry*, 11, 2, 212, Feb. 2019, doi: 10.3390/sym11020212.
40. Kumar, B.R.S., Varalakshmi, N., Lokeshwari, S.S., Rohit, K., Manjunath, Sahana, D.N., Eco-friendly IOT based waste segregation and management, in: *2017 International Conference on Electrical, Electronics, Communication, Computer, and Optimization Techniques (ICEECCOT)*, IEEE, Mysuru, pp. 297–299, Dec. 2017, doi: 10.1109/ICEECCOT.2017.8284686.
41. Zhao, H. and Magoulès, F., A review on the prediction of building energy consumption. *Renew. Sustain. Energy Rev.*, 16, 6, 3586–3592, Aug. 2012, doi: 10.1016/j.rser.2012.02.049.
42. Kalogirou, S.A., Artificial neural networks in renewable energy systems applications: a review. *Renew. Sustain. Energy Rev.*, 5, 4, 373–401, Dec. 2001, doi: 10.1016/S1364-0321(01)00006-5.
43. Mellit, A. and Pavan, A.M., A 24-h forecast of solar irradiance using artificial neural network: Application for performance prediction of a grid-connected PV plant at Trieste, Italy. *Sol. Energy*, 84, 5, 807–821, May 2010, doi: 10.1016/j.solener.2010.02.006.

44. Brancucci Martinez-Anido, C., *et al.*, The value of day-ahead solar power forecasting improvement. *Sol. Energy*, 129, 192–203, May 2016, doi: 10.1016/j.solener.2016.01.049.
45. Gawer, A., Bridging differing perspectives on technological platforms: Toward an integrative framework. *Res. Policy*, vol, 1239–1249, Sep. 2014, doi: 10.1016/j.respol.2014.03.006.
46. Kharlamova, N., Hashemi, S., Træholt, C., The cyber security of battery energy storage systems and adoption of data-driven methods, in: *2020 IEEE Third International Conference on Artificial Intelligence and Knowledge Engineering (AIKE)*, IEEE, pp. 188–192, 2020, Accessed: Nov. 12, 2024, [Online], Available: https://ieeexplore.ieee.org/abstract/document/9355460/.
47. Kumbhar, S., Faika, T., Makwana, D., Kim, T., Lee, Y., Cybersecurity for battery management systems in cyber-physical environments, in: *2018 IEEE Transportation Electrification Conference and Expo (ITEC)*, IEEE, pp. 934–938, 2018, Accessed: Nov. 12, 2024, [Online], Available: https://ieeexplore.ieee.org/abstract/document/8450159/.
48. Li, Y. and Han, Y., A new perspective on battery cell balancing: Thermal balancing and relative temperature control, in: *2016 IEEE Energy Conversion Congress and Exposition (ECCE)*, IEEE, pp. 1–5, 2016, Accessed: Nov. 12, 2024, [Online], Available: https://ieeexplore.ieee.org/abstract/document/7854719/.
49. Pattnayak, R.A. and Vijay, T., Thermal analysis of cell balancing for battery management system in electric vehicle, in: *2020 26th International Workshop on Thermal Investigations of ICs and Systems (THERMINIC)*, IEEE, pp. 1–7, 2020, Accessed: Nov. 12, 2024, [Online], Available: https://ieeexplore.ieee.org/abstract/document/9420510/.
50. Bhuvaneswari, S., Mohan, A., Sundaram, A., Improved BMS: A Smart Electric Vehicle Design based on an Intelligent Battery Management System, in: *2024 International Conference on Inventive Computation Technologies (ICICT)*, IEEE, pp. 1999–2006, 2024, Accessed: Nov. 12, 2024, [Online], Available: https://ieeexplore.ieee.org/abstract/document/10544810/.
51. Ghalkhani, M. and Habibi, S., Review of the Li-ion battery, thermal management, and AI-based battery management system for EV application. *Energies*, 16, 1, 185, 2022.
52. Cuadra, A.G., Battery self-management with Artificial Intelligence in an IoT Architecture, Accessed: Nov. 12, 2024, [Online], Available: https://oa.upm.es/82867/1/TFM_ANDREI_GARCIA_CUADRA.pdf.
53. Ekatpure, R., Optimizing Battery Lifespan and Performance in Electric Vehicles through Intelligent Battery Management Systems. *J. Sustain. Urban Futures*, 14, 5, 11–28, 2024.
54. Zhao, J., Qu, X., Wu, Y., Fowler, M., Burke, A.F., Artificial intelligence-driven real-world battery diagnostics. *Energy AI*, 18, 100419, 2024.

Part 3

INFRASTRUCTURE OPTIMIZATION IN EV

14

Insisting for Electric Vehicle (EV) Charging Infrastructure Management Systems

Jasgurpreet Singh Chohan[1]*, Ashish Singh[2], Jaspreet Kaur[3], Ruby Pant[4] and Kassem AL-Attabi[5]

[1]School of Mechanical Engineering, Rayat Bahra University, Mohali, India
[2]Department of Electrical, Electronics & Electric Vehicle Engineering, NIMS Institute of Engineering & Technology, NIMS University Rajasthan, Jaipur, India
[3]Department of Computer Science Engineering, Chandigarh Engineering College, Chandigarh Group of Colleges, Jhanjeri, Mohali, Punjab, India
[4]Department of Mechanical Engineering, Uttaranchal Institute of Technology, Uttaranchal University, Dehradun, Uttarakhand, India
[5]Department of Computer Science and Engineering, The Islamic University, Najaf, Iraq

Abstract

The rapid adoption of electric vehicles (EVs) around the world presents a promising opportunity to reduce pollution, fossil fuel dependency, and greenhouse gas emissions. To achieve this transition, charging infrastructure for EVs must be expanded substantially. Charging infrastructure has fallen behind the growing number of EVs, and limited public charging facilities have prevented widespread EV use. An emphasis is placed in this chapter on the critical role that EV-CIMS can play in addressing these issues. Managing distributed charging stations efficiently and intelligently enhances user convenience, optimizes load management, and supports grid stability. Based on an analysis of markets, government policies, and consumer preferences, this chapter proposes strategies for increasing electric vehicle charging infrastructure. A budget-balanced subsidy policy, joint public-private partnerships, and dynamic carbon tax are also analyzed in the chapter to foster private-public cooperation. Finally, it discusses future challenges and advances in EV charging infrastructure, stressing the need for flexible, scalable,

**Corresponding author*: jaskhera@gmail.com

Pethuru Raj, Sudesh Yadav, Manas Kumar Mishra, Satya Prakash Yadav and Victor Hugo C. de Albuquerque (eds.) Artificial Intelligence Technologies for Smart and Sustainable Urban Transportation: Integrated Platforms and Use Cases, (235–252) © 2026 Scrivener Publishing LLC

and interoperable management systems that can adapt to rapidly evolving market conditions.

***Keywords*:** Electric vehicles, grid stability, load management, scalable management systems, sustainable transportation, user convenience

14.1 Introduction

Automobiles and energy are changing due to the rise of electric vehicles (EVs). It is an EV's ability to reduce air pollution, fossil fuel dependency, and greenhouse gas emissions that makes them so appealing. Electric vehicles are gaining popularity worldwide, and EV charging infrastructure is one of the driving forces behind this growth. Charge stations for electric vehicles should be highly managed and distributed, equipped with advanced software and communication protocols, unlike conventional fuel stations. The purpose of this chapter is to discuss why EV Charging Infrastructure Management Systems (EV-CIMS) are crucial to the growth, stability, and accessibility of EV charging networks globally [1].

Due to rapid developments in battery and vehicle technologies, electric vehicles (EVs) are becoming increasingly popular worldwide as public awareness about sustainable societies grows. While EVs are becoming more popular, public charging facilities are often limited in number. As an example, Norway's EV market share was 21% in 2017 and is expected to reach 100% by 2025 [2]. However, only 1052 fast chargers were available in the Norwegian network in 2016, compared to 1403 chargers in Germany and 523 in Sweden [3]. Additionally, the Netherlands has 91,000 electric vehicles and aims to reach one million EVs in 2025 (RVO, 2016). By the end of 2016, there were only 7844 charging stations available in the Netherlands [4]. As of 2018, China has sold 2 million EVs, and will sell 5 million by 2020, representing a further increase since 2015 [5]. There are, however, not enough charging facilities to meet the demand. As of 2018, there will be 330 thousand public charging stations in China, according to the China Electric Vehicle Charging Infrastructure Promotion Alliance (CEVCIPA, 2018). There are fewer than 20% of charging infrastructures in relation to EVs. According to government estimates, China will build 12 thousand charging stations and 4.8 million charging piles by 2020 to meet this enormous demand. As a result, charging infrastructure construction needs to be promoted urgently.

The ratio of EV charging stations should be addressed from two perspectives: the market's role and the role of government. EV charging is driven by consumer demand, which is determined by the demand-supply relationship, and the connections between EV charging stations determine supply. Government influences, such as subsidies and taxes, directly affect the income of gas stations and EV charging stations. (1) How can investment companies cooperate more effectively to invest in EV accusing positions? (2) The purpose of this paper is to propose a model for addressing these issues. Charge stations for EVs will be developed based on consumers' preferences. Studies have recently focused on public-private partnerships (PPPs) as a mode of cooperation between governments and investment companies [6] policy incentives for subsidy recipients. To promote public infrastructure development, the government and private companies are collaborating through public-private partnerships. In contrast, PPP projects remain hampered by high risks, such as too little policy coordination and insufficient systematization [7]. It appears that subsidies will promote EV charging infrastructure construction in the short term; however, in the long term, they could create financial problems for the government [8].

Assuming that policies can change over time is a key to achieving a target more effectively. Taxation and subsidy standards influence the willingness to invest in charging infrastructure, as does EV penetration. Dynamic carbon taxes and subsidies are also considered, as well as their advantages. Subsidies and taxes do not balance themselves, though. When a budget-balanced subsidy policy is applied to electric vehicle charging infrastructure, an evolutionary game model can be used to study the problem. As part of the model, static and dynamic policies are categorized and compared between two scenarios. In a recent study [9, 10], it was found that consumers are more inclined to accept electric vehicles (EVs) if charging stations are readily accessible. However, the US does not have enough charging stations for consumers to ease their range concerns. There has been a call for more public charging from governments, utilities, and industry. Electric vehicles offer a market opportunity and can reduce greenhouse gas emissions (GHG) by reducing greenhouse gas emissions. Consequently, more public charging infrastructure is being proposed and built. During the period from January 1, 2011, to December 31, 2013, the US Department of Energy constructed over 17,000 charging stations as part of its EV Project and Charge Point America initiative [11]. Idaho National Laboratory studied the charging network and participating electric vehicles to better understand electric vehicle usage and charging. EVs are being phased in, according to a new plan announced by President Obama in July 2016. As part of the President's plan, charging infrastructure will be improved, and investments in vehicles and manufacturing will be made [12]. States, cities, and utilities are also making investments. According to Caltrans, 1.5 million zero-emission vehicles are expected to be on the road by 2025. CPUC approved Southern California Edison's plan to install 1500 electric vehicle charging stations as part of a pilot program.

Furthermore, San Diego Gas & Electric Company has been approved to install 3500 chargers at 350 locations in its service area [13], whereas PG&E is negotiating infrastructure agreements [14]. Kansas City Power & Light built 1000 electric vehicle charging stations ahead of demand, which helped make Missouri and Kansas a small EV mecca [15, 16]. Furthermore, automakers are building charging stations in conjunction with these aforementioned efforts. With this entire infrastructure being built, the question naturally arises about what everyone hopes to achieve by doing so. Government initiatives should focus more on the impacts of greenhouse gas emissions. Researchers found that electric vehicles are cleaner than conventional vehicles in most US cities, even after accounting for manufacturing emissions. As electric grids evolve, EVs will continue to be cleaner [17]. Further research is needed to determine the correlation between installing electric vehicle charging infrastructure and reducing GHG emissions, especially if it replaces other alternative energy vehicle technologies [18]. Cost is the next issue to be addressed. Utility companies that construct infrastructure must eventually recoup losses incurred in the construction of charging networks [19]. Suppose an increase in electricity prices is required for EV infrastructure because of construction costs or grid improvements to handle increased loads. How would that impact EV sales? Will a reduction in greenhouse gases be achieved by combining infrastructure deployment with electricity price changes? Last is the question of the prioritization [20].

14.2 The Need for EV Charging Infrastructure Management Systems

Charge infrastructure has been lagging behind the exponential growth of the electric vehicle market. Fast-charging stations must be accessible, reliable, and fast, but placing, managing, and integrating them into the power grid is challenging. Through EV Charging Infrastructure Management Systems, these challenges can be addressed through comprehensive, user-friendly, and operationally efficient solutions [7, 21].

14.2.1 User Demand for Convenience

Charging points, rather than fuel pumps, are the main source of power for electric vehicles. For EV users to have a reliable, easy-to-use, and widely available network of chargers, charger management is vital. It can be significantly more convenient for users to monitor, book, and even perform predictive maintenance on stations in real-time.

14.2.2 Utility and Energy Load Management

A large amount of electricity is consumed by charging an EV, especially when several cars are charging at once. To prevent overload and ensure efficient energy distribution, charging infrastructure must be managed effectively, to reduce peak loads and prevent outages.

14.2.3 Integration with Renewable Energy Sources

To achieve sustainable transport, EV charging must be integrated with renewable energy sources. The use of charging stations can be coordinated with renewable generation patterns through the use of intelligent systems, thereby reducing the reliance on fossil fuels for their energy supply.

14.3 Overview of the Charging Infrastructure for Electric Vehicles

There are many places where electric vehicles can be charged, depending on their geographic location, type of connection, and specifications. As a result of this, charging infrastructures for EVs have evolved into multiple types and designs. Electric vehicle chargers, also called electric vehicle supply equipment, are subject to a variety of standards and requirements. In consequence, electric grids vary across locations, as do the EV models they are powered by.

14.3.1 Equipment Specifications for Electric Vehicles

For electric vehicle charging infrastructure to work, electric vehicle supply equipment must be present. To charge electric cars safely and securely, control schemes and cable connections are used with local electrical grid-connected electric vehicle supply equipment. In an EVSE, the control system can perform many functions, such as identifying the user, authorizing charging, recording and sharing data, and protecting data. A minimum set of basic control and management features should be available on EVSEs for all charging needs. Chargers are most commonly charged *via* conductivity, sometimes called plug-in (wired) charges. Electric vehicles with different power ratings, battery sizes, charging techniques, and vehicle types require different conductive charging requirements.

- **Battery characteristics of different EV segments**

A light electric vehicle (LEV), which can be found in 2-wheelers and 3-wheelers, is predicted to accelerate the electrification of transportation worldwide in the next decade. China and India have a big influence on this industry, which is largely the result of their influence [22]. The automotive and light commercial vehicle (LCV) industries are also electrifying. To enable charging in an electric vehicle, the battery specification determines the charging criteria. E-2Ws and E-3Ws are powered by low-voltage batteries. A low-voltage battery also powers the first generation of electric cars. Despite their continued use, especially in taxis, they will eventually disappear. Impending e-car models demonstrate the power of high-voltage batteries in the second generation of electric vehicles. Load-carrying capabilities for electric LCVs will range from low voltage to high voltage [23].

- **Methods of charging and power ratings**

Battery packs in electric vehicles are charged with direct current (DC). For batteries to receive direct current power, a converter is required since most electrical distribution networks provide alternating current (AC). A direct current or an alternating current may be used to conduct charging. DC power is generated by the electric vehicle's onboard charger when an AC EVSE is used. A DC EVSE converts external power to direct current and delivers that current to the batteries through an onboard charger. AC and DC charging can further be divided into four distinct modes: AC charging, DC charging, and hybrid charging as shown in Figure 14.1. Through a cable and plug, electric vehicles can be connected to conventional socket outlets *via* modes 1 and 2. A single EVSE, commonly known as a dumb charger, cannot communicate with the EV, which is why it is not recommended to use it. Mode 2 portable cables are equipped with built-in controls and protection, making them ideal for charging at home. Electric vehicles are charged using modes 3 and 4, respectively, at commercial and public charging stations [24].

The size and power rating of an EVSE varies based on the charging conditions, which in turn affects how much power is required for the

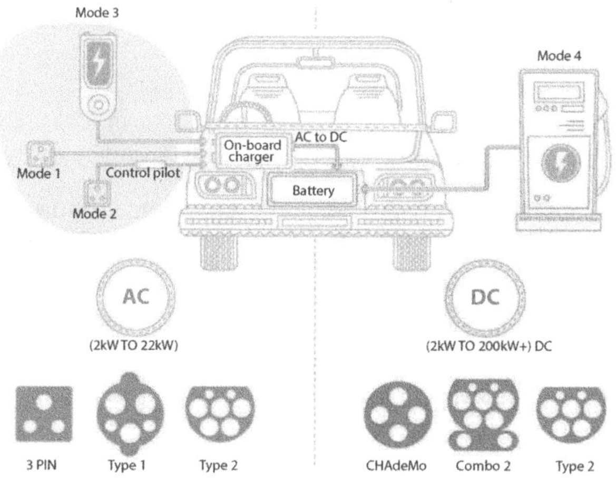

Figure 14.1 Electric vehicle layout.

charging infrastructure. A regular power charge for an electric vehicle can reach 22kW, while a high-power charge can reach 200kW. In spite of the fact that up to 500kW EVSEs are available globally, most of them are suited to large vehicles like buses and trucks. A standard AC outlet can be used to charge an e-2W, an e-3W, or an e-car using standard AC power. With single-phase onboard chargers in LEVs and automobiles, the most efficient chargers have maximum outputs of 7kW [25]. For larger electric vehicles with onboard chargers of 22kW or more, three-phase AC chargers are required. Public energy grids can be used to supply power for typical power charging. Electric vehicles with battery capacities of 30 to 80 kWh use high-power direct current chargers of 50 kW to charge their batteries. A range of power outputs of 25kW to 65kW is available in the market for DC chargers. E-cars can be fast-charged with high-power DC, but they require an external power source and extra infrastructure. Standard power outlets can, therefore, be used to charge most electric cars, even overnight or at slow rates [26].

14.3.2 Standards for Interoperable EV Charging

Network-to-network charging: As a result, the providers of electrical vehicle services operate independent networks in isolation, unable to cooperate or connect with one another. According to current industry jargon, interoperability refers to the ability of electric vehicle owners to use any public charging station by using a single network membership or

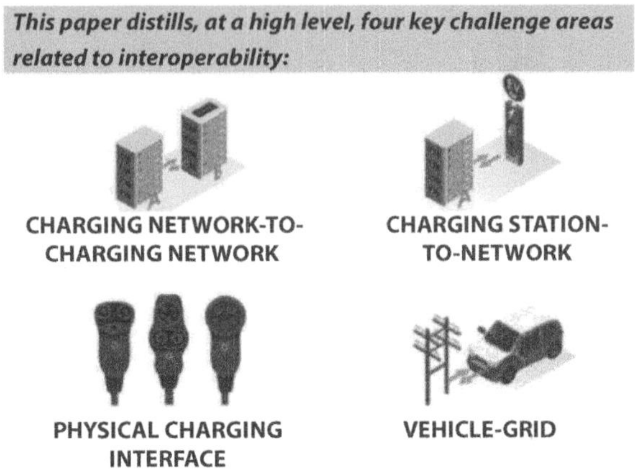

Figure 14.2 Charging network layout.

agreement and a unified platform as mention in Figure 14.2. This is sometimes called "e-roaming". There have been several bilateral agreements between EVSP systems in the past year to establish international partnerships, which means public charging will continue to expand.

14.4 Model Overview

In ParaChoice, fuels such as oil, coal, natural gas, biomass, and zero-carbon fuels are dynamically modeled against light-duty vehicle fleets. Fuel, including electricity, connects the fleet to energy. Prices and demands are shared between the sub-models of energy, fleet, and fuel in the model. We use Energy Information Administration data to provide an initial estimate of energy, electricity, and fuel prices. Federal Highway Administration (2014) collected state motor-vehicle registration data and model availability by powertrain (2014) obtained state motor-vehicle registrations and model availability by powertrain [27] compiled by SRA International, Inc. to determine the composition of LDV fleets. Once the fleet, fuel, and infrastructure are established in the first quarter of 2015, fuel station availability will be determined. As shown in Figure 14.3, the fuel demand for a pump is calculated using the initialized fleet model on the right, which starts with a submodel for the vehicle fleet. Submodels that model fuel production and electricity production pass these values along to models that model energy and electricity consumption. By updating the demand values,

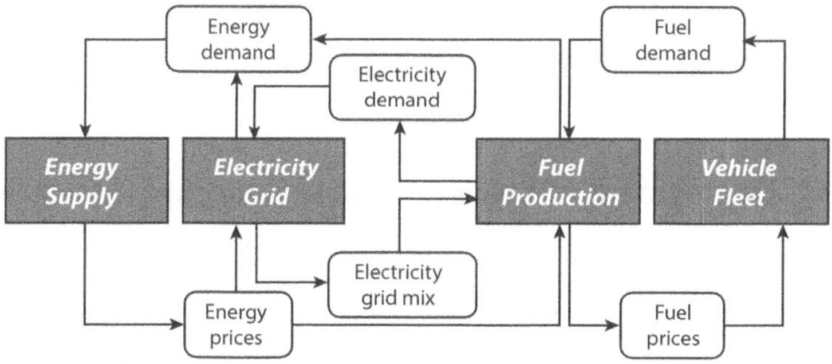

Figure 14.3 Electricity demand and grid mix at a high level.

the energy and electricity sub-models are able to update their prices and the mix of electricity grids. Pump fuel prices are calculated using the fuel production sub-model. A vehicle sub-model evolves the fleet by retiring old cars and selling new ones. Fuel prices are influenced by a variety of factors, including the sub-model for fuel production, vehicle costs, and infrastructure availability. Afterwards, each quarter is simulated until 2050, when the simulation concludes.

14.4.1 Vehicle Fleet

We analyzed EV infrastructure using ParaChoice, in which conventional gasoline spark ignition was compared with diesel compression ignition, and flex-fuel spark ignition was compared with conventional gasoline spark ignition. In addition to the CI and flex-fuel powertrains, hybrid electric vehicle (HEV) and plug-in hybrid electric vehicle (PHEV40) configurations are also modeled with ranges of 10 and 40 miles all-electric, respectively. This model also incorporates a fuel cell electric vehicle (FCEV) and two battery electric vehicles with ranges of 75, 100, 200, and 300 miles. It is necessary to treat dual-fuel vehicles, such as PHEVs, differently when it comes to infrastructure charges. A single-family residence with a PHEV assumes that it is driven serially, with recharging occurring every night. If a vehicle loses charge during the day, it is driven on its liquid tanks. Due to the abundance of gasoline stations and the speed of gasoline refueling, consumers are expected to refuel outside of the home using the most convenient and quickest refueling option, which is liquid refueling. Despite this, this assumption disregards the potential impact of an extremely plentiful DC fast charging infrastructure as well as well-located level 2 charging stations (e.g., in the workplace). For the latter, a future study will examine its impact, but for now, this study aims to examine the impact of broader public charging.

14.4.2 Deployment of Electric Vehicle Charging Infrastructure

In nations that have led the way in electric vehicles, expanding public charging infrastructure will become increasingly important as owners ask for more simplicity and autonomy. Work interruptions in key markets resulted in a 43% decrease in the number of publicly available chargers deployed in 2019, compared to an 83% rise in 2019. Figure 14.4 shows the public chargers with slow and fast charging speeds are most accessible in China. In this region, slow-charging infrastructure accounts for almost half of the world's supply. By 2020, Europe will deploy a third more slow chargers than it does today, with over 245,000 slow chargers deployed. Almost 62,000 slow chargers are located in the Netherlands, the largest number in Europe. Sweden, Finland, and Iceland doubled their slow charger supply in 2020. According to the US Department of Energy, slow chargers installed in 2020 increased by 29% compared to the previous year. In 2020, Korea produced 55,000 slow chargers, taking second place with 44% more.

According to the Chinese government, fast chargers with more than 23 kW of charging power were constructed by 43% in 2020, a slower rate than in 2019 when the number of fast chargers grew by 93%. Despite the limited private charging options in China, the country will be able to achieve its rapid EV deployment targets due to the availability of fast chargers for public use. The installation of fast chargers is occurring at a faster rate than

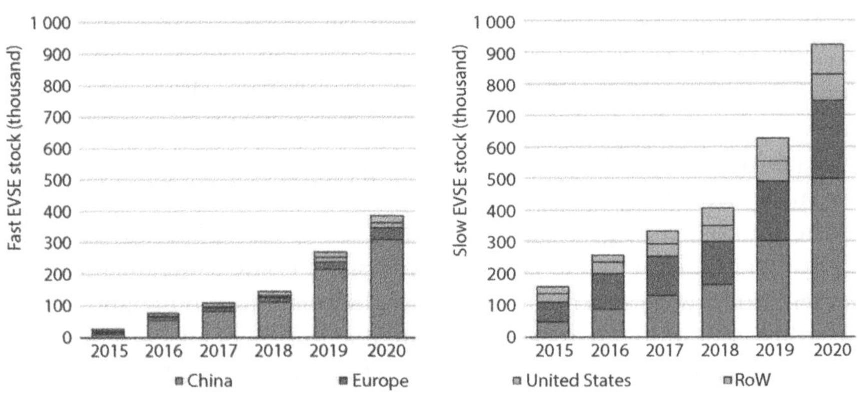

Figure 14.4 Fast and slow EVSE stock.

the installation of slow chargers. There are already over 8000 fast chargers in Germany, 6000 in the UK, 4000 in France, and 1950 in the Netherlands. By 2020, that number is expected to grow by 55%. Fast chargers are available in the US at 18000, of which 62% are Tesla superchargers. In Korea, there are more than 10000 rapid chargers. A fast charger in a public place encourages longer journeys. With them, travellers will be able to travel further and acquire electric cars sooner, as they will be able to charge their cars privately.

Sustainability in transportation is a new concept for mitigating the negative impacts of current transportation systems [28]. Sustainable transportation is characterized by many modes and trends, including electric vehicles (EVs), which have low emissions and are highly efficient [29]. EV usage and EV charging station availability are directly related and influence the popularity of EVs [30]. According to the International Energy Agency (2017), China registered 648,770 electric vehicles in 2017 and 141,254 electric vehicle charging stations, which represented 22% of the total number of electric vehicles. Electric vehicles accounted for 43.3% of all vehicle sales in 2019, with 2,250,000 electric vehicles and 976,500 electric car charging stations sold. More than 75.3% of China's charging infrastructure is distributed across many provinces. China is the most popular developing country for electric vehicles. It is estimated that India does not have sufficient charging stations throughout the country, particularly in the Western province, where most electric vehicles are registered and used. There are more than 5500 electric vehicles registered in Sri Lanka, but only 70 electric vehicle charging stations have been established, accounting for 1.3% of the total [31].

Because of the inconvenience caused to EV users by inadequacies in EVCSs, the existing EVCSs are underutilized. To resolve these issues, it is imperative to locate existing EVCS without taking into account people's travel patterns [32]. As a result of this mismatch between the locations of EVCS and the travel patterns of EVCS users, there are adverse effects on them. So, EVCS should be located in places where charging demand occurs, and charging stations should be located to minimize distances between them [33]. Aside from that, EV owners prefer to charge their EVs after a trip, rather than during the trip, since it will take time for the battery to charge [31] fully. According to researchers, public hotspots, where many people loiter, are the best places to set up EVCS because of these considerations. By eliminating waiting times and maximizing investors' profits, these EVCSs provide solutions to the above-mentioned issues. Although EVCSs have been extensively studied [15] and public hotspots should be considered as bases for EVCSs, no comprehensive study on EVCSs at public hotspots has been conducted. Research has determined that EVCSs are most effective in public hotspots with a high level of loitering. The purpose of this study is to determine which Sri Lankan hotspots are suitable for establishing electric vehicle charging stations, what factors should be considered when establishing these systems, and whether public hotspot-based EV charging stations are technically, economically, legally, culturally, environmentally, and economically feasible.

14.5 Hotspot-Based EVCS

The Wilmington Area Planning Council and PUCSL (2014) research shows that EV users typically commute from home to work, making workplace EVCSs convenient. When workplaces are equipped with charging facilities, it is possible to charge vehicles during the day so that commuters can continue to travel during the day. Aside from that, [34] cited Princeton University and Rutgers University of California as examples of university-based electric vehicle charging stations. Besides shopping centers and amusement parks, fuel stations have also been considered for use as public hotspots. Public hotspots such as supermarkets, amusement parks, and gas stations have also been considered. A public hotspot-based EVCS in an apartment complex has been identified as Fairway Holding Residence by the EV Club Sri Lanka. Further, Sri Lanka's EV green destination hotels provide their guests with charging stations for their electric vehicles. A variety of EVCSs have also been installed at fuel stations by Lanka-Indian Oil Cooperation PLC [35]. EVCS providers in Sri Lanka that operate in shopping centers include 'Kzone', 'Katubedda,' and 'Crescat'.

14.6 Key Features of EV Charging Infrastructure Management Systems

To achieve effective EV-CIMS, a variety of features must be included to enhance grid stability, energy efficiency, and service quality.

14.6.1 Smart Charging and Load Balancing

With smart charging, charger speed and timing are adjusted in real time according to the user's preferences, grid availability, and demand.

As a result of load balancing, the demand is evenly distributed among multiple stations, which eliminates strain on any one of them.

14.6.2 Data Collection and Predictive Maintenance

In an EV-CIMS, information is collected about charging station use, energy consumption, and charging station health. As a result of this information, predictive maintenance can be conducted, reducing downtime and ensuring station availability at all times.

14.6.3 Dynamic Pricing and User Management

Charging stations can be optimized by dynamic pricing based on demand, time of day, or grid conditions, and even encouraged to charge at times that are conducive to renewable energy production.

14.6.4 Integration with Mobile Applications

A mobile app integrated with an EV charging station locater, a charging station availability checker, and a reservation app are critical components of the EV user experience. Also, these apps may display wait times, costs, and alternative options near you in real time.

14.6.5 Grid Interaction and Energy Storage

Using advanced management systems, power can be drawn at low rates and stored for times when demand is high. In some cases, EVs can even feed energy back into the grid when they are not in use through the use of Vehicle-to-Grid (V2G) technology.

14.6.6 Scalability and Flexibility

Flexible, scalable infrastructure management systems are essential as EV markets grow. New charging stations need to be supported, new regulations need to be adapted, and different types of EVs need different charging options.

14.7 Challenges in Implementing EV Charging Infrastructure Management Systems

Successful deployment of EV-CIMS requires addressing several challenges.

14.7.1 High Initial Investment Costs

To implement a robust charging infrastructure, it is necessary to invest significantly in hardware, software, and connectivity. Sharing these costs is often a collaborative effort between governments, utilities, and private companies.

14.7.2 Data Security and Privacy

To manage EV charging infrastructure, numerous data points are required, such as location information, energy usage patterns, and payment information. A critical part of protecting this data is ensuring that it is protected from cyber threats.

14.7.3 Interoperability and Standardization

Charging station equipment, software, and protocols lack universal standards in the industry. Standardization avoids compatibility issues between different manufacturers or providers, making unified management difficult.

14.7.4 Grid Reliability and Capacity

Chargers for electric vehicles may overload power grids. It is imperative to implement management systems that consider grid limitations so that outages and instability are prevented.

14.8 Future Directions and Innovations in EV Charging Infrastructure Management

To enhance efficiency and security, EV-CIMS will incorporate artificial intelligence, machine learning, and blockchain as technology advances. A blockchain-based payment solution may provide secure, transparent charging times and load balancing, while artificial intelligence can optimize charging times. To meet these evolving needs, an adaptive management system is required to handle wireless technologies and fast charging technologies.

14.8.1 AI and Machine Learning for Predictive Optimization

Integrated management systems can predict peak times and dynamically adjust prices based on machine learning algorithms. This ensures a smooth and optimized user experience.

14.8.2 Blockchain for Secure Transactions

In the case of multi-user networks or peer-to-peer energy trading, blockchain technology could make payments both secure and transparent, ensuring both privacy and ease of use.

14.8.3 Ultra-Fast and Wireless Charging

In the future, electric vehicles may even support wireless charging, so adaptive infrastructure management systems will be needed to integrate these emerging technologies.

14.9 Conclusion

Climate change and fossil fuel reliance can be reduced by transitioning to electric vehicles. This transition requires effective EV Charging Infrastructure Management Systems (EV-CIMS), which are able to balance utility demand, grid stability, and energy efficiency. According to the chapter, a combination of market incentives and government policies is needed to overcome the current shortage of EV charging infrastructure. In addition to public-private partnerships and policy interventions, such as dynamic taxation and budget-balanced subsidies, it is vital to overcome challenges such as high investment costs, data security, and grid limitations. EV charging will become even more seamless, efficient, and accessible as AI, machine learning, and blockchain develop and become more prevalent in the market. For a sustainable future of transportation, we need a coordinated approach to building EV infrastructure.

References

1. Hussain, N., Rani, P., Kumar, N., Chaudhary, M.G., A deep comprehensive research architecture, characteristics, challenges, issues, and benefits of routing protocol for vehicular ad-hoc networks. *Int. J. Distrib. Syst. Technol. IJDST*, 13, 8, 1–23, 2022.
2. Figenbaum, E., Electromobility status in Norway: mastering long distances–the last hurdle to mass adoption. *TØI Rep.*, 1627/2018, 2018. Accessed: November 13, 2024, [Online], Available: https://trid.trb.org/View/1509475.
3. Gnann, T., Funke, S., Jakobsson, N., Plötz, P., Sprei, F., Bennehag, A., Fast charging infrastructure for electric vehicles: Today's situation and future needs. *Transp. Res. Part Transp. Environ.*, 62, 314–329, Jul. 2018, doi: 10.1016/j.trd.2018.03.004.
4. Helmus, J.R., Spoelstra, J.C., Refa, N., Lees, M., Van Den Hoed, R., Assessment of public charging infrastructure push and pull rollout strategies: The case of the Netherlands. *Energy Policy*, 121, 35–47, Oct. 2018, doi: 10.1016/j.enpol.2018.06.011.
5. Peterson, S.B. and Michalek, J.J., Cost-effectiveness of plug-in hybrid electric vehicle battery capacity and charging infrastructure investment for reducing US gasoline consumption. *Energy Policy*, 52, 429–438, Jan. 2013, doi: 10.1016/j.enpol.2012.09.059.
6. Zhang, X., Zhou, Y., Han, Q., Game theory-based environmental LCC control behavior analysis. *J. Clean. Prod.*, 211, 1527–1533, 2019.
7. Yang, T., Long, R., Li, W., Suggestion on tax policy for promoting the PPP projects of charging infrastructure in China. *J. Clean. Prod.*, 174, 133–138, Feb. 2018, doi: 10.1016/j.jclepro.2017.10.197.
8. Zhu, L., *et al.*, Study on crowdfunding's promoting effect on the expansion of electric vehicle charging piles based on game theory analysis. *Appl. Energy*, 196, 238–248, Jun. 2017, doi: 10.1016/j.apenergy.2016.11.060.
9. Singer, M., Consumer views on plug-in electric vehicles–National Benchmark Report. *National Renewable Energy Lab.(NREL)*, Golden, CO United States, 2016, Accessed: November 14, 2024, [Online], Available: https://www.osti.gov/biblio/1335806.
10. Charged, F., *The Fully Charged Guide to Electric Vehicles & Clean Energy*, Unbound Publishing, 2020, Accessed: November 14, 2024, [Online], Available: https://books.google.com/books?hl=en&lr=&id=aF6vDwAAQBAJ&oi=fnd&pg=PT31&dq=Davies,+A.,+December+2014.+A+Two+Day+Battle+to+Charge+My+Car+Convinced+Me+We%E2%80%99re+Not+Ready+for+EVs.+Wired.&ots=IRAzihafCl&sig=blJxN8YKE6APPD2wblzVpcf_nJk.
11. Francfort, J., *et al.*, Plug-in electric vehicle and infrastructure analysis. *Ida. Natl. Lab. INLEXT-15-35708*, 359, 1–527, 2015.
12. Sivaram, V., Norris, T., McCormick, C., Hart, D., Energy Innovation Policy: Priorities for the Trump Administration and Congress. *Wash. Inf. Technol.*

Innov. Found., 2016. Accessed: November 14, 2024, [Online], Available: https://legacy-assets.eenews.net/open_files/assets/2016/12/13/document_gw_03.pdf.

13. Lipu, M.S.H., *et al.*, Battery management, key technologies, methods, issues, and future trends of electric vehicles: A pathway toward achieving sustainable development goals. *Batteries*, 8, 9, 119, 2022.
14. John, J.S., Will PG&E Finally get Approval for its EV Charging Pilot? *Greentech Media*, 2016.
15. Levinson, R.S. and West, T.H., Impact of public electric vehicle charging infrastructure. *Transp. Res. Part Transp. Environ.*, 64, 158–177, 2018.
16. Bouallegue, M., *et al.*, Kansas City, Missouri, Streetlight Electric Vehicle Charging, in: *Strategies and challenges for site selection of streetlight electric vehicle infrastructure in Kansas City, Missouri*, Metropolitan Energy Center, United States, 2024, Accessed: November 14, 2024, [Online], Available: https://www.osti.gov/biblio/2459282.
17. Nealer, R., Cleaner cars from cradle to grave: How electric cars beat gasoline cars on lifetime global warming emissions. *JSTOR Sustain. Collect.*, 2015.
18. Singh, A., *et al.*, Smart Traffic Monitoring Through Real-Time Moving Vehicle Detection Using Deep Learning *via* Aerial Images for Consumer Application. *IEEE Trans. Consum. Electron.*, 70, 4, 7302–7309, 2024, doi: 10.1109/TCE.2024.3445728.
19. Berman, B., The ultimate guide to electric car charging networks. *Plugincars Com. Online Httpswww Plugincars Comultimateguide-Electr.-Car-Charg.-Netw.-126530 Html*, 2018.
20. Rani, P., Verma, S., Yadav, S.P., Rai, B.K., Naruka, M.S., Kumar, D., Simulation of the lightweight blockchain technique based on privacy and security for healthcare data for the cloud system. *Int. J. E-Health Med. Commun. IJEHMC*, 13, 4, 1–15, 2022.
21. Liu, J. and Wei, Q., Risk evaluation of electric vehicle charging infrastructure public-private partnership projects in China using fuzzy TOPSIS. *J. Clean. Prod.*, 189, 211–222, 2018.
22. Rani, P. and Sharma, R., IMFOCA-IOV: Intelligent Moth Flame Optimization based Clustering Algorithm for Internet of Vehicle, in: *2023 14th International Conference on Computing Communication and Networking Technologies (ICCCNT)*, IEEE, pp. 1–6, 2023, Accessed: November 09, 2024, [Online], Available: https://ieeexplore.ieee.org/abstract/document/10307646/.
23. Rani, P. and Sharma, R., Intelligent transportation system for internet of vehicles based vehicular networks for smart cities. *Comput. Electr. Eng.*, 105, 108543, 2023.
24. Zhang, L., Zhao, Z., Xin, H., Chai, J., Wang, G., Charge pricing model for electric vehicle charging infrastructure public-private partnership projects in China: A system dynamics analysis. *J. Clean. Prod.*, 199, 321–333, Oct. 2018, doi: 10.1016/j.jclepro.2018.07.169.

25. Rani, S., Talwar, R., Malhotra, J., Ahmed, S.H., Sarkar, M., Song, H., A novel scheme for an energy efficient Internet of Things based on wireless sensor networks. *Sensors*, 15, 11, 28603–28626, 2015.
26. House, M., Forderung für elektroautos und ladestationen (in Chinese), 2017, Accessed: November 13, 2024, [Online], Available: http://mobility-house.com/de/foerderung-fuer-elektroautos-und-ladestationen/%20-%20 muenchen-foerderung.
27. Polk, R., *Co.,* US Vehicle Registration Data MY 2005-2016, Jan. 2015, Compiled by SRA International. Inc.
28. Ryan, S. and Throgmorton, J.A., Sustainable transportation and land development on the periphery: a case study of Freiburg, Germany and Chula Vista, California. *Transp. Res. Part Transp. Environ.*, 8, 1, 37–52, 2003.
29. Poullikkas, A., Sustainable options for electric vehicle technologies. *Renew. Sustain. Energy Rev.*, 41, 1277–1287, 2015.
30. Wu, H. and Niu, D., Study on influence factors of electric vehicles charging station location based on ISM and FMICMAC. *Sustainability*, 9, 4, 484, 2017.
31. Wickramasinghe, P., Electric vehicle charging stations to be regulated. *Dly. Mirror*, 2017.
32. Cai, H., Jia, X., Chiu, A.S., Hu, X., Xu, M., Siting public electric vehicle charging stations in Beijing using big-data informed travel patterns of the taxi fleet. *Transp. Res. Part Transp. Environ.*, 33, 39–46, 2014.
33. Sathaye, N. and Kelley, S., An approach for the optimal planning of electric vehicle infrastructure for highway corridors. *Transp. Res. Part E Logist. Transp. Rev.*, 59, 15–33, 2013.
34. Robinson, J., Brase, G., Griswold, W., Jackson, C., Erickson, L., Business models for solar powered charging stations to develop infrastructure for electric vehicles. *Sustainability*, 6, 10, 7358–7387, 2014.
35. Wettasinghe, C., Lanka IOC to set up electric vehicle charging points at existing filling stations. *Dly. Mirror*, 2017.

15

Illuminating the AIs Role in Shaping Up EV Charging Infrastructures

Jatinder Kumar[1]*, Ashish Singh[2], Hawraa Ali Sabah[3], Yashwant Singh Bisht[4] and Laith H. Jasim[5]

[1]*Department of Computer Science Engineering, Chandigarh Engineering College, Chandigarh Group of Colleges, Jhanjeri, Mohali, Punjab, India*
[2]*Department of Electrical, Electronics & Electric Vehicle Engineering, NIMS Institute of Engineering & Technology, NIMS University Rajasthan, Jaipur, India*
[3]*College of Technical Engineering, National University of Science and Technology, Dhi Qar, Iraq*
[4]*Department of Mechanical Engineering Uttaranchal Institute of Technology, Uttaranchal University, Dehradun, Uttarakhand, India*
[5]*Department of Computer Science and Engineering, The Islamic University, Najaf, Iraq*

Abstract

In response to the increasing adoption of electric vehicles (EVs), charging infrastructure has become increasingly important. Developing innovative solutions is essential to reducing carbon emissions and improving energy efficiency during this shift, which is driven by global commitments. It has become increasingly clear that artificial intelligence (AI) is essential to the development and optimization of EV charging infrastructures. In this chapter, we explore AI's role in designing, placing, and managing charging networks across a number of disciplines. As a result of AI predictive analytics, optimization algorithms, and real-time monitoring, charging stations can be strategically deployed, grid management can be improved, and users have a seamless experience when using EVs. AI insights are used in this chapter to highlight the importance of integrating renewable energy sources, enhancing load balancing, and planning infrastructure. With these developments, AI positions itself as a transformative force in the development of EV charging ecosystems that are sustainable and accessible.

**Corresponding author*: jatinder2090.research@cgcjhanjeri.in

Pethuru Raj, Sudesh Yadav, Manas Kumar Mishra, Satya Prakash Yadav and Victor Hugo C. de Albuquerque (eds.) Artificial Intelligence Technologies for Smart and Sustainable Urban Transportation: Integrated Platforms and Use Cases, (253–268) © 2026 Scrivener Publishing LLC

Keywords: Artificial intelligence, charging infrastructure, electric vehicles, energy efficiency, predictive analytics, real-time monitoring

15.1 Electro Mobility Charging Systems

EVs have become increasingly popular around the world over the past decade, driven by a collective desire to reduce carbon emissions, increase energy efficiency, and move away from fossil fuels. To meet both current demand and future growth, comprehensive and resilient charging infrastructure is urgently needed to support EV adoption. As electric vehicle charging networks expand, artificial intelligence (AI) plays an increasingly important role in their design and management. As AI refines predictive analytics, optimizes algorithms, and monitors real-time data, it is reshaping the EV landscape for improved energy efficiency, lower costs, and better user experience. Sustainable transportation is becoming a focal point of the green energy movement, with electric vehicles (EVs) increasingly prevalent.

Furthermore, as the EV market grows, so does the demand for efficient, reliable charging infrastructure. EV charging is reshaped by artificial intelligence (AI), which transforms how we approach it. Experiencing AI role in the development of EV charging infrastructure allows us to explore its potential for enhancing grid management, forecasting demand, and improving user experiences. The discussion examines the multiple ways that AI is enhancing EV charging and setting the stage for a more affordable, efficient, and accessible future powered by electricity [1].

In place of conventional combustion engines (ICEs), electromobility is becoming more popular as a greener transportation mode. To facilitate the development of electromobility, an accessible and reliable charging infrastructure is crucial [2]. It is crucial to install charging stations in nations that have well-established land and maritime transportation systems to advance electromobility [3]. Charging infrastructure is lacking, which is a major obstacle to electromobility's development. The charging infrastructure and outlets for EVs are not adequate to meet their needs [4]. A lack of infrastructure hinders EV expansion and adoption; charging stations are insufficient, discouraging potential buyers [5]. It is, therefore, imperative that national developments include a charging infrastructure network that supports the widespread adoption of electromobility [6].

EVs are becoming more accepted as transformative and innovative technologies are adopted in the electromobility industry. The dynamic wireless charging system (DWC) and static wireless charging system (SWC) differ fundamentally, according to [7]. The SWC system transfers power from a ground-mounted pad to the vehicle's charging pad while the vehicle is stationary, charging the EV [8]. Alternatively, DWC systems enable EVs to recharge their batteries while in motion by embedding charging pads into the road surface [8]. Recent research has concentrated on stabilizing power by switching segmented coils, especially for scenarios like traffic lights that must manage multiple loads at the same time to improve the efficiency of SWC systems [9]. This article discusses a static wireless charging solution for electromobility. Additionally, studies have examined how wireless charging systems perform under varying operating conditions. A wireless charging technology performs optimally when it is configured with appropriate operating frequencies, coil configurations, and physical spacing between receiving and transmitting coils [10].

A beneficial approach for accelerating the growth of the electromobility charging infrastructure market is to share charging infrastructure with others [5, 11]. Multiple users can share the costs associated with setting up and maintaining charging stations, maximizing the environmental benefits of electromobility [12]. This approach facilitates a charging infrastructure market and promotes the sustainable development of electromobility. Electromobility's development depends on the location of charging stations. Local authorities have decided on a new charging station's location [12, 13]. Using decision-support tools, local authorities can select the best charging stations based on variables such as population density and traffic patterns [12].

A smart charging system enables EVs to seamlessly integrate into the power grid, ensuring the long-term sustainability of public charging stations [14]. Organizing the charging process is crucial to grid integration and to the profitability of charging infrastructure through smart charging strategies [14]. The use of smart charging can make electric vehicles a flexible grid resource and enable electromobility ancillary services to be provided to the grid during times of crisis [14]. As a secondary function and support mechanism, EMAS provides additional services beyond primary mobility needs. The use of augmented reality technology might facilitate grid stabilization, demand response, and energy trading in the context of intelligent electromobility systems [15]. As electric vehicles are developed and their success is demonstrated, as is the expansion of charging infrastructure, they could greatly contribute to sustainable urban development [16, 17]. EV charging and automated driving also benefit safety due to the ability of smart charging to reduce risks and improve communication between vehicles [18]. Chargers and batteries can also be integrated into PV solar panels and electric vehicles (EVs) to reduce grid reliance and increase energy efficiency [18].

In terms of consumer benefits, smart charging has a number of advantages. As a result, it facilitates peak-load shifting and optimizes EV users' economic benefits [19]. The use of smart charging enables consumers to travel more affordably. By contrast, the charging process can be adapted to the power system's needs, ensuring the most efficient use of energy [20]. A smart charger facilitates grid management by allowing EVs to respond rapidly to the energy system's demands. In the future, smart charging will be more important as mass adoption of EVs increases, so intelligent energy systems will be able to schedule energy systems effectively as EV users increase [21]. Further, smart charging may reduce the strain placed on distribution transformers caused by electric vehicles' charging requirements, ensuring a reliable power supply [22].

In recent years, electric vehicles (EVs) have gained momentum as a critical tool in the fight against climate change. EVs have become more affordable and practical, thanks to advances in battery technology, attracting both consumers and policymakers [23]. For electric vehicles to be successfully integrated into mainstream transportation, a robust charging infrastructure and the availability of electric vehicles are critical. As charging infrastructure becomes ubiquitous, EVs will be adopted more widely by providing users with convenient and accessible charging solutions [24, 25]. There are currently two types of EV charging infrastructure: Residential and workplace-appropriate levels 1 and 2. With the increase in the popularity of EVs, fast-charging infrastructure, such as DC fast chargers, has become more important in facilitating long-distance travel [26].

15.2 Literature Review

Several studies have examined how electric vehicles (EVs) are integrated into power grids to gain a better understanding of how they interact with transportation and energy systems [27]. Studies of EV charging strategies and the dynamics of EV operation within power grids have highlighted the need for efficient management to mitigate the adverse impacts on grids. Electricity demand profiles of plug-in hybrid EVs have been studied in aggregate, providing insight into grid stability and capacity planning issues [28]. According to some researchers, increased load variability poses a challenge to power distribution systems caused by EV adoption. In a number of studies, specific ramifications have been examined, such as the impact of plug-in electric vehicles on ageing transformers and rooftop solar photovoltaics, as well as the broader effects of EVs on distribution systems [29]. In addition, discussions extend to enhancing smart grids with microgrids and reviewing key technologies associated with pure electric vehicles. Researchers have studied extremely fast charging technologies for EVs, conducted simulation-based case studies to quantify their impact on the electric grid, and developed smart parking lot management systems to facilitate EV charging [30]. Furthermore, metaheuristics have been used to solve real-world EV charging scheduling problems, behavioral theories have been used to model EV charging behavior, and machine learning approaches have been investigated so far to understand EV charging behavior. Furthermore, a nuanced understanding of the charging infrastructure requirements of private battery electric vehicles has been obtained through meticulous research into factors influencing fast charging behavior [31].

15.3 Electric Vehicle Charging Infrastructure

The wide adoption and use of electric vehicles depend on a stable charging infrastructure, which is a core component of the EV ecosystem. Figure 15.1 illustrates how electric vehicle charging infrastructure is currently designed, along with the challenges, types, and factors that influence its optimization.

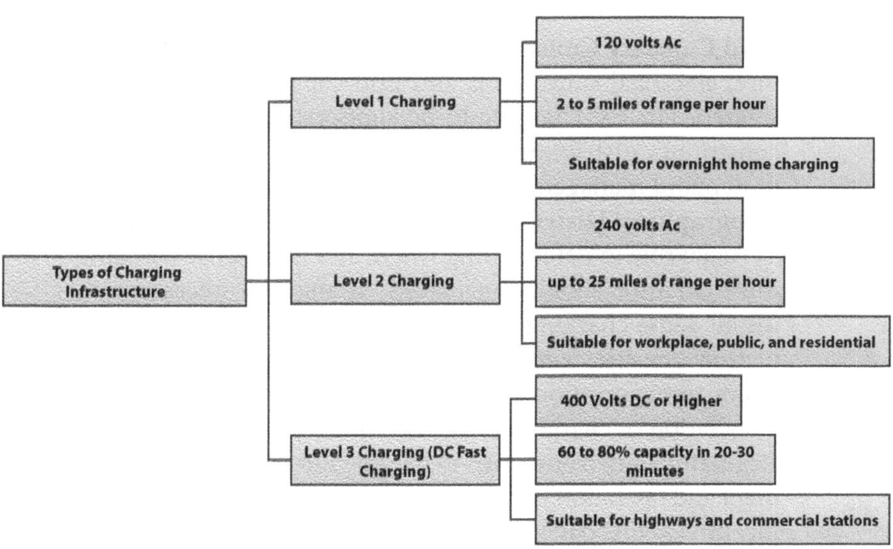

Figure 15.1 Classify charging levels using AI techniques.

15.3.1 Infrastructural Types of Charging

The types of charging infrastructure for electric vehicles are classified according to the power, voltage, and duration of charging:

- **Level 1 Charging:** Level 1 charging is delivered using household outlets (120 volts AC), typically delivering 2 to 5 miles of range per hour. Charging the battery overnight at home with Level 1 is convenient, but it may not be the best method for recharging the battery quickly.
- **Level 2 Charging:** There are two charging stations: The 120-volt Level 1 and the 240-volt Level 2 are typically used. With the Level 2 charger, you can charge anywhere, whether at work, in a public parking lot, or your home.
- **Level 3 Charging (DC Fast Charging):** Battery capacity can be increased to 60% to 80% within 20 to 30 minutes using Level 3 charging or DC fast charging. Highways, major transportation routes, and commercial charging stations with high voltages (typically 400 to 450 volts DC) are common places to deploy Level 3 chargers.
- **Charging Demand Patterns:** Optimizing charging station placements, capacities, and operations requires an understanding of charging demand patterns. EV owners' charging preferences, daily commute patterns, and peak charging hours all influence demand for charging infrastructure.
- **Grid Capacity Constraints:** To deploy charging infrastructure, electrical substations and grid capacity must be considered. For EV charging to be supported by grid capacity constraints, distribution networks may need to be upgraded or demand management strategies implemented.
- **Geographical Distribution of EVs:** Space distribution plays a crucial role in determining where and how dense charging infrastructure should be placed. It may be necessary to install dense charging infrastructure in cities with high population densities and rates of EV adoption. In rural areas, major transportation routes may be the priority area for the installation of charging infrastructure.
- **Regulatory Environment:** A charging infrastructure's deployment and operation may be affected by regulatory policies and incentives. Electric vehicles can be adopted more readily if subsidies are provided, tax incentives are offered, and zoning

Table 15.1 Explains the types of infrastructure used to charge electric vehicles.

Types of charging	Account	Charge speed	Application
Level 1 Charging	AC outlet (120 volts) used in the average household	It is possible to cover 2-5 miles per hour	Charges for residential use
Level 2 Charging	AC 240 volts	Per hour, the range can reach 25 miles	Parking in public areas and charging at work
Level 3 Charging (DC Fast Charging)	The voltage should be high (usually 400 volts DC or higher)	In 20-30 minutes, battery capacity reaches 60-80%	Transport corridors, highways

regulations are enacted to encourage the installation of charging infrastructures.

- **User Preferences and Behavior:** To design user-friendly chargers and services, it is imperative to understand users' preferences and behaviors.

The following Table 15.1 illustrates levels one to three of the electric vehicle charging infrastructure. Charge speeds and typical uses are explained for each charging. As a result of this information, stakeholders will be better able to understand how different charging options can be applied to certain situations based on their characteristics.

15.4 Optimizing the Charging Infrastructure Using Artificial Intelligence

With advanced algorithms and data analytics, artificial intelligence (AI) can optimize electric vehicle charging infrastructure and make real-time decisions in real-time. In this section, we explore how AI can help optimize charging infrastructure in a multitude of ways.

15.4.1 Predicting Charging Demand with Data Analytics

AI is widely used to predict charging demand, which is a key application of AI in charging infrastructure optimization. Using AI algorithms that analyze historical data, such as charging patterns, weather conditions, traffic patterns, and user behavior, it is possible to predict future charging demand with accuracy. In addition to predicting peak demand periods and allocating resources efficiently, these predictive models can also optimize the use of charging infrastructure. The use of predictive models such as regression analysis and time series forecasting is common in the prediction of charging demand. It is important to consider a number of factors when determining charging behavior, which include the time of day, the day of the week, seasonal trends, and special events. A charging demand prediction system powered by artificial intelligence adjusts to changing conditions and improves its accuracy over time. Artificial intelligence algorithms can be used to identify patterns in charging demand on a spatial and temporal scale, allowing stakeholders to optimize charging station placement and capacity. Using clustering algorithms, for example, it is possible to choose locations with high demand and prioritize investments in infrastructure based on proximity and demand similarities between charging stations.

15.4.2 Managing Dynamic Charges with AI

With dynamic charging systems, charging schedules and energy distribution are optimized according to real-time conditions, including energy prices, grid load, and customer preferences. With these systems, charging stations can be intelligently controlled, charging rates can be adjusted, charging sessions can be prioritized, and energy consumption can be balanced across a grid network on a dynamic basis. In dynamic charging management, reinforcement learning algorithms learn optimal charging policies by interacting with the environment in a trial-and-error manner. When reinforcement learning agents maximize cumulative rewards, they are better able to maximize user satisfaction, minimize energy costs, and reduce grid congestion. A dynamic charging management system can optimize electric vehicle fleet charging behaviour based on grid conditions and market dynamics while also accommodating demand response and grid balancing. These systems can help utilities manage the variability of load, reduce peak demand, and improve grid stability by combining demand-side management with EV charging off-peak hours. By integrating artificial intelligence with dynamic charging managers, multiple charging protocols, communication standards, and grid interfaces can be supported, enhancing interoperability and compatibility. Additionally, these systems integrate seamlessly with smart grid technologies and energy management systems, facilitating the coordination and control of charging infrastructure across various environments and stakeholders.

15.4.3 Planned Infrastructure Optimization Algorithms

Optimization algorithms based on artificial intelligence play an important role in planning, deploying, and expanding electric vehicle charging networks. Table 15.2 describes the key applications of artificial intelligence in optimizing electric vehicle charging infrastructure using mathematical optimization techniques such as linear programming, integer programming, and genetic algorithms. Several applications are described, including prediction of charging demand, management of dynamic charging,

Table 15.2 AI Applications for optimizing electric vehicle charging infrastructure.

Application	Description	Key techniques	Benefits
Charging Demand Prediction	Using historical data, we can predict future charging demand based on weather designs and the behavior of users	Predictive analytics, machine learning	Allocation of resources and grid planning should be optimized
Dynamic Charging Management	Optimization of charging rates and schedules in real time	Learning reinforcement, dynamic pricing	Congestion on the grid and energy costs should be minimized
Optimisation Algorithms for Infrastructure Planning	Optimal charging station locations and capacities should be determined	Genetic algorithms, linear programming	Infrastructural reduction

and optimization algorithms for planning infrastructure. A showcase of AI techniques and their benefits can enhance stakeholders' understanding of AI's diverse capabilities to enhance charging infrastructure efficiency.

15.5 Charging Intelligent Infrastructures

Electric vehicle success relies on a well-distributed, accessible, and reliable charging infrastructure. Electric vehicle charging infrastructure must satisfy a variety of needs, including rapid urban recharges and slower, overnight home charging. Energy demand is inherently variable, vehicles are dispersed geographically, and electrical grid capacity limits make traditional infrastructure management methods inadequate. In this regard, AI proves to be a valuable tool, providing a scalable and data-driven solution [32].

15.5.1 The Challenges of Developing EV Charging Infrastructure

Demand Prediction: Achieving accurate forecasts of charging demand.
Grid Load Management: Keeping the electricity grid stable by balancing charging loads.
Resource Allocation: Placing charging stations in strategic locations to ensure maximum coverage and accessibility.
User Experience: Offering real-time information about charging options, minimizing wait times, and ensuring availability.
Energy Sourcing: Charge networks for electric vehicles that integrate renewable energy and energy efficiency.

15.5.2 Predicting Demand and Selecting Sites with AI

EV infrastructure planners need to understand when and where the vehicles will need to be charged. A predictive analytics model that uses artificial intelligence can analyze historical data, traffic patterns, EV adoption trends, and other factors to predict demand accurately. Charge stations can be strategically placed to anticipate demand in areas that may otherwise be underserved with this capability [33].

Case Study: Predictive Modeling for Urban Charging Stations
According to demographic and mobility data, AI models have been used to identify densely populated urban areas that are likely to adopt electric vehicles. A wide range of data is used to refine these models, including the

number of electric vehicle registrations, nearby charging stations, and even the weather. It is possible to predict a spike in charging demand during heat waves when drivers prefer shorter commutes but need to recharge more frequently to keep their air conditioners running [34].

15.5.3 Managing and Balancing Loads in Real Time

Maintaining the electric grid's stability while meeting the charging demand is a critical aspect of EV infrastructure. EV charging generates more load than traditional grids can handle, especially during peak hours. Load-balancing algorithms based on artificial intelligence offer a solution, however [35].

Using AI to balance loads dynamically
With artificial intelligence-driven load balancing systems, charging rates on networks can be automatically adjusted in real-time with information on grid capacity, energy prices, and current demand. During times of low grid demand or abundant renewable energy resources, these systems can prioritize charging. Additionally, they can reduce the rate of charging at low-priority stations or homes during peak grid demand periods, preventing brownouts and overloads on the grid [36].

Application example: Smart Charging Hubs with AI
The use of smart charging hubs in high-density EV areas, such as urban centers and large commercial facilities, is being investigated as a model of the future. Using AI algorithms in these hubs, drivers can be directed to the best charging stations according to grid conditions, EV traffic, and user preferences, minimizing both congestion and energy consumption [37].

15.5.4 The Integration of Renewable Energy Sources with AI

For EV infrastructure to reduce dependence on fossil fuels, renewable energy sources such as solar and wind power must be integrated because renewable energy sources are inherently variable due to factors such as weather, day of the week, etc. The use of artificial intelligence assists in addressing this challenge by optimizing how renewable energy is utilized to charge, store excess energy, and manage peak energy demands [38].

Systems for storing energy using AI
Environments can benefit from storing surplus renewable energy in large batteries during off-peak times. The use of artificial intelligence algorithms

to monitor and control the production and release of renewable energy can help meet peak energy demands. AI, therefore, facilitates the integration of renewables into EV charging infrastructure, enabling charging demand to be decoupled from fossil-fuel grid peaks [39].

Stations for charging electric vehicles with solar power
A mixed-energy system that combines solar and energy storage is being developed for regions with high solar potential. Through AI algorithms, solar energy can be optimized for immediate charging needs, and excess energy can be stored for nighttime usage, making the stations both eco-friendly and economical [40].

15.5.5 Infrastructural Challenges and Considerations in AI-Driven Charging

EV charging infrastructure can achieve many benefits with AI, but there are a number of challenges as well:

Keeping your data private and secure: The data used by AI models includes sensitive grid information. A secure and private environment is essential to harnessing this data effectively.
The cost of energy and the strain on the grid: Using AI-driven charging to reduce grid strain can reduce energy costs, but if it is not managed correctly, it can also cause high energy costs.
Interoperability: For an integrated infrastructure, AI systems must be able to work across a variety of platforms, vehicle types, and charging networks.
Regulatory Challenges: Providing AI-driven systems to energy grids often involves navigating regional regulatory frameworks with varying requirements.

15.5.6 The Future of AI in EV Charging Infrastructure

Increasingly, electric vehicles will require AI-driven, scalable infrastructure. The future may bring the following advancements:

Chargers for autonomous vehicles that are enhanced: As autonomous vehicles become more autonomous, they may be able to locate, charge, and disconnect themselves on their own, which would reduce human intervention and wait times.

V2G Integration: V2G systems will likely rely heavily on artificial intelligence to balance load and reduce grid strain by allowing EVs to give back energy during peak demand.
Incorporation of IoT Devices: By integrating IoT devices with AI, we will be able to gather more precise data on charging habits, allowing us to improve our predictive abilities.

15.6 Conclusion

Therefore, integrating AI into electric vehicle charging infrastructure represents a significant contribution to sustainable transportation. The use of artificial intelligence-driven solutions can manage charging demand, optimize infrastructure placement, and balance grid loads in a scalable and adaptive manner. With AI, renewable energy sources are used efficiently, and energy distribution is dynamic, ensuring an eco-friendly and resilient charging network. EV adoption will continue to rise, increasing the importance of artificial intelligence in infrastructure development and ultimately supporting the widespread adoption of electric vehicles. AI-enabled charging infrastructure will become a cornerstone of the modern transportation landscape as research and innovation aim to enhance AI's interoperability across diverse platforms and improve data privacy.

References

1. Rani, P. and Sharma, R., Intelligent Transportation System Performance Analysis of Indoor and Outdoor Internet of Vehicle (IoV) Applications Towards 5G. *Tsinghua Sci. Technol.*, 29, 6, 1785–1795, Dec. 2024, doi: 10.26599/TST.2023.9010119.
2. Zema, T., Sulich, A., Grzesiak, S., Charging Stations and Electromobility Development: A Cross-Country Comparative Analysis. *Energies*, 16, 1, 32, Dec. 2022, doi: 10.3390/en16010032.
3. Sun, X., Li, Z., Wang, X., Li, C., Technology Development of Electric Vehicles: A Review. *Energies*, 13, 1, 90, Dec. 2019, doi: 10.3390/en13010090.
4. Figura, J. and Gądek-Hawlena, T., The Impact of the COVID-19 Pandemic on the Development of Electromobility in Poland. The Perspective of Companies in the Transport-Shipping-Logistics Sector: A Case Study. *Energies*, 15, 4, 1461, Feb. 2022, doi: 10.3390/en15041461.
5. Melander, L. and Wallström, H., The benefits of green horizontal networks: Lessons learned from sharing charging infrastructure for electric freight

vehicles. *Bus. Strategy Environ.*, 32, 4, 1835–1846, May 2023, doi: 10.1002/bse.3222.

6. Rani, P., Verma, S., Yadav, S.P., Rai, B.K., Naruka, M.S., Kumar, D., Simulation of the lightweight blockchain technique based on privacy and security for healthcare data for the cloud system. *Int. J. E-Health Med. Commun. IJEHMC*, 13, 4, 1–15, 2022.
7. Trung, N.K. and Diep, N.T., A Maximum Transfer Efficiency Tracking Method for Dynamic Wireless Charging Systems of Electric Vehicles. *J. Electr. Comput. Eng.*, 2021, 1–10, Nov. 2021, doi: 10.1155/2021/5562125.
8. Yang, Y., El Baghdadi, M., Lan, Y., Benomar, Y., Van Mierlo, J., Hegazy, O., Design Methodology, Modeling, and Comparative Study of Wireless Power Transfer Systems for Electric Vehicles. *Energies*, 11, 7, 1716, Jul. 2018, doi: 10.3390/en11071716.
9. Liu, H., Tan, L., Huang, X., Zhang, M., Zhang, Z., Li, J., Power Stabilization based on Switching Control of Segmented Transmitting Coils for Multi Loads in Static-Dynamic Hybrid Wireless Charging System at Traffic Lights. *Energies*, 12, 4, 607, Feb. 2019, doi: 10.3390/en12040607.
10. Sahany, S., Biswal, S.S., Kar, D.P., Sahoo, P.K., Bhuyan, S., Impact of Functioning Parameters on the Wireless Power Transfer System Used for Electric Vehicle Charging. *Prog. Electromagn. Res. M*, 79, 187–197, 2019, doi: 10.2528/PIERM18092610.
11. Baraniak, J. and Starzyński, J., Modeling the Impact of Electric Vehicle Charging Systems on Electric Power Quality. *Energies*, 13, 15, 3951, Aug. 2020, doi: 10.3390/en13153951.
12. Pan, A., Zhao, T., Yu, H., Zhang, Y., Deploying Public Charging Stations for Electric Taxis: A Charging Demand Simulation Embedded Approach. *IEEE Access*, 7, 17412–17424, 2019, doi: 10.1109/ACCESS.2019.2894780.
13. Panchal, C., Stegen, S., Lu, J., Review of static and dynamic wireless electric vehicle charging system. *Eng. Sci. Technol. Int. J.*, 21, 5, 922–937, Oct. 2018, doi: 10.1016/j.jestch.2018.06.015.
14. Moghaddam, Z., Ahmad, I., Habibi, D., Phung, Q.V., Smart Charging Strategy for Electric Vehicle Charging Stations. *IEEE Trans. Transp. Electrif.*, 4, 1, 76–88, Mar. 2018, doi: 10.1109/TTE.2017.2753403.
15. Khayyer, P. and Izadian, A., Power management strategies for hybrid electric trucks in smart-grids, in: *2012 IEEE PES Innovative Smart Grid Technologies (ISGT)*, IEEE, Washington, DC, USA, pp. 1–3, Jan. 2012, doi: 10.1109/ISGT.2012.6175767.
16. Rus, C., Leba, M., Negru, N., Marcuş, R., Rîsteiu, M., Electric vehicles in smart grid and smart city for Petroşani case. *MATEC Web Conf.*, 342, 05002, 2021, doi: 10.1051/matecconf/202134205002.
17. Rani, P. and Sharma, R., An experimental study of IEEE 802.11 n devices for vehicular networks with various propagation loss models, in: *International Conference on Signal Processing and Integrated Networks*, Springer, pp. 125–135, 2022.

18. Vitale, C., *et al.*, CARAMEL: results on a secure architecture for connected and autonomous vehicles detecting GPS spoofing attacks. *EURASIP J. Wirel. Commun. Netw.*, 2021, 1, 115, Dec. 2021, doi: 10.1186/s13638-021-01971-x.
19. Zhang, Y., He, Y., Su, F., Wang, X., Zhang, D., Characterization of interaction between electric vehicles and smart grid. *E3S Web Conf.*, 237, 02004, 2021, doi: 10.1051/e3sconf/202123702004.
20. Hildermeier, J., Kolokathis, C., Rosenow, J., Hogan, M., Wiese, C., Jahn, A., Smart EV Charging: A Global Review of Promising Practices. *World Electr. Veh. J.*, 10, 4, 80, Nov. 2019, doi: 10.3390/wevj10040080.
21. Han, X., Wang, H., Liang, D., Master-slave game optimization method of smart energy systems considering the uncertainty of renewable energy. *Int. J. Energy Res.*, 45, 1, 642–660, Jan. 2021, doi: 10.1002/er.5798.
22. Masoum, M.A.S., Moses, P.S., Hajforoosh, S., Distribution transformer stress in smart grid with coordinated charging of Plug-In Electric Vehicles, in: *2012 IEEE PES Innovative Smart Grid Technologies (ISGT)*, IEEE, Washington, DC, USA, pp. 1–8, Jan. 2012, doi: 10.1109/ISGT.2012.6175685.
23. Faddel, S., Al-Awami, A.T., Mohammed, O.A., Charge control and operation of electric vehicles in power grids: A review. *Energies*, 11, 4, 701, 2018.
24. Darabi, Z., Ferdowsi, M., Keyhani, Marwali, Impact of Plug-In Hybrid Electric Vehicles on Electricity Demand Profile, in: *Smart Power Grids 2011Power Systems*, A. and M. (Eds.), pp. 319–349, Springer Berlin Heidelberg, Berlin, Heidelberg, 2012, doi: 10.1007/978-3-642-21578-0_11.
25. Yadav, S.P., Jindal, M., Rani, P., de Albuquerque, V.H.C., dos Santos Nascimento, C., Kumar, M., An improved deep learning-based optimal object detection system from images. *Multimed. Tools Appl.*, 83, 10, 30045–30072, 2024.
26. Shafiee, S., Fotuhi-Firuzabad, M., Rastegar, M., Investigating the impacts of plug-in hybrid electric vehicles on power distribution systems. *IEEE Trans. Smart Grid*, 4, 3, 1351–1360, 2013.
27. Papadopoulos, P., Skarvelis-Kazakos, S., Grau, I., Cipcigan, L.M., Jenkins, N., Sep. 2012, Electric vehicles' impact on British distribution networks. *IET Electr. Syst. Transp.*, 2, 3, 91–102, Sep. 2012, doi: 10.1049/iet-est.2011.0023.
28. Etezadi-Amoli, M., Choma, K., Stefani, J., Rapid-charge electric-vehicle stations. *IEEE Trans. Power Deliv.*, 25, 3, 1883–1887, 2010.
29. Yoldaş, Y., Önen, A., Muyeen, S.M., Vasilakos, A.V., Alan, I., Enhancing smart grid with microgrids: Challenges and opportunities. *Renew. Sustain. Energy Rev.*, 72, 205–214, 2017.
30. Sujitha, S. and Srinivasa, G., Design and Performance of High Frequency Temperature Compensated Crystal Oscillator. *J. Adv. Res. Dyn. Control Syst.*, 2019.
31. Li, Z., Khajepour, A., Song, J., A comprehensive review of the key technologies for pure electric vehicles. *Energy*, 11, 182, 552–561, 2019.

32. Ronanki, D., Kelkar, A., Williamson, S.S., Extreme fast charging technology—Prospects to enhance sustainable electric transportation. *Energies*, 12, 19, 3721, 2019.
33. Ramanujam, A., Sankaranarayanan, P., Vasan, A., Jayaprakash, R., Sarangan, V., Sivasubramaniam, A., Qantifying The Impact of Electric Vehicles On The Electric Grid: A Simulation-Based Case-Study, in: *Proceedings of the Eighth International Conference on Future Energy Systems*, ACM, Shatin Hong Kong, pp. 228–233, May 2017, doi: 10.1145/3077839.3077854.
34. Kuran, M.S., Viana, A.C., Iannone, L., Kofman, D., Mermoud, G., Vasseur, J.P., A smart parking lot management system for scheduling the recharging of electric vehicles. *IEEE Trans. Smart Grid*, 6, 6, 2942–2953, 2015.
35. García-Álvarez, J., González, M.A., Vela, C.R., Metaheuristics for solving a real-world electric vehicle charging scheduling problem. *Appl. Soft Comput.*, 65, 292–306, 2018.
36. Hu, L., Dong, J., Lin, Z., Modeling charging behaviour of battery electric vehicle drivers: A cumulative prospect theory-based approach. *Transp. Res. Part C Emerg. Technol.*, 102, 474–489, 2019.
37. Spoelstra, J.C., Charging behaviour of Dutch EV drivers, 2014, Master's Thesis, Accessed: Nov. 14, 2024. [Online]. Available: https://studenttheses.uu.nl/handle/20.500.12932/17792.
38. Shahriar, S., Al-Ali, A.-R., Osman, A.H., Dhou, S., Nijim, M., Machine learning approaches for EV charging behaviour: A review. *IEEE Access*, 8, 168980–168993, 2020.
39. Rani, P. and Sharma, R., Intelligent transportation system for internet of vehicles based vehicular networks for smart cities. *Comput. Electr. Eng.*, 105, 108543, 2023.
40. Rani, P. and Sharma, R., IMFOCA-IOV: Intelligent Moth Flame Optimization based Clustering Algorithm for Internet of Vehicle, in: *2023 14th International Conference on Computing Communication and Networking Technologies (ICCCNT)*, IEEE, pp. 1–6, 2023, Accessed: Nov. 09, 2024, [Online], Available: https://ieeexplore.ieee.org/abstract/document/10307646/.

16

Deciphering Smart Grid Integration and Energy Management

Jatinder Kumar[1], Protyay Dey[2], Jasgurpreet Singh Chohan[3*], Sanjeev Kumar Shah[4] and Laith Jasim[5]

[1]*Department of Computer Science Engineering, Chandigarh Engineering College, Chandigarh Group of Colleges, Jhanjeri, Mohali, Punjab, India*
[2]*Department of Computing Science and Artificial Intelligence, NIMS Institute of Engineering and Technology, NIMS University Rajasthan, Jaipur, India*
[3]*School of Mechanical Engineering, Rayat Bahra University, Mohali, India*
[4]*Department of Electronics & Communication Engineering, Uttaranchal Institute of Technology, Uttaranchal University, Dehradun, Uttarakhand, India*
[5]*Department of Computer Science and Engineering, The Islamic University, Najaf, Iraq*

Abstract

The smart grid plays an important role in transforming energy distribution and management to become sustainable. With smart grids, decentralized renewable energy, electric vehicles, and battery energy storage systems can all work together using advanced communication, control, and digital technologies. This chapter examines how green energy is integrated into smart grids, as well as how energy management systems (EMS) increase grid reliability and reduce costs. With layered structures, grid stability and resilience can be ensured. This chapter examines the integration of renewable energy with smart grids, energy management systems (EMS), and smart grids layered to maximize stability and resilience. The future energy landscape will also require concepts like demand response, distributed energy resources management, and prosumer involvement.

***Keywords*:** Battery energy, storage systems, demand response, electric vehicles (EVs), grid stability, green energy integration, smart grids, sustainable

**Corresponding author*: jaskhera@gmail.com

Pethuru Raj, Sudesh Yadav, Manas Kumar Mishra, Satya Prakash Yadav and Victor Hugo C. de Albuquerque (eds.) Artificial Intelligence Technologies for Smart and Sustainable Urban Transportation: Integrated Platforms and Use Cases, (269–286) © 2026 Scrivener Publishing LLC

16.1 Introduction

Energy landscapes around the world are changing. With smart grids, we are transforming energy distribution by integrating advanced communication, control, and digital technologies. In the future, renewable energy, decentralized power sources, and electrified transportation will replace traditional grids. A smart grid's role in sustainable energy management is explored in this chapter, as well as the challenges that arise during its integration. Whenever possible, we should concentrate on reducing global warming's emissions in the sectors that are responsible for most of its emissions, namely transportation and electricity production. Electric vehicles should be powered by renewable energy sources (RESs) to achieve sustainable development.

The intermittent nature of a power grid can, however, adversely impact voltage profiles and conventional on-load tap changer control systems. Furthermore, these technologies also have challenges, such as uncoordinated grid parameters, increased complexity of systems, intermittent renewable power, and high costs of electric vehicles. Consequently, this causes serious issues such as power quality problems, energy imbalances, and reliability issues [1]. Increasing access to renewable energy has led to the emergence of active distribution systems. As a result, energy flows between sources must be controlled when there are multiple sources and storage systems. A smart grid system is essential for maximizing the potential of new sources of energy and minimizing their negative effects on electricity grids. Load continuity is ensured in all weather conditions, improving network stability. As defined in IEC 61970, an EMS is a set of computer systems consisting of a software platform, a set of applications, and a software platform that provides essential support services. In accordance with the International Electrotechnical Commission's standard, energy generation and transmission facilities should operate efficiently and at the lowest cost possible. Energy management (EM) keeps the smart grid (SG) balanced between supply and demand while managing all system constraints that ensure efficient, safe, and reliable operation. A reduction in power generation costs is also achieved through optimization. EMSs are therefore focused on managing and reducing the cost and amount of energy for a specific application to an absolute minimum. A distribution system can benefit from energy management, but it comes with a number of limitations, such as protecting customer privacy, operating in a large system, and upgrading it regularly.

16.1.1 Smart Grid Systems

Integrated with the smart grid, information and communication technologies (ICT) enable electricity to be transported from production points to consumers smartly and efficiently since they can automatically balance production, feasting, and delivery. Several components must be incorporated, including renewable energy sources, distributed microprocessor rooms, and electricity storage systems, to achieve this: adjusting the flow of electricity from suppliers to consumers, making grids flexible and reliable, and integrating renewable energy sources, distributed microprocessor rooms, and electricity storage systems [2]. As shown in Figure 16.1, the SG consists of several components. In addition to providing real-time, two-way communication, SG technology also allows power to be restored more quickly after a blackout. It can have a detrimental domino effect on banking, communication, manufacturing, traffic, and security when power outages occur on a regular basis [3]. Automated, controlled, and monitored bidirectional flow of electricity between plugs and outlets is handled using smart meters, digital sensors, and control systems.

A smart meter is an important tool for utility companies to communicate with their customers, which makes them an attractive target for criminals [4]. An example of a cyber-attack is a data integrity attack. A hacker could use the supposedly protected grid data to inject false information that is not readily detectable by current operational methods.

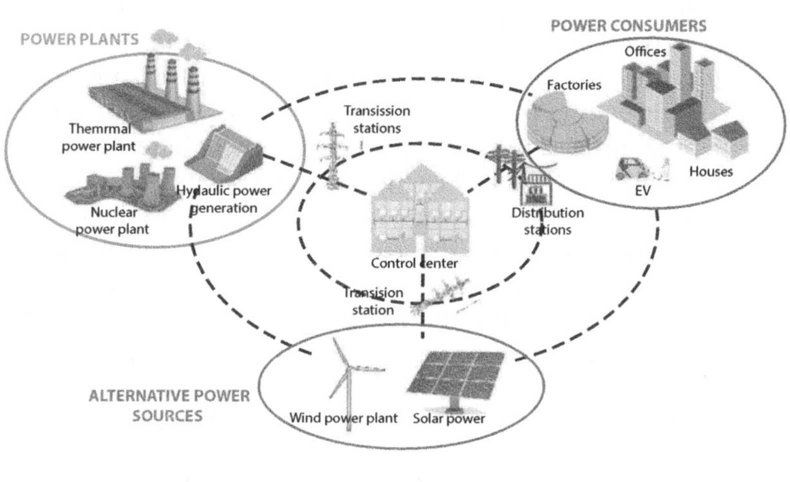

Figure 16.1 An overview of the SG's concept.

According to a study, several data integrity attacks have been demonstrated to affect four different load forecasting models negatively. This study found that the SVR model has the highest level of robustness, whereas the fuzzy interaction regression model has the lowest level of robustness. When data integrity is breached significantly, all models fail to perform satisfactorily. A serious challenge exists in this respect for load forecasters as well as the forecasting community. Recent case studies have shown that SVRs are the most resilient load prediction models when data integrity is compromised [5, 6].

It is much more effective to use a number of small, discrete plants rather than a single large one to minimize the effect of data attacks or catastrophic events. The SG is self-regenerating, so if this occurs, it will quickly recover after isolating the affected line and reconnecting its power. The process cannot be achieved without a smart switch, such as an electronic system, for protecting transformer windings against rapid short-circuiting [7]. There are three layers of SG systems: the control layer, the application layer, and the physical power layer. Our main focus is on control and management. The SG must improve energy efficiency and renewable energy consumption through energy management.

16.1.2 Energy Management System

An EMS is a system for managing demand, load, and distribution in real-time, using a variety of control techniques [8]. As a result of the EMS, different forms of energy are distributed between consumers optimally while sustainable energy sources are introduced without compromising the network's reliability, safety, or security. It can monitor, supervise, control, and optimize consumers in addition to transmitting, distributing, and generating power. The system can be used to control, dispatch power, program, and transmit safety management in real-time [9]. Plug-in electric vehicles (PEVs) and energy storage systems (ESS) are increasingly integrating into the grid, along with renewable energy, high-performance buildings, and other factors that require elaborate energy management systems. According to Figure 16.2, there has been an increase in publication rates between 2012 and 2020. Even though the number of publications has remained relatively constant in recent years, we have seen a significant increase as well as improvements and development towards a common management and control system programming, with an emphasis on improving the network's efficiency and quality.

There are a number of objectives in an EMS: technical, economic, techno-economic, environmental, and social-economic. Economic objectives are the most common focus of EMS research. In addition, these

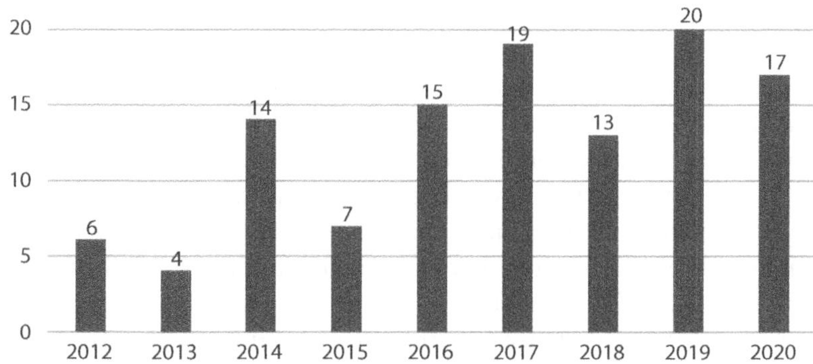

Figure 16.2 A smart grid publication list versus year.

objectives include maximizing profits for aggregators, royalties, and other costs involved in operating energy. In addition to economic objectives, technical aspects must also be taken into account. EMSs may be optimized to achieve optimal economic performance without considering these factors but at the cost of power outages, power failures, or damage to distribution network appliances [10].

16.1.3 System for Managing Transmission Energy

There is still a need for central optimization of energy management in transmission systems. A centrally controlled EMS uses high-performance computing systems that are connected to dedicated, secure networks to manage energy usage [11]. Controllers can either be aggregators or utilities that collect data about each node, such as load/consumer energy consumption patterns and DER energy production patterns, to optimize energy consumption [12]. DR loads (demand-responsive flexible loads) for large-scale energy use are emphasized in this area. In the proposal, DR is taken into consideration when planning wind power on a large scale, for example. As a result of this methodology, wind energy production and load fluctuations can be limited by combining demand resources and energy. An analysis of how EVs and RES integrate in SG has also been conducted. For efficient engagement and programming of thermal units, EVs, and RES, the fireworks algorithm is implemented as a binary version.

There is an increasing demand for energy around the world, which is largely met by non-renewable sources at present. At the moment, a majority of electricity is generated by coal-fired plants, nuclear plants, petroleum plants, and natural gas plants. Non-renewable energy sources, however, should be used only in moderation when excessive use can adversely affect the environment [13, 14]. Unlike conventional grids, conventional grids rely on transmission and distribution interactions, not consumer interactions. Smart grid systems have recently been proposed in response to dynamic power grid demands to ensure that users are involved effectively [15–17]. The smart grid allows data and energy to flow bi-directionally through a network of integrated communication infrastructure [18, 19]. Throughout the world, consumers are participating actively in the development of sustainable, reliable, and economically viable power supplies. Management of the increased demand can be improved with AMI and HEMS. To help utilities understand consumers' consumption patterns, AMI enables HEMSs and smart meters to communicate with them.

Furthermore, consumers receive information about energy availability, prices, and incentives from the grid [20, 21]. Demand response programs, which are provided by utilities based on the requirements of different demand periods, are one way to overcome the energy shortage. When consumers produce and share renewable energy, they become prosumers. Authors are the producers and sharers of energy, converting them from consumers to prosumers. Consumers who use renewable energy sources share any surplus energy generated with grids and/or other consumers instead of using it themselves [22]. An illustration of the prosumer concept can be found in Figure 16.3. The increasing demand for energy results in negative impacts on the environment, the economy, and society. An energy-sharing

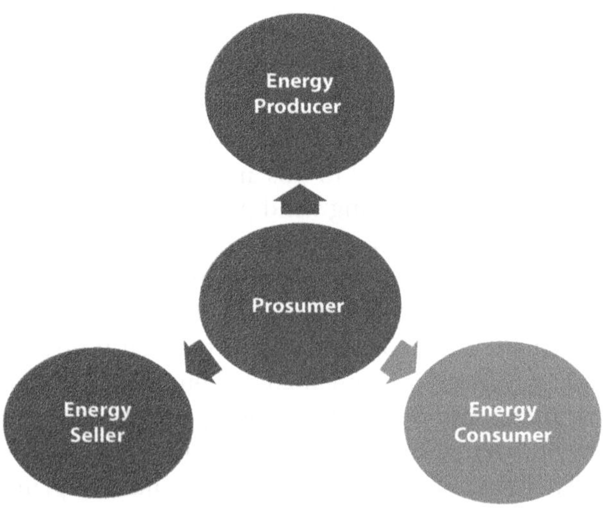

Figure 16.3 A concept of prosumers.

network uses smart grids to manage prosumers in groups based on a variety of criteria, including behavior related to energy consumption.

The future smart grids will rely heavily on consumers, but they also play a crucial role in managing peak demand. As a result, PEMS and its associated challenges need to be investigated and reviewed. In future smart grids, prosumers' impact will be better understood and analyzed. Due to this, this paper discusses PEMS within a smart grid context. It provides a deeper understanding of prosumers, their challenges, and the required communication techniques and optimization techniques. This project has the following objectives:

- An overview of PEMS will be provided,
- PEMS system communications infrastructure and optimization techniques are examined to give you a better understanding of how they work,
- A review of existing PEMS research is required,
- The challenges of PEMS are outlined.

16.2 A Smart Grid EMS Based on Communication Technologies

A communication infrastructure is needed to manage consumer energy. In addition to smart grid communication technologies, PEMS may benefit from them. According to the literature, communication technologies can facilitate a variety of smart grid applications. In this section, we discuss a few key technologies.

16.2.1 GPRS

Packet switching technology is what gives GPRS its name. IP (Internet Protocol) technology is used. GSM (Global System for Mobile) is used as the communication network in GPRS [23]. A GSM network operates at frequencies of 900MHz and 1800MHz. A theoretical maximum data rate of 171.2Kbps is available with GPRS in comparison to GSM [24]. Many applications of GPRS have been developed for managing and monitoring energy. An example is the use of GPRS for remote monitoring of substations [24]. The system uses GPRS to read meters automatically. Transmission lines are monitored using GPRS and ZigBee technologies [25]. The GPRS technology in smart grid applications allows long-distance data transmission, making PEMS a viable alternative. Data can be transferred among utilities, microgrids, and prosumers using GPRS. Conversely, applications requiring high data rates can be severely hampered by low data rates.

16.2.2 WiMAX (IEEE 802.16)

Additionally, wireless technologies such as Worldwide Interoperability for Microwave Access (WiMAX) are available, and they operate according to IEEE 802.16 standards and have a range of 50 kilometers [26]. This device provides up to 70Mbps of data rate [27]. WiMAX operates in two frequencies, 11-66 GHz and 2-11 GHz, both in line of sight [27].

For PEMS implementation, WIMAX provides long-distance data transfer capability, therefore connecting consumers, utilities, and control centers. Smart meters, AMI, feeder status monitoring, etc., are among the smart grid applications that have already been implemented with WiMAX [28, 29].

16.2.3 Bluetooth (IEEE 802.15)

According to IEEE 802.15.1, Bluetooth is a technology that is rising rapidly. In addition to its energy efficiency, low power consumption, and short range, the technology can transmit data at speeds up to 1 Mbps and cover distances of up to 10 meters [30]. An ISM (Industrial, Scientific, and Medical) frequency of 2.4 GHz is used [31, 32]. There are already several smart grid applications that use Bluetooth technology. A user vehicle-charging system based on Bluetooth connects electric vehicles to charging stations *via* wired plugs and wirelessly to drivers' mobile phones *via* Bluetooth so that they are informed about charging status. Power systems have been monitored and managed using Bluetooth applications. Wireless short-range communication is used in smart grid applications, as discussed above. Moreover, it can be used for short-distance communications in PEMS, such as sending data from smart meters to prosumers.

16.2.4 Power Line Communication (PLC)

In PLC, power lines are used as a medium for bidirectional communication. Two categories of frequencies are recognized here: narrow bands (NB) and wide bands (BB). NBs with a 3 km range use frequencies between 3 and 500 kHz and offer data rates between 10 and 500 kbps [33, 34]. There is a possibility of achieving 500 Mbps using the BB band with a relatively high-frequency band between 2 and 250 MHz. Implementation of this technology is difficult due to channel noise. The existing infrastructure has prompted researchers to explore the feasibility of implementing PLCs for various applications related to smart grids. The Author [35], studied the application of PLCs in low-voltage and medium-voltage systems in a smart grid environment through simulations. This paper proposes an architecture for PLC communication for smart meters [36]. Based on simulations, measurements, and tests [37], describes the performance of an electric vehicle charging system based on PLCs.

16.3 Smart Grids: An Overview

As smart grids become more prevalent, images of interconnected electrical grids, electric vehicles, and rooftop solar and wind generators often accompany the term. Using smart grid technology, energy distribution is more efficient, reliable, and sustainable. While conventional grids originate from a central source and deliver electricity to end-users in one direction, smart grids are bidirectional. By monitoring and maintaining systems in real-time and responding to demand and supply changes, real-time monitoring is possible [38].

16.3.1 An Overview of Smart Grid Components

Smart grids are composed of multiple layers and technologies. A smart grid's functionality depends on a number of key components:

Sensors and Monitoring Devices: Various grid sections contain these sensors to capture information on energy production, consumption, and transmission.
Communication Networks: Grid operators, energy producers, and consumers can exchange information in real-time, thanks to a robust communication infrastructure.
Advanced Metering Infrastructure (AMI): Utilities and consumers can leverage AMI's energy usage data to develop dynamic pricing models and demand-response strategies.
Distributed Energy Resources (DERs): A decentralized generator, storage system, and controllable load are examples of resources that improve flexibility and resilience.

Automation and Control Systems: Automation systems make energy flow management, improve performance, and make it possible to respond to disruptions.

16.3.2 Goals of a Smart Grid

There are several objectives that smart grids aim to achieve:

Improving Energy Efficiency: Using smart grids to monitor and automate usage patterns can help reduce energy waste.
Enhancing Reliability: Proactively maintaining the system and detecting faults as soon as they arise minimizes downtime and improves power quality.
Facilitating Renewable Energy Integration: The flexibility of renewable energies like wind and solar energy is supported by smart grids that balance supply and demand dynamically.
Empowering Consumers: The flexibility of renewable energies like wind and solar energy is supported by smart grids that balance supply and demand dynamically.

16.4 Integrating Smart Grids with Existing Infrastructure

Smart grids must be integrated into existing infrastructure, which is a complex but critical process. Conventional grid infrastructures are built with centralized power plants and unidirectional flows, which cannot meet the demands of smart grids [39]. Grid systems need to be upgraded and modernized in several phases:

16.4.1 Upgrading Infrastructure

Transformations, sensors, and control systems must be upgraded as part of the transition to smart grids. Retrofitting or replacing traditional grid components will be needed for bidirectional power flow, as well as communication with distributed energy resources and consumers [40].

16.4.2 Synchronizing with Renewable Sources

Renewable energy sources have an inherent intermittent nature. For a smart grid to be successful, variable energy sources, including wind and solar power, must be integrated without compromising grid stability. A smart grid achieves this by utilizing energy storage solutions, demand response mechanisms, and flexible generation resources [41].

16.4.3 Digitalizing the Grid

For the smart grid to work, digitalization is required. SCADA systems, for instance, allow utility companies to monitor the performance of the grid and detect problems before they spiral out of control. By creating virtual replicas of physical infrastructure, digital twins allow scenarios to be tested, and weaknesses can be identified without impacting actual grid operations [42].

16.4.4 Cybersecurity Measures

Smart grids are increasingly connected and digitalized, posing significant cybersecurity risks. Interconnected systems must have robust cybersecurity frameworks so that they can be protected from hackers, data breaches, and other possible threats.

16.5 Energy Management in the Smart Grid

Intelligent grids plan, monitor, and optimize energy production, storage, and consumption. A well-managed energy supply improves efficiency and cuts carbon emissions and costs at the same time.

16.5.1 Demand Response

A critical part of energy management, demand response (DR) shifts or reduces power usage during peak hours, improving grid stability and minimizing the need for expensive infrastructure upgrades. Through smart appliances and devices, utilities can automate and manage loads through DR programs by rewarding consumers for reducing their load during peak hours [43].

16.5.2 Distributed Energy Resources Management (DERM)

The DERM system coordinates the operation of DERs, which include solar panels, wind turbines, and batteries. With DERM, supply and demand are balanced so DERs can operate efficiently and renewable energy can be seamlessly integrated into the grid [44].

16.5.3 Energy Storage Solutions

Energy storage can be achieved through a variety of technologies, including lithium-ion batteries, flow batteries, and pumped hydro storage. Sunlight and wind energy are stored in storage systems and then released when peak demand is at its highest. A reliable energy supply can be ensured with the help of smart grids that incorporate storage solutions [45].

16.5.4 Rates and Pricing for Real-Time Usage

With smart grids, electricity prices can be dynamically adjusted according to supply and demand. As a result of real-time energy pricing, consumers are able to make cost-effective energy decisions, resulting in more balanced grid loading. Customers can benefit from time-of-use rates by reducing their energy costs and strain on the grid [46].

16.5.5 Electric Vehicle (EV) Integration

As electric vehicles become more prevalent, smart grid management faces a range of challenges as well as opportunities. Although electric vehicles consume significant amounts of energy, they can also be used to store energy. Electric vehicles can feed stored energy back into the grid during peak hours as part of V2G technology, which provides additional support for grid stability [47].

16.6 Managing Energy and Integrating Smart Grids

A smart grid is difficult to integrate and manage efficiently because there are a variety of complexities involved:

High Initial Costs: The development of communication networks, the deployment of sensors, and the upgrade of grid infrastructure require major investments.

Interoperability Issues: It is possible that different equipment and systems do not work together seamlessly, posing obstacles to the creation of a seamless, integrated network.
Regulatory and Policy Hurdles: Energy systems are often surrounded by a complex regulatory environment characterized by varying policies that can impede their efficient integration.
Data Privacy Concerns: Detailed consumption data collected by smart grids raises privacy concerns in terms of handling, storing, and sharing the data.
Public Acceptance: It is crucial that consumers are aware of and accept energy management, as active consumer participation is crucial to its success.

Electric power ensures a high quality of life and productivity; therefore, electric power plays a critical role in economic growth. However, economic growth and electricity demand are depleting energy resources due to current global economic growth. As part of Smart Grid technology, energy management systems are a fundamental part of preventing resource depletion and promoting economic growth.

Energy management systems, in general, only monitor and collect data about consumer electronics' energy consumption. Each device in these systems must be manually actuated to reduce energy consumption. With the development of smart grid technologies, management systems will need to become smarter and capable of responding to the needs of charges, energy management, and timing [48].

The following concepts are being developed to integrate architectural solutions for energy management systems into a smart grid. To interoperate with the smart grid, Chang-Sci Choi developed an architecture for energy management based on AMI solutions called EMM (Energy Monitoring and Management). Several assumptions have been adopted, but the most important is the division of streams of operations and building settings [48]. Sizing was first based on complex apartments and homes in South Korea, which had advanced home networks and several features, including electricity, gas, water, and cogeneration.

16.7 Conclusion

In the future, smart grids will fundamentally change how energy is managed, distributed, and consumed. A smart grid not only provides reliability and flexibility but it also incorporates renewable energy sources and energy management systems to give consumers access to real-time information and dynamic pricing. Even though smart grids can be beneficial, there are several challenges to their integration, including high initial costs, regulatory barriers, cybersecurity risks, and public acceptance. However, smart grid technology has the potential to reduce greenhouse gas emissions, increase grid stability, and boost economic growth through improved energy efficiency. For sustainable energy systems to reach their full potential, researchers and developers must keep researching about smart grid technologies.

References

1. Zhang, W.J. and Van Luttervelt, C.A., Toward a resilient manufacturing system. *CIRP Ann.*, 60, 1, 469–472, 2023, doi: 10.1016/j.cirp.2011.03.041.
2. Koutsopoulos, I. and Tassiulas, L., Challenges in demand load control for the smart grid. *IEEE Netw.*, 25, 5, 16–21, Sep. 2011, doi: 10.1109/MNET.2011.6033031.
3. Rathor, S.K. and Saxena, D., Energy management system for smart grid: An overview and key issues. *Int. J. Energy Res.*, 44, 6, 4067–4109, May 2020, doi: 10.1002/er.4883.
4. Shen, H., Liu, Y., Xia, Z., Zhang, M., An efficient aggregation scheme resisting on malicious data mining attacks for smart grid. *Inf. Sci.*, 526, 289–300, Jul. 2020, doi: 10.1016/j.ins.2020.03.107.
5. Luo, J., Hong, T., Fang, S.-C., Benchmarking robustness of load forecasting models under data integrity attacks. *Int. J. Forecast.*, 34, 1, 89–104, Jan. 2018, doi: 10.1016/j.ijforecast.2017.08.004.
6. Luo, J., Hong, T., Fang, S.-C., Robust Regression Models for Load Forecasting. *IEEE Trans. Smart Grid*, 10, 5, 5397–5404, Sep. 2019, doi: 10.1109/TSG.2018.2881562.
7. Energy Management Systems: Operation and Control of Electric Energy, [Online]. Available: https://books.google.co.in/books?hl=en&lr=&id=lcbuCAAAQBAJ&oi=fnd&pg=PA2&ots=EmVnrmD9bk&sig=_lSY6OleF0YbjbdPwHrfRhlQ1-A&redir_esc=y#v=onepage&q&f=false.
8. Ovalle, A., Hably, A., Bacha, S., Ahmed, M., Voltage support by optimal integration of plug-in hybrid electric vehicles to a residential grid, in: *IECON 2014 - 40th Annual Conference of the IEEE Industrial Electronics Society*, IEEE, Dallas, TX, USA, pp. 4430–4436, Oct. 2014, doi: 10.1109/IECON.2014.7049169.
9. Yadav, S.P., Jindal, M., Rani, P., de Albuquerque, V.H.C., dos Santos Nascimento, C., Kumar, M., An improved deep learning-based optimal object detection system from images. *Multimed. Tools Appl.*, 83, 10, 30045–30072, 2024.

10. Singh, A., *et al.*, Smart Traffic Monitoring Through Real-Time Moving Vehicle Detection Using Deep Learning *via* Aerial Images for Consumer Application. *IEEE Trans. Consum. Electron.*, 70, 4, 7302–7309, 2024, doi: 10.1109/TCE.2024.3445728.
11. Yan, X., Zhang, L., Wu, Y., Luo, Y., Zhang, X., Secure smart grid communications and information integration based on digital watermarking in wireless sensor networks. *Enterp. Inf. Syst.*, 11, 2, 223–249, Feb. 2017, doi: 10.1080/17517575.2015.1033767.
12. Verma, P.P., Srinivasan, D., Swarup, K.S., Mehta, R., A Review of Uncertainty Handling Techniques in Smart Grid. *Int. J. Unc. Fuzz. Knowl. Based Syst.*, 26, 03, 345–378, Jun. 2018, doi: 10.1142/S0218488518500186.
13. Rathnayaka, A.J.D., Potdar, V.M., Dillon, T., Hussain, O., Kuruppu, S., Goal-Oriented Prosumer Community Groups for the Smart Grid. *IEEE Technol. Soc. Mag.*, 33, 1, 41–48, 2014, doi: 10.1109/MTS.2014.2301859.
14. Steinheimer, M., Trick, U., Ruhrig, P., Energy communities in Smart Markets for optimization of peer-to-peer interconnected Smart Homes, in: *2012 8th International Symposium on Communication Systems, Networks & Digital Signal Processing (CSNDSP)*, IEEE, Poznan, Poland, pp. 1–6, Jul. 2012, doi: 10.1109/CSNDSP.2012.6292732.
15. Mahmood, A., Javaid, N., Razzaq, S., A review of wireless communications for smart grid. *Renew. Sustain. Energy Rev.*, 41, 248–260, Jan. 2015, doi: 10.1016/j.rser.2014.08.036.
16. Ciuciu, I.G., Meersman, R., Dillon, T., Social network of smart-metered homes and SMEs for grid-based renewable energy exchange, in: *2012 6th IEEE International Conference on Digital Ecosystems and Technologies (DEST)*, IEEE, Campione d'Italia, Italy, pp. 1–6, Jun. 2012, doi: 10.1109/DEST.2012.6227922.
17. Skopik, F. and Wagner, C., Novel Energy Saving Opportunities in Smart Grids Using a Secure Social Networking Layer, in: *2012 IEEE 36th Annual Computer Software and Applications Conference*, IEEE, Izmir, Turkey, pp. 557–566, Jul. 2012, doi: 10.1109/COMPSAC.2012.75.
18. Khan, A.R., Mahmood, A., Safdar, A., Khan, Z.A., Khan, N.A., Load forecasting, dynamic pricing and DSM in smart grid: A review. *Renew. Sustain. Energy Rev.*, 54, 1311–1322, Feb. 2016, doi: 10.1016/j.rser.2015.10.117.
19. Usman, A. and Shami, S.H., Evolution of Communication Technologies for Smart Grid applications. *Renew. Sustain. Energy Rev.*, 19, 191–199, Mar. 2013, doi: 10.1016/j.rser.2012.11.002.
20. Opris, I. and Caracasian, L., On the implementation of the functionalities of smart metering systems, in: *2013 8th International Symposium on Advanced Topics in Electrical Engineering (ATEE)*, Bucharest, Romania, IEEE, pp. 1–6, May 2013, doi: 10.1109/ATEE.2013.6563366.
21. Zhou, L., Xu, F., Ma, Y.-N., Impact of smart metering on energy efficiency, in: *2010 International Conference on Machine Learning and*

Cybernetics, IEEE, Qingdao, China, pp. 3213–3218, Jul. 2010, doi: 10.1109/ICMLC.2010.5580715.

22. Razzaq, S., Zafar, R., Khan, N., Butt, A., Mahmood, A., A Novel Prosumer-Based Energy Sharing and Management (PESM) Approach for Cooperative Demand Side Management (DSM) in Smart Grid. *Appl. Sci.*, 6, 10, 275, Oct. 2016, doi: 10.3390/app6100275.
23. Xenakis, C., Apostolopoulou, D., Panou, A., Stavrakakis, I., A Qualitative Risk Analysis for the GPRS Technology, in: *2008 IEEE/IFIP International Conference on Embedded and Ubiquitous Computing*, IEEE, Shanghai, China, pp. 61–68, Dec. 2008, doi: 10.1109/EUC.2008.123.
24. Kong, L., Jin, J., Cheng, J., Introducing GPRS technology into remote monitoring system for prefabricated substations in China, in: *2005 2nd Asia Pacific Conference on Mobile Technology, Applications and Systems*, IEEE, Guangzhou, p. 6, 2005, doi: 10.1109/MTAS.2005.243794.
25. Yang, Y., Xie, G., Xu, X., Jiang, Y., A Monitoring System Design in Transmission Lines based on Wireless Sensor Networks. *Energy Procedia*, 12, 192–199, 2011, doi: 10.1016/j.egypro.2011.10.027.
26. Kuzlu, M. and Pipattanasomporn, M., Assessment of communication technologies and network requirements for different smart grid applications, in: *2013 IEEE PES Innovative Smart Grid Technologies Conference (ISGT)*, IEEE, Washington, DC, pp. 1–6, Feb. 2013, doi: 10.1109/ISGT.2013.6497873.
27. Habib, M. and Ahmad, M., A Review of Some Security Aspects of WiMAX and Converged Network, in: *2010 Second International Conference on Communication Software and Networks*, IEEE, Singapore, pp. 372–376, 2010, doi: 10.1109/ICCSN.2010.93.
28. Aalamifar, F., Lampe, L., Bavarian, S., Crozier, E., WiMAX technology in smart distribution networks: Architecture, modeling, and applications, in: *2014 IEEE PES T&D Conference and Exposition*, IEEE, Chicago, IL, USA, pp. 1–5, Apr. 2014, doi: 10.1109/TDC.2014.6863432.
29. Mao, R. and Julka, V., WiMAX for advanced metering infrastructure, in: *2012 International Conference on Green and Ubiquitous Technology*, IEEE, Bandung, Indonesia, pp. 15–19, Jul. 2012, doi: 10.1109/GUT.2012.6344171.
30. Rashid, R.A. and Yusoff, R., Bluetooth Performance Analysis in Personal Area Network (PAN), in: *2006 International RF and Microwave Conference*, IEEE, Putra Jaya, Malaysia, pp. 393–397, Sep. 2006, doi: 10.1109/RFM.2006.331112.
31. McDermott-Wells, P., Bluetooth scatternet models. *IEEE Potentials*, 23, 5, 36–39, Jan. 2005, doi: 10.1109/MP.2005.1368914.
32. Cordeiro, C.D.M., Abhyankar, S., Toshiwal, R., Agrawal, D.P., A novel architecture and coexistence method to provide global access to/from Bluetooth WPANs by IEEE 802.11 WLANs, in: *Conference Proceedings of the 2003 IEEE International*, IEEE, Phoenix, AZ, USA, pp. 23–30, 2003, doi: 10.1109/PCCC.2003.1203680.
33. Galli, S., Scaglione, A., Wang, Z., Power Line Communications and the Smart Grid, in: *2010 First IEEE International Conference on Smart Grid*

Communications, IEEE, Gaithersburg, MD, USA, pp. 303–308, Oct. 2010, doi: 10.1109/SMARTGRID.2010.5622060.

34. Galli, S., Scaglione, A., Wang, Z., For the Grid and Through the Grid: The Role of Power Line Communications in the Smart Grid. *Proc. IEEE*, 99, 6, 998–1027, Jun. 2011, doi: 10.1109/JPROC.2011.2109670.
35. Cataliotti, A., Cosentino, V., Di Cara, D., Russotto, P., Tine, G., On the use of narrow band power line as communication technology for medium and low voltage smart grids, in: *2012 IEEE International Instrumentation and Measurement Technology Conference Proceedings*, IEEE, Graz, Austria, pp. 619–623, May 2012, doi: 10.1109/I2MTC.2012.6229503.
36. Sendin, A., Simon, J., Urrutia, I., Berganza, I., PLC deployment and architecture for Smart Grid applications in Iberdrola, in: *18th IEEE International Symposium on Power Line Communications and Its Applications*, IEEE, Glasgow, United Kingdom, pp. 173–178, Mar. 2014, doi: 10.1109/ISPLC.2014.6812374.
37. Barmada, S., Raugi, M., Tucci, M., Maryanka, Y., Amrani, O., PLC systems for electric vehicles and Smart Grid applications, in: *2013 IEEE 17th International Symposium on Power Line Communications and Its Applications*, IEEE, Johannesburg, pp. 23–28, Mar. 2013, doi: 10.1109/ISPLC.2013.6525819.
38. Rani, P. and Sharma, R., Intelligent transportation system for internet of vehicles based vehicular networks for smart cities. *Comput. Electr. Eng.*, 105, 108543, 2023.
39. Rani, P. and Sharma, R., IMFOCA-IOV: Intelligent Moth Flame Optimization based Clustering Algorithm for Internet of Vehicle, in: *2023 14th International Conference on Computing Communication and Networking Technologies (ICCCNT)*, IEEE, pp. 1–6, 2023, Accessed: Nov. 09, 2024. [Online]. Available: https://ieeexplore.ieee.org/abstract/document/10307646/.
40. Rani, S., Talwar, R., Malhotra, J., Ahmed, S.H., Sarkar, M., Song, H., A novel scheme for an energy efficient Internet of Things based on wireless sensor networks. *Sensors*, 15, 11, 28603–28626, 2015.
41. Rani, P., Verma, S., Yadav, S.P., Rai, B.K., Naruka, M.S., Kumar, D., Simulation of the lightweight blockchain technique based on privacy and security for healthcare data for the cloud system. *Int. J. E-Health Med. Commun. (IJEHMC)*, 13, 4, 1–15, 2022.
42. Rani, P., Hussain, N., Khan, R.A.H., Sharma, Y., Shukla, P.K., Vehicular Intelligence System: Time-Based Vehicle Next Location Prediction in Software-Defined Internet of Vehicles (SDN-IOV) for the Smart Cities, in: *Intelligence of Things: AI-IoT Based Critical-Applications and Innovations*, Springer International Publishing, Cham, pp. 35–54, 2021, doi: 10.1007/978-3-030-82800-4_2.
43. Pereira, R., Figueiredo, J., Melicio, R., Mendes, V.M.F., Martins, J., Quadrado, J.C., Consumer energy management system with integration of smart meters. *Energy Rep.*, 1, 22–29, Nov. 2015, doi: 10.1016/j.egyr.2014.10.001.

44. Adika, C.O. and Wang, L., Autonomous Appliance Scheduling for Household Energy Management. *IEEE Trans. Smart Grid*, 5, 2, 673–682, Mar. 2014, doi: 10.1109/TSG.2013.2271427.
45. Heymann, B., Bonnans, J.F., Martinon, P., Silva, F.J., Lanas, F., Jiménez-Estévez, G., Continuous optimal control approaches to microgrid energy management. *Energy Syst.*, 9, 1, 59–77, Feb. 2018, doi: 10.1007/s12667-016-0228-2.
46. Zhang, Y., Meng, F., Wang, R., Zhu, W., Zeng, X.-J., A stochastic MPC based approach to integrated energy management in microgrids. *Sustain. Cities Soc.*, 41, 349–362, Aug. 2018, doi: 10.1016/j.scs.2018.05.044.
47. Sousa, T., Morais, H., Vale, Z., Faria, P., Soares, J., Intelligent Energy Resource Management Considering Vehicle-to-Grid: A Simulated Annealing Approach. *IEEE Trans. Smart Grid*, 3, 1, 535–542, Mar. 2012, doi: 10.1109/TSG.2011.2165303.
48. Choi, C.-S., Han, J., Park, W.-K., Jeong, Y.-K., Lee, I.-W., Proactive energy management system architecture interworking with Smart Grid, in: *2011 IEEE 15th International Symposium on Consumer Electronics (ISCE)*, IEEE, Singapore, Singapore, pp. 621–624, Jun. 2011, doi: 10.1109/ISCE.2011.5973905.

17

Decoding the Aspects of Intelligent Traffic Management

Preeti Rani[1]*, Jatinder Kumar[2], Sandeep Singh[3], Protyay Dey[4] and Laith H. Jasim[5]

[1]*College of Computing Sciences & IT (CCSIT), Teerthanker Mahaveer University, Moradabad (U.P.), India*
[2]*Department of Computer Science Engineering, Chandigarh Engineering College, Chandigarh Group of Colleges, Jhanjeri, Mohali, Punjab, India*
[3]*Department of Civil Engineering, School of Engineering and Technology, Bahra University, Waknaghat, Distt. Solan, HP, India*
[4]*Department of Computing Science and Artificial Intelligence, NIMS Institute of Engineering and Technology, NIMS University Rajasthan, Jaipur, India*
[5]*Department of Computer Science and Engineering, The Islamic University, Najaf, Iraq*

Abstract

The objective of this chapter is to address the escalating problems of urban traffic congestion by designing and developing an intelligent traffic management system (ITMS). It includes innovative solutions for traffic monitoring, route optimization, and accident detection based on advanced technologies, including wireless sensor networks, machine learning, and fuzzy logic. Real-time vehicle data collection is enabled through magnetometers, and urban traffic tracking is achieved through scalable architecture and predictive modeling. As a result of the system's integration of cost-effective methodologies, accurate traffic flow analysis and efficient emergency response mechanisms are ensured, which enhance urban mobility, reduce pollution, and improve overall road safety. It is evident from a comparison of this approach with existing solutions that it promotes an efficient and sustainable management of urban traffic.

**Corresponding author*: preetiresearcher1@gmail.com

Pethuru Raj, Sudesh Yadav, Manas Kumar Mishra, Satya Prakash Yadav and Victor Hugo C. de Albuquerque (eds.) Artificial Intelligence Technologies for Smart and Sustainable Urban Transportation: Integrated Platforms and Use Cases, (287–300) © 2026 Scrivener Publishing LLC

***Keywords*:** Accident detection, machine learning, road safety, route optimization, traffic flow analysis, traffic monitoring, urban mobility, urban traffic congestion

17.1 Introduction

The term smart cities refers to a paradigm shift in thinking about technologies that can be used to make cities 'smarter' so that people can enjoy a better quality of life. The European Commission launched the European Initiative on Smart Cities [1], focusing on four key aspects of urban living: buildings, heating and cooling systems, electricity, and transportation [2, 3]. Participating in this program will allow us to identify and promote sustainable transportation methods, develop intelligent public transportation systems that utilize real-time data, and develop traffic management systems (TMS) for reducing congestion, improving safety, and developing green applications (for example, reducing fuel consumption, greenhouse gas emissions, and energy consumption) [4, 5].

Observations by Soufiene Djahel and John Murphy at UCD's Performance Engineering Laboratory indicate an increase in the use of road infrastructure. At Dublin City University, Gabriel-Miro Muntean and Ronan Doolan participated in the Performance Engineering Laboratory, where they received tremendous growth in their manuscript in November 2013. There are many implications associated with this increase, including congestion management, road safety, and environmental impacts. As a result of the use of wireless sensing technology, communication technology, simulation tools, and modeling tools by researchers from academia and industry in recent years, the efficiency of existing road traffic management systems has improved, so future smart cities will be able to address these issues. There is a delay in medical services, police operations, and fire and rescue operations due to traffic congestion. Emergency vehicles must respond quickly and efficiently to incidents, robberies, and criminal attacks in order to protect individuals and the public. In addition, recent road traffic statistics reveal that vehicles are being involved in more crashes than ever before. Crash rates are higher around congested roads due to drivers' tendency to drive faster when experiencing congestion to compensate for the experience of delays. Accidents involving emergency vehicles can have considerable negative repercussions on individuals, groups, and society.

As urbanization continues, city roads are becoming increasingly congested because of rapid growth. The result is that vehicles can stand for an extended period. When a vehicle stands for a long time, it pollutes the environment, causing breathing problems and delaying emergencies such as car accidents that could result in death. A number of factors, aside from development, contribute to traffic congestion, including development [6]. Increasingly, vehicles can be repaired, in part due to the growing number of repairable vehicles. To alleviate traffic congestion, an ITMS should be developed that considers the number of vehicles in the area. As technology advances, various types of advancements are being made. Microcontrollers,

Table 17.1 Developing ITMS: challenges and opportunities.

Challenges	Description
Vehicle shadow	Vehicles block sunlight and skylights, thereby blocking ambient light. As a result, shadows are cast below the vehicle. Shadows on vehicles provide both necessary and unnecessary information. It makes locating and recognizing moving vehicles difficult, as well as extracting important information about them.
Vehicle occlusion	It happens when images of three-dimensional traffic scenes are converted into two-dimensional images, thereby occluding visual details regarding the vehicles. The complexity of traffic scenes makes it difficult to analyze them accurately.
Resolution Change	Cameras collect real-time images of moving vehicles in different pixel sizes as they are acquired in real-time by the camera. It makes it difficult to analyze a moving vehicle accurately.
Camera coordination	Using multiple cameras to coordinate traffic is meant to increase the quality of images captured from a scene. Combining the features of multiple cameras is often the easiest way to accomplish this.
wind and weather changes	ITMS developers may have difficulty developing ITMS when lighting conditions are changing (twilight, night, day, and sunny days) and when the camera shakes due to wind.
Intra variations of vehicles	Vehicles have many internal variances, such as length, width, size, and color, which make it very difficult to analyze them properly.

WSNs, and fuzzy logic (FL) are among the technologies used in traffic control. Among its many applications are environmental impact assessments, electronic tolling, anomaly detection, crime detection, and traffic signal management [7, 8]. As ITMS applications become more common, they are being used to improve people's quality of life. ITMS systems are more robust when they are aware of these challenges in Table 17.1.

In UTMC systems and for road users, wireless communication is one of the similar research studies that has been conducted. All traffic management centers, traffic junctions, and vehicles must be equipped with wireless devices. It was proposed that data be transmitted using very high frequencies on public broadcasting. GSM (Global System for Mobile Communications) supports a variety of technologies, such as general packet radio services (GPRS), radio data systems (RDS) modulated by VHF/FM, as well as digital audio broadcasting and private mobile radio systems (VHF/UHF). There are several public terrestrial radio systems available today, including TETRA, UMTS, DSRC, 5.8GHz, Wireless LAN, and paging services (VHF) [9]. Compared to previous work, the Intelligent Urban Traffic Management System utilizes only one system, which consists of 5.8GHz integrated antennas and a single software program. In this way, different wireless hardware and software applications can be integrated and compatible without causing a lot of problems. In addition to being more practical and economical, it also has the advantage of not requiring vehicles to have devices [8].

17.2 Related Work

In addition to reducing traffic congestion and improving response times to incidents, traffic management systems (TMS) can also enhance commuter experiences [10]. Figure 17.1 illustrates how a TMS typically consists of a number of complementary phases, each of which controls traffic flow in a city. As part of a TMS, heterogeneous monitoring equipment gathers and analyzes traffic parameters and reports this information to a central server. Mobile sensors, wireless networks, and cellular networks can all be used for this purpose. It includes detecting random incidents and reporting them as soon as possible by using these monitoring tools. In the data fusion, processing, and aggregation (DFPA) stage, traffic information is extracted from these data feeds by combining and aggregating them. It will be followed by data exploitation (DE), which involves calculating short-term traffic forecasts or optimal routes for vehicles based on the processed data.

A TMS's service delivery consists of delivering this knowledge through a mobile device or on-board unit to end users.

Additionally, TMS contributes to a country's economic growth, to the safety of its citizens, and to the nation's security beyond helping drivers and road authorities. It is still difficult to accurately measure traffic parameters or report in real-time events occurring on the roads with most road traffic surveillance technologies, especially in developing countries. Furthermore, gathering traffic data requires a process that reduces noise and improves quality. In the absence of sufficient financial resources and because of the lack of flexibility of such equipment, it may not be possible to deploy highly sophisticated equipment that accurately estimates traffic flows and detects emergency events. Both developed and developing countries need to implement cost-effective and flexible alternatives to ensure better road traffic management.

In recent years, a growing body of research has been conducted on the management and development of smart cities. Urban smartness has been measured by a variety of indicators developed by researchers over the years. Using smart city rankings, one can compare smart cities. Artificial intelligence (AI), as applied to urban planning, was examined in a thorough literature review [11–13]. A review of 91 publications was conducted from Scopus and Web of Science databases to identify key applications of artificial intelligence in urban data analytics, infrastructure management, disaster management, and development control. The governance of smart cities also relies heavily on content management and collaboration tools. Citizens and city authorities benefit from better communication, resulting in better public service delivery. Smart cities need these tools to ensure effective governance. In addition to promoting transparency, they also encourage active citizen participation. In cities without a robust system for managing content and streamlining workflows, meeting increased demands for fast and efficient services may become increasingly difficult. A smart city vision would not be complete without the IoT, which links devices and sensors to generate real-time data. As a result of this interconnectedness, cities are also vulnerable to cyberattacks. In sectors such as transportation and healthcare, urban systems must be well-protected in order to function efficiently.

According to [3], smart city services must be aligned with end-user requirements and expectations. The perceptions and expectations of smart city residents were analyzed through a pilot survey. According to the authors, technology providers and policymakers should focus on ICT-enabled services to meet citizen needs. An assessment toolkit is presented that allows businesses and policymakers to adapt services to citizens' needs, improving smart city services. Several case studies aimed at enhancing sustainability and resilience were analyzed by the author [4], of smart city software applications, and the benefits were grouped into six possible areas for action: governance, economy, environment, people, mobility, and living. Developing smart city software that can be integrated into multiple domains is crucial to addressing future challenges.

17.3 Proposed Methodology

17.3.1 Design Objectives

The following subsection outlines some of the important objectives our project will achieve.

Traffic Monitoring: In a smart city, it is one of the most important components. Local authorities can monitor an area, route, or street through traffic monitoring. In this way, traffic from neighboring cities can be tracked during a particular period or day. Planning smart cities and building city infrastructure can be made more efficient by using historical traffic monitoring data.

Pollution Avoidance: In addition to posing a threat to the environment, rising pollution levels are harmful to human health and well-being. Pollution levels in cities are directly proportional to traffic congestion. A long queue of vehicles results in enormous emissions of pollutants, increasing temperatures, reducing rainfall, and causing respiratory problems.

Route Optimization: The shortest route appears to have problems with commute times, fuel consumption, and average waiting times. In these scenarios, optimum routes are the most economical route, considering factors like traffic congestion, distance, time, and fuel consumption. All these parameters must be balanced to determine an optimal route for a traveller, taking into account the amount of time and money spent on travel.

Green Corridor: Several years have passed since the concept of green corridors was introduced. There are many traffic signals on this corridor, all of which are green, which is, in fact, a route between two sources. To facilitate emergency vehicle access to their destinations quickly and without waiting, the green corridor is used.

Accident Detection: In today's world, there are a lot of accidents due to overcrowded streets. Traffic management involves detecting accidents since they inform medical services to treat accident victims, and it also affects traffic flow and congestion levels in specific areas.

Jamming: Efficient traffic management reduces average waiting times by preventing traffic jams and preventing traffic jams.
Vehicle Tracking: Local administrations can monitor vehicles by tracking their travel routes, travel times, speeds, visited locations, and vehicle types. These parameters are crucial for maintaining law and order within a city.

17.3.2 Method and Materials

Counting the cars and determining their speeds is done with a magnetometer. The magnetic field normally ranges from 0.25 to 0.65 gauss and is basically uniform throughout. Due to their high permeability, sensors utilizing highly permeable ferrous materials disturb the earth's magnetic field in the targeted area [14]. There is typically a 10-meter range between a magnetic sensor and a vehicle. Figure 17.1 illustrates the architecture of the system.

The system collects information about vehicles' movements through each data collection node. To communicate <i>*via*</i> mobile internet, these nodes transmit data to the aggregation nodes using the ALOHA protocol. In addition to its high scalability and low construction and

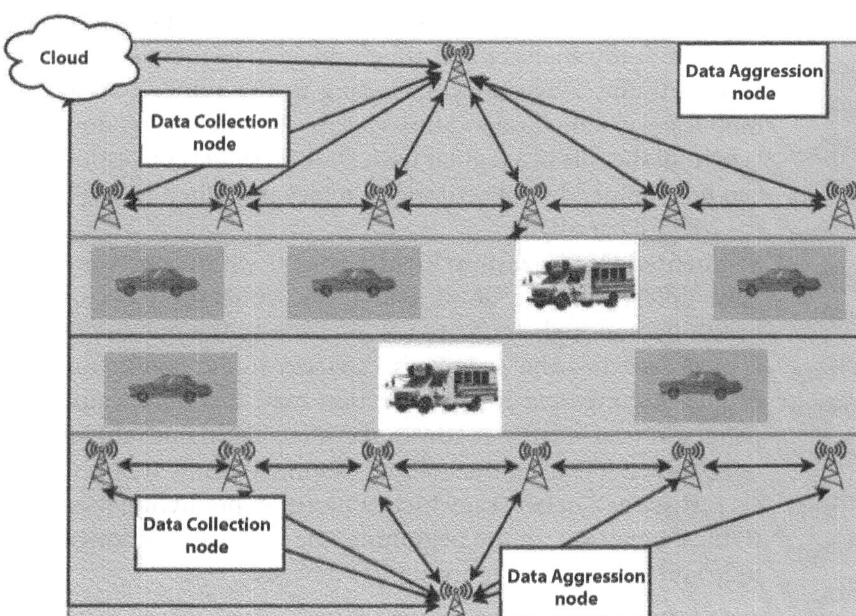

Figure 17.1 The architecture of the traffic system.

maintenance costs, the system is also cost-effective to operate. There is a description of the proposed road congestion monitoring system, which includes data collection, traffic flow simulation, segmented organization companies, time- and place publishing/subscribing processes, gridlock forecasting for local sections, and a service for drivers experiencing traffic jams from origin to destination.

- **Data Gathering:** The development of a realistic system requires accurate data collection and analysis. A number of topics, including real-time traffic and vehicle data, are discussed in this chapter. Based on the information gathered, segments are created, congestion is forecasted, and essential elements are formalized.
- **Vehicle Detector Information Gathering:** In this study, traffic conditions were assessed using VD data. For a better understanding of traffic flow distribution, we used a macroscopic traffic flow model. VD data was analyzed. The VD system was constructed after 184 million traffic records had been recorded on Taichung City's major roads in the previous system.
- **Vehicle Data Collection in Real-Time:** In the absence of just-in-time information, traffic congestion is a major problem. The study analyzes real-time vehicle movement data to control traffic congestion. According to the authors, automobiles play a dual role in this system. A vehicle's first duty is to fulfil the role of a regular user. Data on traffic congestion can be obtained by subscribing to topics published through the MQTT paradigm.
- **Segmented Organization:** Road segments within a network have different traffic flows, and traffic congestion occurs most often on high-traffic roads. As soon as the number of vehicles reaches the segment's capacity, traffic congestion begins. A congested roadway section may, therefore, cause congestion on nearby roadway sections. Taking advantage of historical VD data, the authors examined traffic flow levels across road networks, which is useful in predicting local traffic congestion. In the beginning, the authors used longitude and latitude data to locate VD devices.

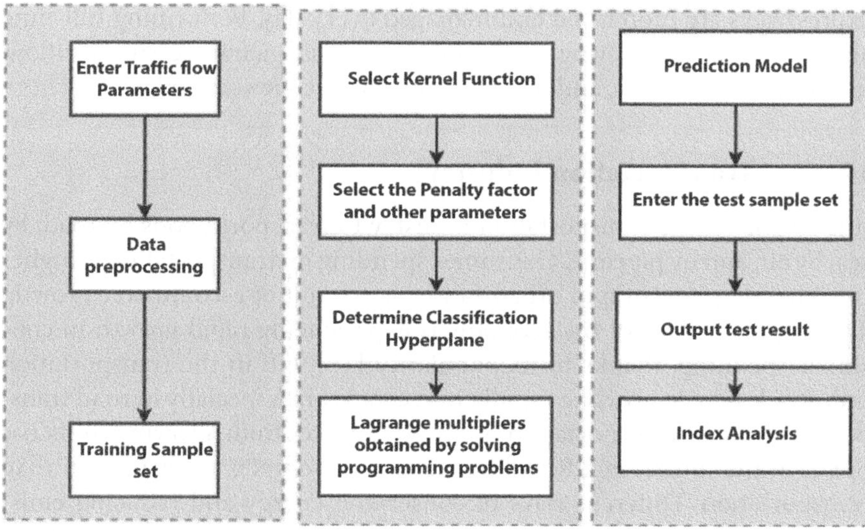

Figure 17.2 Modeling of short-term traffic flow.

Computation models should be able to receive feedback from anomalous flows so that they can be adjusted accordingly. Figure 17.2 shows the process of modeling short-term traffic flows.

17.4 ITS Applications in Various Transport Sectors

There have been numerous implementations of integrated rail transit monitoring and management as part of ITS management and control in the transportation field [15]. This article demonstrates an example of how it can be used to manage and control transportation operations. Many parties need to cooperate in order to manage and control the traffic system holistically. Today, the era of rapid development of information in expressway construction is transforming expressways. In many countries, transmission backbones and provincial comprehensive management platforms have been completed. Moreover, information technology has advanced considerably in comprehensive provincial management. There is a slight backwardness in information management levels in comparison with road networks, and monitoring capabilities need to be enhanced. Traffic control and management systems on a sectional level focus on ensuring expressways are monitored and managed every day. Performing full-time and all-round monitoring, the system can also guarantee smooth flow, improve management, and increase efficiency at the section level.

17.4.1 Transportation Industry

It is all part of the transportation industry to transport goods by road, by sea, by air, and by pipeline. Consumer spending on transportation is higher in China despite its rapid urbanization as a result of e-commerce growth. This popular sector of the economy is experiencing rapid growth in consumer spending, thanks to its popularity [16]. CE in the transportation industry has been growing rapidly in recent years, especially in road transport [17]. Road transportation is expected to continue growing in its use of CE in the future. ECER will be faced with complex challenges in road transportation. Different ways of conserving energy and reducing emissions. Two aspects of ECER are discussed in relation to different services in the transportation sector: road and maritime.

17.4.2 Low CE of Urban Transportation

Scholars are interested in using ITS to schedule and monitor social vehicles and UPT. Sharing economy business models have been transforming how domestic residents travel. DiDi travel data, along with other statistical data, were analyzed by the author [18], to determine how shared travel affects CO2 emissions and transportation structure nationwide. It is estimated that shared travel increased annual CO2 emissions by 518,000 tons due to the substitution effect, resulting in 371,000 tons of ECER. A shared travel platform that prioritizes traffic structure would also promote the widespread use of electric vehicles and result in a greater return on investment [16].

17.4.3 Road Traffic Transportation Infrastructure

As part of the transportation infrastructure, there are facilities for safety, roads, intelligent transportation systems, etc. This industry uses a wide range of equipment, such as signal lights, road markings, guardrails, isolation fences, isolation piers, noise barriers, lighting equipment, parking facilities, traffic gantries, protective nets, guardrails, isolation fences, signs, markings, delineators, antiglare boards, anti-fall nets, and raised spikes.

To build smart cities, a digitalization of highway infrastructure is an essential part of creating road-vehicle coordination and the direction of future intelligent highway development.

17.5 Result and Discussion

iFogSim framework is used in the implementation of the proposal [17]. In addition to providing an internet of things (IOT) simulation environment, there are also cloud computing simulation classes available. Eclipse or NetBeans can be used to implement this Java-based simulation toolkit. To implement machine learning algorithms, Weka [15] was used, which is a popular open-source tool for this purpose.

Based on the Random Forest algorithm, the traffic estimation below illustrates the accuracy of the feature selection, as shown in Figure 17.3. Vehicle counts are depicted in the graph alongside their actual and estimated values. Though estimated traffic counts may differ from actual counts, in a larger perspective, both estimates and actual traffic counts show the same increases and decreases in traffic over all time intervals.

This system is beneficial for reducing average waiting times on given roads as well as making accurate predictions based on the varying levels of

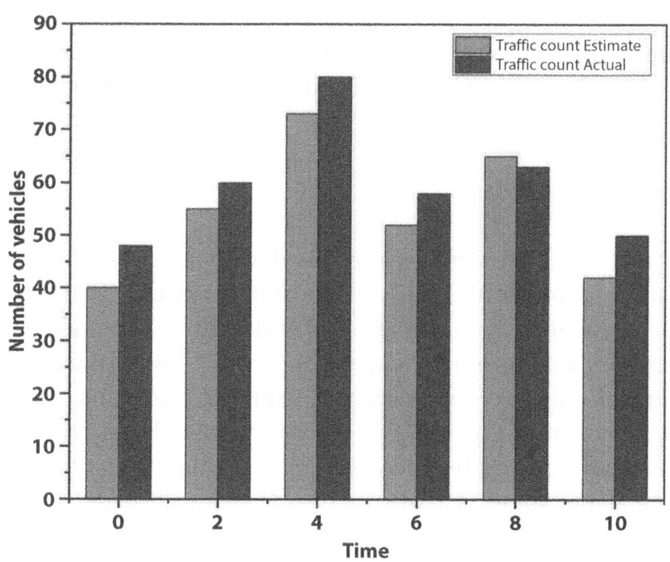

Figure 17.3 Traffic estimation.

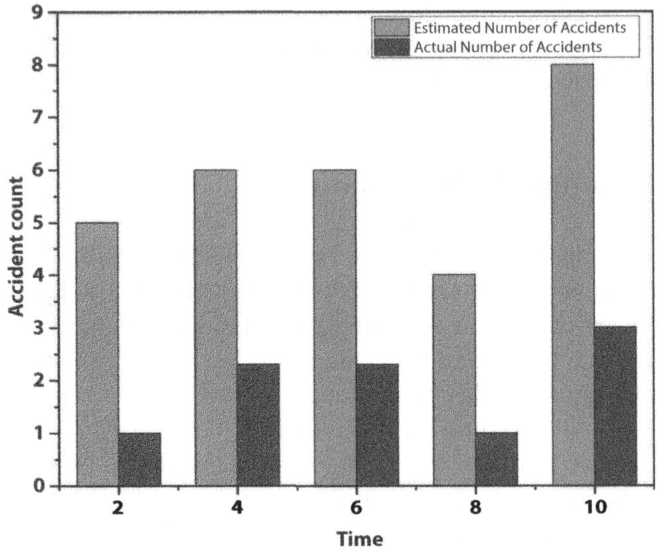

Figure 17.4 Accident detection.

traffic, as shown in Figure 17.4. Our accident detection mechanism is illustrated in the following graphs, which compare the actual and estimated accidents occurring on the road over time.

Estimated accident numbers always exceed actual accident numbers. Accordingly, we were able to detect actual accidents, as well as other scenarios that exhibited similar characteristics to those of accidents. As a result, accidents are more likely to occur than actual accidents.

17.6 Conclusion

A smart traffic management system was developed by analyzing real-time traffic data and performing advanced computations. There is a significant improvement in the estimation of traffic flow, congestion reduction, and emergency response efficiency. There are a few minor limitations to the system, like the overestimation of accident scenarios, but its overall effect on urban mobility and the environment is huge. The ITMS integrates advanced sensing technologies with predictive analytics for a cost-effective, scalable solution that can be adapted to a wide range of urban settings. It is possible that the system will be refined in the future and may be applied to a wider range of transportation sectors to increase its utility and performance.

References

1. Bayashot, Z.A., The Contribution of AI-Powered Mobile Apps to Smart City Ecosystems. *J. Softw. Eng. Appl.*, 17, 03, 143–154, 2024, doi: 10.4236/jsea.2024.173008.
2. Rani, P. and Sharma, R., IMFOCA-IOV: Intelligent Moth Flame Optimization based Clustering Algorithm for Internet of Vehicle, in: *2023 14th International Conference on Computing Communication and Networking Technologies (ICCCNT)*, IEEE, pp. 1–6, 2023.
3. Ahmed, M., *et al.*, A Blockchain-Based Emergency Message Transmission Protocol for Cooperative VANET. *IEEE Trans. Intell. Transp. Syst.*, 23, 10, 19624–19633, Oct. 2022, doi: 10.1109/TITS.2021.3115245.
4. Alkadi, S., Al-Ahmadi, S., Ben Ismail, M.M., Toward Improved Machine Learning-Based Intrusion Detection for Internet of Things Traffic. *Computers*, 12, 8, 148, Jul. 2023, doi: 10.3390/computers12080148.
5. Ban, X., Herring, R., Margulici, J.D., Bayen, A.M., Optimal Sensor Placement for Freeway Travel Time Estimation, in: *Transportation and Traffic Theory 2009: Golden Jubilee*, W.H.K. Lam, S.C. Wong, H.K. Lo (Eds.), pp. 697–721, Springer US, Boston, MA, 2009, doi: 10.1007/978-1-4419-0820-9_34.
6. Rani, P., Verma, S., Yadav, S.P., Rai, B.K., Naruka, M.S., Kumar, D., Simulation of the lightweight blockchain technique based on privacy and security for healthcare data for the cloud system. *Int. J. E-Health Med. Commun. IJEHMC*, 13, 4, 1–15, 2022.
7. Rani, P. and Sharma, R., Intelligent transportation system for internet of vehicles based vehicular networks for smart cities. *Comput. Electr. Eng.*, 105, 108543, 2023.
8. Rani, P., *et al.*, Federated Learning-Based Misbehaviour Detection for the 5G-Enabled Internet of Vehicles. *IEEE Trans. Consum. Electron.*, 70, 2, 4656–4664, 2023.
9. Son, T.H., Weedon, Z., Yigitcanlar, T., Sanchez, T., Corchado, J.M., Mehmood, R., Algorithmic urban planning for smart and sustainable development: Systematic review of the literature. *Sustain. Cities Soc.*, 94, 104562, Jul. 2023, doi: 10.1016/j.scs.2023.104562.
10. Singh, A., *et al.*, Smart Traffic Monitoring Through Real-Time Moving Vehicle Detection Using Deep Learning *via* Aerial Images for Consumer Application. *IEEE Trans. Consum. Electron.*, 70, 4, 1–1, 2024, doi: 10.1109/TCE.2024.3445728.
11. Rani, P. and Sharma, R., An experimental study of IEEE 802.11 n devices for vehicular networks with various propagation loss models, in: *International Conference on Signal Processing and Integrated Networks*, Springer, pp. 125–135, 2022.
12. Chhajer, P., Shah, M., Kshirsagar, A., The applications of artificial neural networks, support vector machines, and long–short term memory for stock market prediction. *Decis. Anal. J.*, 2, 100015, 2022.

13. Bartin, B., Ozbay, K., Iyigun, C., Clustering-Based Methodology for Determining Optimal Roadway Configuration of Detectors for Travel Time Estimation. *Transp. Res. Rec. J. Transp. Res. Board*, 2000, 1, 98–105, Jan. 2007, doi: 10.3141/2000-12.
14. Majid, M., *et al.*, Applications of wireless sensor networks and internet of things frameworks in the industry revolution 4.0: A systematic literature review. *Sensors*, 22, 6, 2087, 2022.
15. Yang, C., Zha, M., Wang, W., Liu, K., Xiang, C., Efficient energy management strategy for hybrid electric vehicles/plug-in hybrid electric vehicles: review and recent advances under intelligent transportation system. *IET Intell. Transp. Syst.*, 14, 7, 702–711, Jul. 2020, doi: 10.1049/iet-its.2019.0606.
16. Alrawi, F., The importance of intelligent transport systems in the preservation of the environment and reduction of harmful gases. *Transp. Res. Procedia*, 24, 197–203, 2017.
17. Wang, Y., Wen, Z., Cao, X., Dinga, C.D., Is information and communications technology effective for industrial energy conservation and emission reduction? Evidence from three energy-intensive industries in China. *Renew. Sustain. Energy Rev.*, 160, 112344, 2022.
18. Arun, K., Ishan, G., Sanmeet, K., Loan approval prediction based on machine learning approach. *IOSR J. Comput. Eng.*, 18, 3, 18–21, 2016.

18

Exploring the Impact of Computer Vision in Smart Transportation

Umesh Chandra Garjola[1]*, Sandeep Singh[2], Kamaljeet Kaur[3], Protyay Dey[4] and Laith Hussein[5]

[1]*College of Computing Sciences & IT (CCSIT), Teerthanker Mahaveer University, Moradabad (U.P.), India*
[2]*Department of Civil Engineering, School of Engineering and Technology, Bahra University, Waknaghat, Distt. Solan, HP, India*
[3]*Department of Computer Science Engineering, Chandigarh Engineering College, Chandigarh Group of Colleges, Jhanjeri, Mohali, Punjab, India*
[4]*Department of Computing Science and Artificial Intelligence, NIMS Institute of Engineering and Technology, NIMS University Rajasthan, Jaipur, India*
[5]*College of Technical Engineering, The Islamic University, Najaf, Iraq*

Abstract

The need for sustainable urban development has led to the adoption of connected lighting systems to reduce energy consumption and minimize environmental impact. Municipalities, which invest heavily in traditional lighting systems, are now transitioning to Solid-State Lighting (SSL) technologies like LEDs due to their energy efficiency and adaptability. This paper examines the integration of advanced roadside surveillance systems and smart lighting to enhance urban sustainability. This chapter addresses challenges such as traffic congestion, accident detection, and intelligent transportation. Innovative methods like ConvLSTM2D for capturing spatiotemporal data and DenseNet-transformer for feature extraction and sequence processing are proposed using leveraging computer vision (CV) and machine learning (ML). The findings highlight the potential of integrating connected lighting systems with intelligent technologies to create sustainable, energy-efficient urban environments.

**Corresponding author*: umeshgarjola6@gmail.com

Pethuru Raj, Sudesh Yadav, Manas Kumar Mishra, Satya Prakash Yadav and Victor Hugo C. de Albuquerque (eds.) Artificial Intelligence Technologies for Smart and Sustainable Urban Transportation: Integrated Platforms and Use Cases, (301–322) © 2026 Scrivener Publishing LLC

***Keywords*:** Accident detection, computer vision, connected lighting, intelligent transportation systems, machine learning, traffic management, smart cities

18.1 Introduction

Artificial lighting contributes to approximately 1900 Mt of CO2 emissions, which equates to about 19% of total global electricity production [1]. As outdoor lighting requires a high wattage and long operating hours to ensure traffic visibility, it has a high energy consumption. Municipalities may consume 60% of their electricity for street lighting, for instance [2]. Lighting maintenance often consumes a substantial portion of a municipality's budget [3]. Street lighting systems typically last 30–40 years, so they are usually inefficient and expensive because they are long-lasting [4]. It will be necessary to replace these lighting systems in the near future, preferably with one that will last for a long time and be cost-effective. SSL is developing rapidly, allowing for energy-efficient, long-lasting, and eco-friendly products, such as LED [5]. In the upcoming decades, LED lighting is expected to significantly reduce costs and CO2 emissions compared with mercury vapor, metal halides and high-pressure sodium (HPS) [6]. In addition to being adaptable, LED street lighting can also be intelligent, further reducing energy consumption [7].

18.1.1 Surveillance Systems Along Roadsides: An Overview

Roadside surveillance systems are becoming more accurate and reliable as technology advances by detecting objects, tracking them, classifying them and analyzing their behavior [8, 9]. The process of observing certain human behaviors with the aim of influencing, managing, protecting, or directing them is called surveillance [10]. An example would be the observation of a person from a distance by CCTV or the capture of electronic transmissions like internet traffic or phone calls. In addition to gathering intelligence, preventing and investigating crimes, and protecting individuals, groups, and objects, governments commonly use surveillance. Typical traffic surveillance involves video cameras installed at various locations with real-time traffic monitoring, e.g., along the highways, to control traffic flow and facilitate government traffic management. A roadside surveillance system detects, recognizes, classifies, tracks, analyses, and retrieves traffic behavior and activities, according to [11]. Objects that are persistent or transient in an environment can be monitored in real time as a result of this technology. A roadside surveillance system can assist in solving transportation issues, such as traffic congestion and road hazards, by making use of advanced information and communication technologies [12]. The public and road users alike depend on smooth traffic conditions, and traffic congestion can negatively affect them. In [13], it points out that traffic congestion is extremely expensive and negatively impacts productivity, fuel consumption, sustainability, and quality of life for everyone. As a result of traffic congestion, journeys are delayed, air pollution and CO_2 emissions increase, emergency vehicles are blocked, and spill-over effects occur on congested transportation. During traffic congestion, truck drivers might have to delay their loading and unloading activities, which would have an impact on logistics companies' management. Parking spots are sometimes found by driving around the road network. This way of wasting fuel and energy is wasteful. The purpose of roadside surveillance systems is to improve fleet management and traffic flow due to traffic congestion, accidents, and low traffic efficiency [14].

18.2 Related Work

18.2.1 Computer Vision Functions

A significant amount of visual data is emerging in the ITS field [15]. CV studies analyze images and videos and present detailed traffic information. CV techniques play a number of fundamental roles in ITS, as presented in Figure 18.1. A CV method can perform a wide variety of basic

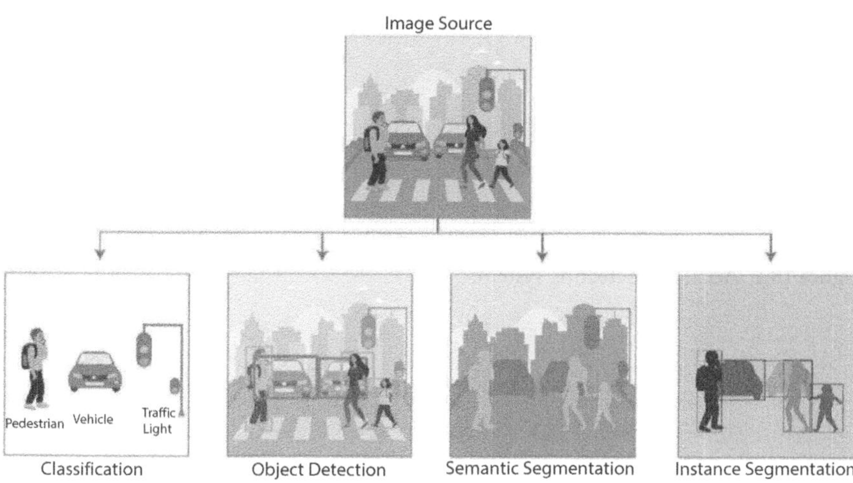

Figure 18.1 Computer vision functions for ITS.

functions, such as classification, object detection, semantic segmentation, and instance segmentation [16].

A CV technique can be used to classify objects using images or videos captured by cameras. Images can automatically assign labels to their sub-objects. Models are created by dividing objects into parts. CV techniques are also used to detect objects [17]. Vehicles and pedestrians can be detected in images, which is crucial for the development of numerous applications. Traffic density can be calculated using deep learning models, pedestrians can be detected, and other vehicles can also be detected with deep learning models. Classification works by determining the location of the object, while object detection analyses each relevant object in the image and categorizes it [18]. This type of artificial intelligence can be used for classification as well as regression. In the image, the corner coordinate points allow the machine to position the object. Semantic segmentation classifies any pixel that belongs to an object as shown in Figure 18.2. Pixels of an object need to be grouped and categorized accordingly. Objects are grouped according to semantic segmentation models. Sometimes, however, it's necessary to group pedestrians and vehicles separately in traffic.

In some cases, however, it is necessary to separate vehicles and pedestrians in traffic. In these cases, instance segmentation is employed.

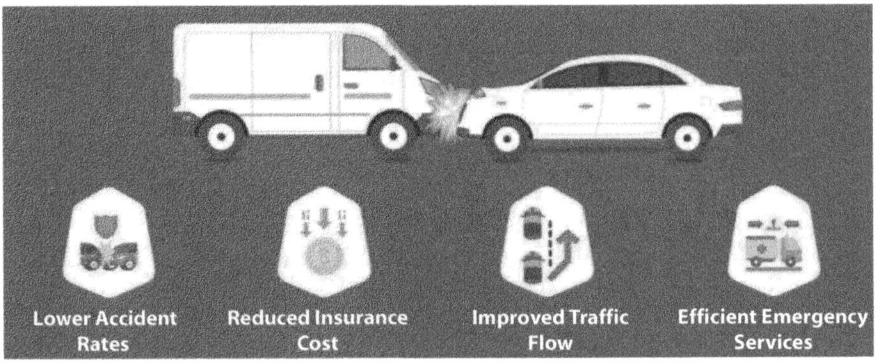

Figure 18.2 Accidents detecting system.

Pixels are assigned classes through instance segmentation and semantic segmentation. Even overlapping objects can be segmented separately by instance segmentation. As shown in Table 18.1, computer vision techniques can provide solutions to several problems in the field of ITS.

Table 18.1 Comparative analysis of computer vision techniques for ITS.

Computer vision	**Application areas**	**Sample datasets**	**Performance metrics**
Detection of objects	An image/video is boxed, and its coordinates are found • Light detection • A method of detecting and classifying traffic signs • A system for detecting pedestrians • Detecting and counting types of vehicles	CIFAR-10, 100, and Cityscape are just a few of the additional algorithms that are available, including ImageNet, LISA, GTSDB, Cityscape and Pascal VOC	Accuracy, precision, recall, average precision, mAP (mean average precision) and RMSE
Object Segmentation	An image must be classified into its pixels, and the individual masks of the objects should be obtained • Estimating speed • Tracking and determining the road lines • Optimizing routes	LISA, KITTI, COCO, Cityscape, BD100K	mAP

(*Continued*)

Table 18.1 Comparative analysis of computer vision techniques for ITS. (*Continued*)

Computer vision	Application areas	Sample datasets	Performance metrics
Image Enhancement	Light pollution, fog, haze, rain, and low lighting can corrupt images • Raindrops are removed from images captured by cameras • Increasing the resolution of low-resolution objects • Blurred images can be sharpened • Cartesian coordinates for fish-eye cameras	REDS	PSNR
Object Tracking	In video tracking, there is a problem • Pedestrian and vehicle tracking • A vehicle's speed can be detected • Route extraction	MOT19	MOTA
Event Identification/ Prediction	Interpreting the video: Understanding what happened • Accident recognition and prediction • Obstruction estimation • A hazardous situation or route is detected	AUC 101, Kinetics 600	Accuracy mAP

(*Continued*)

Table 18.1 Comparative analysis of computer vision techniques for ITS. (*Continued*)

Computer vision	Application areas	Sample datasets	Performance metrics
Anomaly Detection	Transport system abnormality detection • A sudden appearance of pedestrians or objects on the road • In rail systems, there may be anomalies • Drivers who are drowsy or drunk are detected, as are drivers who are texting, talking on their cell phones, etc. • A license plate recognition system is used to detect traffic violations and suspicious vehicles	The CIFAR-10/ CIFAR-100 is a component of the ShanghaiTech Model	AUC
Density Analysis	Public transport density analysis for pedestrians, passengers, or vehicles • Traffic can be detected automatically • Analysis of parking lot vehicle densities • Analyzing pedestrian density at specific locations	A 5K race at Oxford, a mall run at UCSD, a Chinese technology show at UCF, and the World Expo	MAE and MSE

(*Continued*)

Table 18.1 Comparative analysis of computer vision techniques for ITS. (*Continued*)

Computer vision	Application areas	Sample datasets	Performance metrics
Image/Event Search	Video archives cannot be used to extract license plates, vehicles, or pedestrians • Traffic camera archive searches for target plates by law enforcement • Finding people or vehicles in digital archives for security reasons • A technique for detecting similar objects within a given category	Pascal VOC, Oxford 5K	Accuracy

18.3 Proposed Methodology

Prioritizing a vehicle starts with detecting it. Detecting vehicles is an essential component of intelligent transportation system. In the study, two algorithms are compared based on different parameters for detecting, tracking, and prioritizing vehicles. Tracking and categorizing vehicles are done based on their type. Train aggregated channel feature (ACF) in MATLAB is used to train video data [19, 20]. The ACF and point tracker algorithms are used to detect accidents and manage emergencies.

ACF and point-tracking algorithms were used to detect all vehicles [21]. A training algorithm is applied to all vehicles to detect and train them. This method involves the following steps:

- A vehicle is detected.
- Keeping track of vehicles.
- Labeling vehicles according to their types.
- Determine the vehicle's priority.
- Keeping track of vehicles.

18.3.1 ACF Object Detection System

The procedure first detects vehicles, ambulances, bicycles, and pedestrians. In the first step, an identical input image is altered numerous times to gain the feature of varying frequencies at the same resolution. After noise suppression is removed, smoothing filters are applied [22]. Channel feature labels are assigned to each picture element. According to the ACF object detection algorithm, it is more precise compared to algorithms using single-channel features. An algorithm that detects pedestrians and non-vehicles is proposed. As part of this algorithm, the "vehicle detector ACF" function is used. This function detects pedestrians using "people detector ACF" [23, 24]. A multi-type vehicle, pedestrian, and non-vehicle can be detected using this algorithm without sacrificing image quality. As soon as the vehicles have been detected and labeled, the proposed study will assign priorities based on their categories. A channel feature extraction method called "Aggregated Channel feature" is used to represent object features [21, 25]. Using it, you can detect fractures in moving objects uniquely based on the image. As a first step, an input image is converted into a pyramid of multiple-resolution images to extract data from the ACF algorithm as quickly as possible. The algorithm extracts similar and unique features from high- and low-resolution images.

18.3.2 Point Tracker Algorithm

The algorithm first develops a sequence of images. Every image is detected, and the distance between two tracking points is minimized. Images must be sequenced to capture the scale of the vehicles' numerous geographic shapes. Vehicle detection is handled by a "Vision Point Tracker" [26]. This algorithm uses the Kanada Lucas-Tornasi (KLT), which is a highlighting algorithm. As input to the algorithm, video data is detected for motion estimation and object tracking, and data stability is provided. In comparison to the ACF algorithm, it didn't modify the shape of images in cases where visual texture is displayed [27–29]. Tracking and the video frame are initialized before detecting a vehicle, and every point is specified at its initial position [30].

18.3.3 Intelligent Transportation Systems: Computer Vision Applications

With ITSs, transport safety, efficiency, energy conservation, law enforcement, and emissions decrease have greatly improved. In this context, CVs play a crucial role, so researchers should be interested in CV applications. ITS CV studies have been published in the WoS database since 2000, with more than a thousand studies included in that database. In light of the multidisciplinary nature of ITS, these publications covered a variety of scientific disciplines, including engineering, transportation science, and telecommunications.

Several CV-based technologies have been applied to road transport systems [31, 32], as well as a complete overview of CV systems for traffic analysis in urban areas [32]. In contrast, these studies lack the most recent CV methods. Machine learning techniques to improve the efficiency of transportation systems have been widely used over the last few years. Research indicates that new learning techniques are replacing traditional machine learning techniques, and transport systems are widely adopting DL techniques. A presentation by the University of Rochester focused on deep learning models' potential for improving transportation systems' intelligence [33]. A number of transportation applications were examined with DL models, such as traffic sign recognition, flow estimation, and speed estimation. As part of the discussion, the applicability and weaknesses of DL models were also examined in relation to ITS and evolving future trends.

According to studies, advanced computing power, big data analysis, and machine learning will enable transportation systems to become smarter [33]. Presented below are summaries and categorizations the studies based on different CV applications in ITS. At the end of each section, a table presents the key findings in each category. There is a wide variety of CV applications in ITS, as well as different datasets and metrics used to evaluate the methods' effectiveness. A method's performance can't be compared to that of other methods since different datasets and metrics are used. Furthermore, this may make it seem as if the performance of methods decreased in some categories over time. It is, therefore, necessary to include datasets and metrics for each method that is utilized in the literature to assess its performance.

18.3.4 Intelligent Transportation Systems and Machine Learning (ML)

As a rule of thumb, machine learning is described as the ability of a system to make intelligent decisions without explicit programming. With ML, computer systems learn to classify, group, predict, identify patterns, and perform other tasks based on data. In this process, samples are analyzed with algorithms and statistical models to archive learning. By categorizing

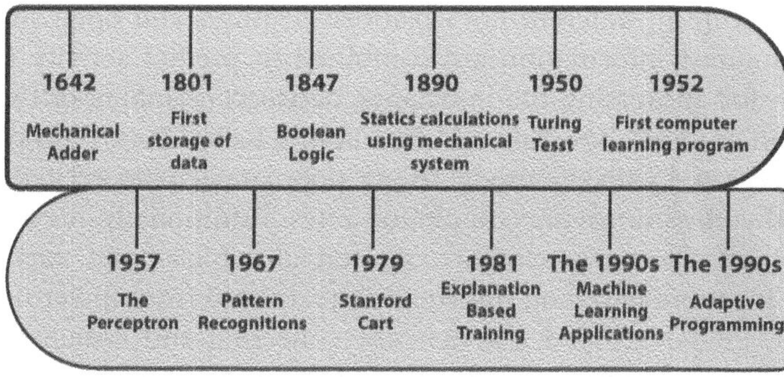

Figure 18.3 Adapting machine learning to the changing world.

data based on quantifiable properties, ML algorithms analyze data by finding relationships between quantifiable features and specific output values, or labels [14]. In the training phase, new data is analyzed to discover patterns and make judgments. A recent surge in the popularity of machine learning has been attributed to exponential data collection and technological advances that enable it, but its roots date back nearly 400 years [34]. Data has always been analyzed to gain immediate insights since the dawn of time. Figure 18.3 above shows how machine learning has evolved.

18.3.4.1 Machine Learning: The Evolution

Data training, processing, and prediction have become faster and more accurate, thanks to machine learning. Today, machine learning algorithms are increasingly being used to solve transportation-related problems. A smartphone-based travel survey employing location-aware technology helps researchers and city planners gather comprehensive mobility data. By leveraging data mining and machine learning, we can better understand the demands on road and parking infrastructure. This goal can be achieved through FMS, which tracks and analyzes mobile devices and traditional GPS trackers to identify movement patterns. Moreover, machine learning algorithms can extract additional trip details, such as travel time, mode of transportation, and stopping points [35].

In Singapore, an FMS platform monitored the movements of heavy goods vehicles, capturing their stops and routes. By merging raw mobility data with contextual information, machine learning algorithms can identify further trip details, including travel time, transportation mode, and stopping points. However, a large percentage of recorded stops were unverified, and the types of reported activities were unequal.

According to [35], different FMS platform features and open street map point-of-interest information are combined to predict activity type. An algorithm for addressing the imbalance between members of the classes is based on gradient-boosting and resampling techniques. FMS platforms can be integrated with the model to pre-populate activity-related fields of surveys, thereby improving completion rates. Additionally, it can be used to investigate the movement of commercial vehicles and parking patterns in Singapore based on information recovered from unverified stops. Moreover, [36] developed a system for collecting and visualizing travel data to understand mobility patterns and travel habits better. Data is collected and merged from a variety of sources by the FMS system, which then visually presents the results. A mobile sensing device, such as a smartphone, a GPS logger, and machine learning algorithms are used to compile multi-day travel data using the FMS Data Collection Platform. A third feature of the FMS Data Fusion and Visualization Platform is that it integrates data from a variety of sources into knowledge so that users can view their mobility diary, provide additional input, and explore their patterns of mobility. The research aimed to combat distracted driving, a major cause of auto accidents, by developing a device similar to an eye blink sensor using AdaBoost algorithms [37]. Several distractions can cause a driver to be distracted, potentially leading to an accident if not eliminated by this alarm system. Their study compared the use of four models in anticipating traffic flow for planned work zones. Nonparametric regression, quadratic regression, regression trees, and multilayer feedforward neural networks are the four models. A look-back traffic flow at upstream, downstream, and present locations of the latest interval had the greatest impact on highway data. The data from the arterials were influenced by traffic flows only during a look-back at the current site [38].

An investigation was conducted to compare RF with SVM and ANN to detect road accidents [39]. Traffic simulation scenarios utilize SUMO [40] for data collection. A vehicle simulation was used to obtain information about cars' speeds and positions. A 10-fold cross-validation technique is used to train and verify the above classifiers. There is no doubt that the RF algorithm performs better than the two other classifiers in terms of accuracy and sensitivity. The author [41], presents a comparison was made between the proposed system and several machine learning approaches, including LightGBM, KNN, SVM, and RF. Tests showed that the proposed system had an average F1-score of 0.98, a higher average accuracy, and a higher recall rate than the competition. In addition, the concept can be easily applied to a variety of drivers and locations. This system works in real-time because it processes the smartphone's sensor input directly.

The author [42], developed a predictive algorithm to assist drivers when performing emergency manoeuvres using advanced driver assistance systems (ADAS). The solution provided a real-time solution to the planning and control issues. Through the combination of a model predictive control and an environmental model, no reference signal is required. Crudely discretizing inputs obtain real-time capabilities so that state trajectory can be predicted for all possible combinations of inputs. According to [43], clustered neural networks (CNNs) are employed as input variables to predict online journey times based on raw detector data from traffic vectors. Firstly, there is a base term and, subsequently, a variation term. In this study, the effectiveness of the suggested method was evaluated using CORSIM, the microscopic simulator.

18.3.4.2 Challenges

Increasing automobile traffic is outpacing population growth, causing highways to become clogged and certain roads to become unsafe to travel on [44]. On the other hand, road data can be analyzed for traffic regulation. Since transportation sector data is so large, standard data analytics methods may not be as effective.

For a machine-learning solution to be effective and efficient, data type, features, and algorithms must be taken into account. Although today's internet allows massive amounts of data to be generated on a regular basis, collecting data in domains like cybersecurity and smart transportation is not easy. In order to analyze smart city applications in the future, it is imperative to gather usable data and manage it effectively. To analyze real-world data effectively, it is necessary to analyze data collection methods in greater detail. History may also contain many unclear, missing, and outlier values, as well as nonsensical data [45].

As a result, machine learning algorithms have a significant impact on the final model's quality and availability of data for training. The data obtained from multiple sources must, therefore, be cleaned and pre-processed appropriately. The current pre-processing procedures must be altered or improved, or new data preparation methods must be used effectively [44]. Data analysis and insight extraction can be achieved with machine learning algorithms. Due to these reasons, it may not be easy to choose the right learning algorithm. The data characteristics may affect the result of a learning algorithm, which increases complexity [16]. By selecting the incorrect ML algorithm, there may be unanticipated consequences, such as a loss of effort and a reduction in model efficacy and accuracy. A machine learning-based solution's success is heavily dependent on data and learning algorithms.

Insufficient data, poor quality, irrelevant, or non-representative data might make learning more difficult. A scenario such as this could render the model ineffective or inaccurate.

18.4 Result and Discussion

The main objective of this research is to extract both RGB frames and optical flow information from video sequences using the CONVLSTM2D architecture. LSTMs and convolutional neural networks are integrated in this model, which is pivotal for our research. As opposed to traditional CNNs, RNNs, or LSTMs, our approach captures the dynamic spatio-temporal characteristics of accidents effectively. In accident scenarios, ConvLSTM2D is particularly well suited to detecting motion patterns. The analysis involves looking at the intricate interplay between spatial patterns and temporal patterns in individual frames. In our experiments, we evaluate various architectural and parameter configurations to determine whether they are effective at identifying features indicative of accidents. As a result of the model, accident detection precision has been significantly improved, demonstrating its effectiveness in enhancing autonomous accident detection systems within smart city infrastructures, as well as providing a more nuanced and accurate way of detecting traffic accidents.

A. I3D-CONVLSTM2D RGB ONLY

In the I3D-CONVLSTM2D RGB Only model, the input video was analyzed exclusively for RGB frames to identify accidents. In this model, [20, 32], and 32 filters were used within the ConvLSTM2D layers. Overfitting was also mitigated using dropouts. As a result of the model, accident-related features were more easily distinguished. A mean average precision of 78% was achieved, as well as accuracy, precision, recall, and an F1 score of 72%.

B. I3D-CONVLSTM2D NON-TRAINABLE RGB + OPTICAL FLOW

This second iteration used an explicit, non-trainable setting for all layers of the I3D component, turning it into a fixed feature extractor. In this research, we use the CONVLSTM2D architecture as a means of extracting both RGB frames and optical flow information from video sequences. Multiple layers of CONVLSTM2D were added to the classifying head of the I3D model to process RGB frames and optical flow simultaneously as

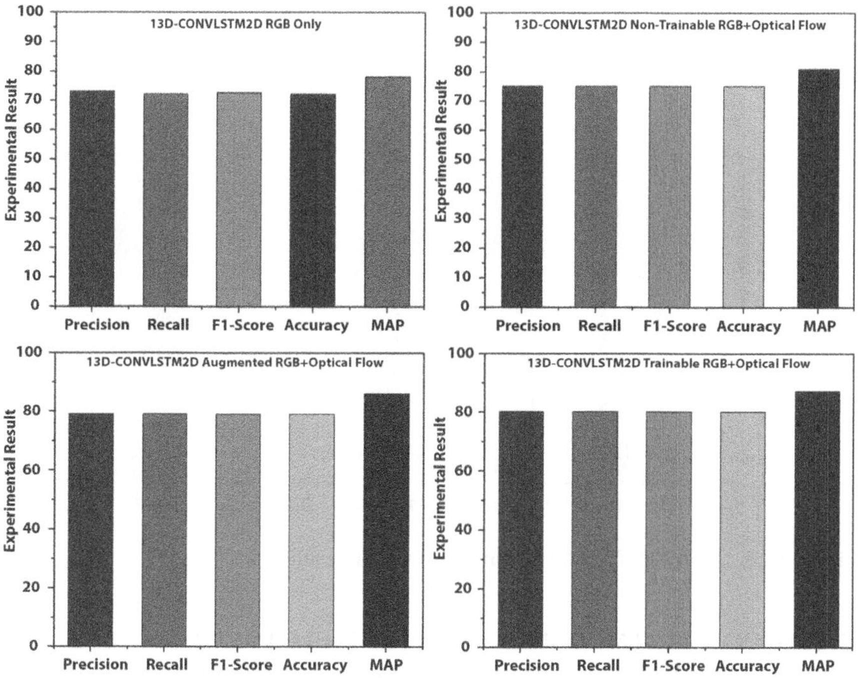

Figure 18.4 Experimental results of I3D-CONVLSTM2D.

shown in Figure 18.4. An I3D component that does not allow training that incorporates a fixed feature extraction approach. There is a 75% accuracy, a 95% precision, a 98% recall and an 81% mean average precision for the key metrics.

C. I3D-CONVLSTM2D AUGMENTED RGB + OPTICAL FLOW

The third model examined how data augmentation techniques affected accident detection with I3D-CONVLSTM2D Augmented RGB + Optical Flow. In particular, cropping, zooming, and careful rotation were key components of our augmentation strategy. Rotation was limited to ensure realistic representations and avoid producing upside-down videos. With video augmentation, we were able to improve the recognition of accident-related features in the dataset and learn more about temporal features. This integration did not lead to the anticipated improvement in accident detection, contrary to our initial hypothesis. A MAP of 86%, an accuracy score of 86%, a precision score of 84%, and a recall score of 79% were recorded for the model as shown in Figure 18.5.

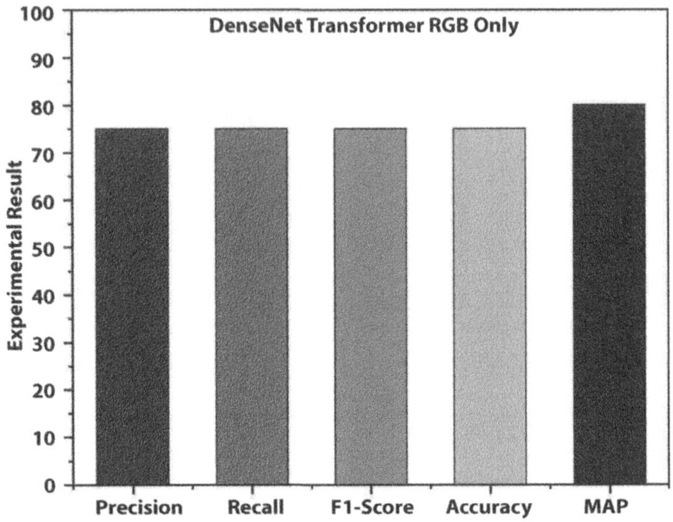

Figure 18.5 DenseNet-Transformer of experimental result.

D. I3D-CONVLSTM2D RGB + OPTICAL-FLOW (TRAINABLE)

By integrating the I3D model, which comprises at least 15 trainable layers, we rely on this principle to generate the I3D-CONVLSTM2D RGB + Optical-Flow (Trainable) model. The density layers were enriched with 512, 256, and 256 neurons to improve the model's ability to extract and represent features. The ConvLSTM2D layer was modified to include extra filters, which significantly improved the model's ability to process spatio-temporal information. Performance metrics have significantly increased due to these strategic enhancements. An 80% F1 score, 87% accuracy, 87% precision, and 87% recall were achieved by the model, with a mean average precision (MAP) of 87%.

E. DENSENET-TRANSFORMER RGB ONLY

The DenseNet-Transformer model was trained, first extracting features from DenseNet, followed by sequence processing using transformer layers. As part of the architecture, additional dense layers were added, as well as a fully connected classification layer. The model achieved 75% accuracy, recall, and F1 score.

18.5 Conclusion

Connected lighting systems and intelligent technologies play an important role in addressing urban sustainability challenges. Energy consumption and carbon emissions can be significantly reduced when advanced lighting solutions are combined with roadside surveillance. In the fields of accident detection, traffic management, and urban planning, CV and ML techniques have shown improved performance, including models like ConvLSTM2D and DenseNet-Transformer. To solve urban issues efficiently and scalable, data-driven, adaptive approaches are necessary. Achieving sustainable, energy-efficient urban ecosystems will require such integrated solutions as cities grow. These technologies need to be optimized for broader implementation and explored in conjunction with other frameworks for smart cities in the future.

References

1. OECD; IEA. Light's Labour's Lost. Policies for Energy-Efficient Lighting, Available online: Accessed: Nov. 08, 2024, [Online], Available: http://www.iea.org/publications/freepublications/publication/name,3644,en.html.
2. Fiaschi, D., Bandinelli, R., Conti, S., A case study for energy issues of public buildings and utilities in a small municipality: Investigation of possible improvements and integration with renewables. *Appl. Energy*, 97, 101–114, Sep. 2012, doi: 10.1016/j.apenergy.2012.03.008.
3. Jägerbrand, A.K. and Robertson, K., Renewal of street and road lighting in Swedish municipalities, in: *CIE conference 15-16 April, 2013, Paris, France*, pp. 1009–1015, 2013, Accessed: Nov. 08, 2024, [Online], Available: https://www.diva-portal.org/smash/record.jsf?pid=diva2:698397.
4. Gaston, K.J., A green light for efficiency. *Nature*, 497, 7451, 560–561, May 2013, doi: 10.1038/497560a.
5. De Almeida, A., Santos, B., Paolo, B., Quicheron, M., Solid state lighting review – Potential and challenges in Europe. *Renew. Sustain. Energy Rev.*, 34, 30–48, Jun. 2014, doi: 10.1016/j.rser.2014.02.029.
6. Miyairi, K., Endo, K., Matsuda, S., Kokuryo, J., Study of CO_2 Emission in Relation to Economic Efficiency in LED Street Lighting. *J. Illum. Eng. Inst. Jpn.*, 97, 2, 57–64, 2013, doi: 10.2150/jieij.97.57.
7. Gong, Y. and Han, P., Research on energy-saving scheme based on LED street lamp management-system. *Appl. Mech. Mater.*, 448, 2850–2855, 2014.

8. Rani, P. and Sharma, R., Intelligent Transportation System Performance Analysis of Indoor and Outdoor Internet of Vehicle (IoV) Applications Towards 5G. *Tsinghua Sci. Technol.*, 29, 6, 1785–1795, Dec. 2024, doi: 10.26599/TST.2023.9010119.
9. Rani, P., *et al.*, Federated Learning-Based Misbehaviour Detection for the 5G-Enabled Internet of Vehicles. *IEEE Trans. Consum. Electron.*, 70, 2, 4656–4664, 2023.
10. Lyon, D., Surveillance studies: An overview, 2007, Accessed: Nov. 06, 2024, [Online], Available: https://books.google.com/books?hl=en&lr=&id=_dTHJgh3-f0C&oi=fnd&pg=PR4&ots=JqtHQqwTMg&sig=-k5k6YD-5hoUu5pNyKUDnSxpl2WI.
11. Valera, M. and Velastin, S.A., Intelligent distributed surveillance systems: a review. *IEE Proc. Vis. Image Signal Process.*, 152, 2, 192, 2005, doi: 10.1049/ip-vis:20041147.
12. Figueiredo, L., Jesus, I., Machado, J.T., Ferreira, J.R., De Carvalho, J.M., Towards the development of intelligent transportation systems, in: *ITSC 2001. 2001 IEEE intelligent transportation systems. Proceedings (Cat. No. 01TH8585)*, IEEE, pp. 1206–1211, 2001, Accessed: Nov. 06, 2024, [Online], Available: https://ieeexplore.ieee.org/abstract/document/948835/.
13. Problems in the Urban Environment: Traffic Congestion and Its Effects, Available online: Accessed: Nov. 06, 2024, [Online], Available: https://ro.uow.edu.au/wollgeo/14/.
14. Qin, M., Wang, Y., Cui, Z., Zhu, Y., The design and realization of an advanced urban traffic surveillance and management system, in: *2006 6th World Congress on Intelligent Control and Automation*, 2006, Accessed: Nov. 06, 2024, [Online], Available: https://www.infona.pl/resource/bwmeta1.element.ieee-art-000001713673.
15. Rani, P. and Sharma, R., IMFOCA-IOV: Intelligent Moth Flame Optimization based Clustering Algorithm for Internet of Vehicle, in: *2023 14th International Conference on Computing Communication and Networking Technologies (ICCCNT)*, IEEE, pp. 1–6, 2023.
16. Sarker, I.H., Kayes, A.S.M., Watters, P., Effectiveness analysis of machine learning classification models for predicting personalized context-aware smartphone usage. *J. Big Data*, 6, 1, 57, Dec. 2019, doi: 10.1186/s40537-019-0219-y.
17. Yadav, S.P., Jindal, M., Rani, P., de Albuquerque, V.H.C., dos Santos Nascimento, C., Kumar, M., An improved deep learning-based optimal object detection system from images. *Multimed. Tools Appl.*, 83, 10, 30045–30072, 2024.
18. Rani, P., Verma, S., Yadav, S.P., Rai, B.K., Naruka, M.S., Kumar, D., Simulation of the lightweight blockchain technique based on privacy and security for healthcare data for the cloud system. *Int. J. E-Health Med. Commun. IJEHMC*, 13, 4, 1–15, 2022.

19. Mathworks, 2019, [Online], Available: Accessed: Nov. 07, 2024, [Online], Available: https://www.mathworks.com/help/vision/ref/trainacfobjectdetector.html.
20. Mahavar, V. and Juremalani, J., Literature Review on Traffic Control Systems Used Worldwide. *J. Emerg. Technol. Innov. Res.*, 5, 5, 77–79, 2018.
21. Khushi, Smart Control of Traffic Light System using Image Processing, in: *2017 International Conference on Current Trends in Computer, Electrical, Electronics and Communication (CTCEEC)*, IEEE, pp. 99–103, 2017, Accessed: Nov. 07, 2024, [Online], Available: https://ieeexplore.ieee.org/abstract/document/8454966/.
22. Rani, P. and Sharma, R., An experimental study of IEEE 802.11 n devices for vehicular networks with various propagation loss models, in: *International Conference on Signal Processing and Integrated Networks*, Springer, pp. 125–135, 2022.
23. Sharma, N., Sultana, H.P., Singh, R., Patil, S., Pareek, S., Image segmentation in Constrained IOT servers. *Procedia Comput. Sci.*, 165, 336–342, 2019.
24. Tahir, W., *et al.*, Enhanced Vehicle Detection Mechanism for Traffic Management in Smart Cities. *Wirel. Pers. Commun.*, 135, 4, 1925–1945, Apr. 2024, doi: 10.1007/s11277-023-10833-2.
25. Karami, Z. and Kashef, R., Smart transportation planning: Data, models, and algorithms. *Transp. Eng.*, 2, 100013, 2020.
26. Trivedi, J., Mandalapu, S., Dhara, D., Review paper on intelligent traffic control system using computer vision for smart city. *Int. J. Sci. Eng. Res.*, 8, 6, 14–17, 2017.
27. Chakraborty, P.S., Nair, P., Sinha, P.R., Behera, I.K., Real time optimized traffic management algorithm. *Int. J. Comput. Sci. Inf. Technol.*, 6, 4, 119, 2014.
28. Merrad, W., Rachedi, A., Busawon, K., Binns, R., A survey on smart traffic network control and optimization, in: *2016 International Conference Multidisciplinary Engineering Design Optimization (MEDO)*, IEEE, pp. 1–6, 2016, Accessed: Nov. 07, 2024, [Online], Available: https://ieeexplore.ieee.org/abstract/document/7746553/.
29. Nwaigwe, D.N., Amiara Chidiebere, A., Okwunze, C.F., Egege, C.C., Analytical Study of Causes, Effects and Remedies of Traffic Congestion in Nigeria: Case Study of Lagos State. *Int. J. Eng. Res. Adv. Technol.*, 5, 9, 11–19, 2019. Accessed: Nov. 07, 2024, [Online], Available: https://pdfs.semanticscholar.org/cb00/0dc5b7987d69c48cdbbed3afffb81a688e1c.pdf.
30. Roychowdhury, P. and Das, S., Automatic road traffic management system in a City. *Trends Transp. Eng. Appl.*, 1, 2, 38, 2014.
31. Loce, R.P., Bernal, E.A., Wu, W., Bala, R., Computer vision in roadway transportation systems: a survey. *J. Electron. Imaging*, 22, 4, 041121–041121, 2013.
32. Buch, N., Velastin, S.A., Orwell, J., A Review of Computer Vision Techniques for the Analysis of Urban Traffic. *IEEE Trans. Intell. Transp. Syst.*, 12, 3, 920–939, Sep. 2011, doi: 10.1109/TITS.2011.2119372.

33. Wang, Y., Zhang, D., Liu, Y., Dai, B., Lee, L.H., Enhancing transportation systems *via* deep learning: A survey. *Transp. Res. Part C Emerg. Technol.*, 99, 144–163, Feb. 2019, doi: 10.1016/j.trc.2018.12.004.
34. Sheth, A., History of Machine Learning, 2018, Available online: Accessed: Nov. 06, 2024, [Online], Available: https://medium.com/bloombench/history-of-machine-learning-7c9dc67857a5.
35. Low, R., Cheah, L., You, L., Commercial Vehicle Activity Prediction With Imbalanced Class Distribution Using a Hybrid Sampling and Gradient Boosting Approach. *IEEE Trans. Intell. Transp. Syst.*, 22, 3, 1401–1410, Mar. 2021, doi: 10.1109/TITS.2020.2970229.
36. You, L., Zhao, F., Cheah, L., Jeong, K., Zegras, C., Ben-Akiva, M., Future mobility sensing: An intelligent mobility data collection and visualization platform, in: *2018 21st International Conference on Intelligent Transportation Systems (ITSC)*, IEEE, pp. 2653–2658, 2018, Accessed: Nov. 06, 2024, [Online], Available: https://ieeexplore.ieee.org/abstract/document/8569697/.
37. Oladimeji, D., Gupta, K., Kose, N.A., Gundogan, K., Ge, L., Liang, F., Smart transportation: an overview of technologies and applications. *Sensors*, 23, 8, 3880, 2023.
38. Hou, Y., Edara, P., Sun, C., Traffic Flow Forecasting for Urban Work Zones. *IEEE Trans. Intell. Transp. Syst.*, 16, 4, 1761–1770, Aug. 2015, doi: 10.1109/TITS.2014.2371993.
39. Dogru, N. and Subasi, A., Traffic accident detection using random forest classifier, in: *2018 15th Learning and Technology Conference (L&T)*, IEEE, pp. 40–45, 2018, Accessed: Nov. 06, 2024, [Online], Available: https://ieeexplore.ieee.org/abstract/document/8368509/.
40. Behrisch, M., Bieker, L., Erdmann, J., Krajzewicz, D., SUMO–simulation of urban mobility: an overview. *Proceedings of SIMUL 2011, The Third International Conference on Advances in System Simulation*, ThinkMind, 2011, Accessed: Nov. 06, 2024, [Online], Available: https://elib.dlr.de/71460.
41. Li, P., Abdel-Aty, M., Cai, Q., Islam, Z., A Deep Learning Approach to Detect Real-Time Vehicle Maneuvers Based on Smartphone Sensors. *IEEE Trans. Intell. Transp. Syst.*, 23, 4, 3148–3157, Apr. 2022, doi: 10.1109/TITS.2020.3032055.
42. Keller, M., Haß, C., Seewald, A., Bertram, T., A model predictive approach to emergency maneuvers in critical traffic situations, in: *2015 IEEE 18th International Conference on Intelligent Transportation Systems*, IEEE, pp. 369–374, 2015, Accessed: Nov. 06, 2024, [Online], Available: https://ieeexplore.ieee.org/abstract/document/7313161/.
43. Yu, J., Chang, G.-L., Ho, H.W., Liu, Y., Variation based online travel time prediction using clustered neural networks, in: *2008 11th International IEEE Conference on Intelligent Transportation Systems*, IEEE, pp. 85–90, 2008,

Accessed: Nov. 06, 2024, [Online], Available: https://ieeexplore.ieee.org/abstract/document/4732594/.

44. Sarker, I.H., Machine Learning: Algorithms, Real-World Applications and Research Directions. *SN Comput. Sci.*, 2, 3, 160, May 2021, doi: 10.1007/s42979-021-00592-x.
45. Mohammadi, M. and Al-Fuqaha, A., Enabling Cognitive Smart Cities Using Big Data and Machine Learning: Approaches and Challenges. *IEEE Commun. Mag.*, 56, 2, 94–101, Feb. 2018, doi: 10.1109/MCOM.2018.1700298.

19

Exposing the Importance of Connected Lighting for Urban Sustainability

Zainab. R. Abdulsada[1], Kamaljeet Kaur[2*], Sapna Singh[3], Devendra Singh[4] and Mohammed H. Al-Farouni[5]

[1]College of Technical Engineering, National University of Science and Technology, Dhi Qar, Iraq
[2]Department of Computer Science Engineering, Chandigarh Engineering College, Chandigarh Group of Colleges, Jhanjeri, Mohali, Punjab, India
[3]Department of Electrical, Electronics & Electric Vehicle Engineering, NIMS Institute of Engineering & Technology, NIMS University Rajasthan, Jaipur, India
[4]Department of Computer Science & Engineering, Uttaranchal Institute of Technology, Uttaranchal University, Dehradun, Uttarakhand, India
[5]Department of Computer Science and Technology, The Islamic University, Najaf, Iraq

Abstract
Increasing urban populations and balancing development and environmental responsibility have made urban sustainability a global concern. Connected lighting is an important element of sustainable cities as they become more interconnected. An intelligent lighting system is digitally networked to continuously adapt in real-time, enhancing life quality, reducing energy consumption, and enhancing safety. In addition to energy savings, connected lighting systems reduce carbon emissions and can be integrated with other smart city technologies, as discussed in the chapter. It also introduces a new framework for measuring sustainability in connected lighting, taking into account economic, ecological, and social factors. The framework aims to guide urban planners and decision-makers towards long-term sustainability-aware lighting solutions.

Keywords: Connected lighting, energy savings, intelligent systems, smart city, sustainability framework, sustainable cities, urban planning, urban sustainability

**Corresponding author*: Kamaljeet2640.research@cgcjhanjeri.in.

Pethuru Raj, Sudesh Yadav, Manas Kumar Mishra, Satya Prakash Yadav and Victor Hugo C. de Albuquerque (eds.) Artificial Intelligence Technologies for Smart and Sustainable Urban Transportation: Integrated Platforms and Use Cases, (323–336) © 2026 Scrivener Publishing LLC

19.1 Introduction

Sustainability in cities has become a top priority on all agendas as cities struggle to accommodate urban populations while maintaining environmental resources. There are several common sustainability solutions, including renewable energy, waste management, and green architecture. Connected lighting is a vital but underemphasized piece of this urban puzzle. In addition to optimizing urban efficiency and energy savings, connected lighting improves the overall quality of life and enhances urban efficiency. In this chapter, we explore connected lighting's transformative impact on urban sustainability and highlight key areas where cities can leverage these systems to become more resilient. Globally, artificial lighting consumes about 19% of electricity, which produces about 1900 metric tons of CO_2 yearly [1]. As outdoor lighting requires high wattages to ensure public safety and traffic visibility, it consumes a substantial amount of energy. It is estimated that street lighting consumes 60% of a municipality's total electricity [2]. Public lighting inevitably represents a significant portion of public finances. A road lighting installation typically lasts 30-40 years, making it inefficient and costly [3]. Lighting systems like this will need to be replaced soon, preferably with long-term, cost-effective, and sustainable alternatives.

LED light emitting diode (LED) lamps, for example, are a new, energy-efficient and environmentally friendly product based on solid-state lighting (SSL) [4]. The energy savings associated with LED lighting could reduce CO_2 emissions in the future by saving lighting owners money and reducing mercury vapor or high-pressure sodium light sources [5]. By using LED lighting for energy-saving schemes, street and road lighting systems can become adaptable and intelligent, which further reduces energy consumption [6]. Based on energy consumption and carbon dioxide emissions, replacing old equipment does not appear to be a disadvantage with LED/SSL installations. Despite these advantages, LED/SSL lighting has several important disadvantages from a sustainable development perspective. Although LED lighting can save energy, it can also cause light pollution, ecological damage, and degradation of the environment [7–9]. Although LPS and HPS lights are different in terms of environmental friendliness and astronomical benefits, low-pressure sodium lights are not dimmable, and they have a low scotopic-to-photopic ratio [10], which implies that they are somewhat less visible. Generally, SSL lighting is more energy efficient and dimmable, which reduces lighting levels and improves S/P ratios and color rendering indexes [11]. LED lighting has been rapidly adopted in outdoor applications, but several important aspects, such as unwanted glare and rebound effects, have been overlooked. Despite this, street lighting remains one of LED technology's fastest-growing applications.

LED/SSL lighting has an important role to play in a long-term and sustainable lighting system, but its impact on outdoor applications needs to be fully evaluated and prioritized. LED lighting can be selected based on energy or economics, but choosing products that are suitable for different sustainability aspects is much more challenging. Several guidelines exist for enhancing social sustainability or reducing environmental impact. There is currently no sustainability indicator-based system for the evaluation of LED/SSL lighting systems, nor can existing transportation indicator systems be used for the evaluation of LED/SSL lighting systems [12, 13]. In addition to being difficult to apply, they are not applicable to lighting systems in general, so lighting systems cannot be used with European transportation and environmental indicator systems [14]. In addition to energy consumption, CO2 emissions, and price, other factors play a significant role in choosing the best LED/SSL product. The environmental and social impacts of outdoor lighting systems, which are designed to last for decades, are extremely concerning when aspects of sustainability are overlooked.

As part of a new framework for sustainability indicators, LED/SSL lighting systems are either used to evaluate a broad range of objectives or to highlight a particular aspect of interest. An ecological, economic, and social perspective is taken into consideration when examining LED/SSL lighting through a literature review. Indicators of sustainability are developed to facilitate prioritization when choosing LED/SSL products and ensure long-term goals are met when making short-term decisions regarding LED lighting systems. It is predicted that outdoor lighting research and development will benefit significantly from this framework since such a system does not exist at present. Moreover, the framework provides a resource for planners, decision-makers, developers, designers, and consumers interested in LED/SSL lighting outdoors. There are a number of types of additional research that should be conducted to gain a comprehensive understanding of the sustainability aspects of LED/SSL lighting, including environmental impacts, external costs, and social sustainability since LED/SSL lighting has not yet been extensively explored or developed.

19.2 Sustainability

To achieve sustainable development, three categories must be considered: normative, scientific, and strategic [15]. It is within this chapter that the normative perspective is defined as representing sustainable development's dimensions and key principles. Sustainability indicators are identified through science and are used to select and suggest indicators in this paper while planning and decision-making perspectives are used in the analysis. According to the normative perspective, sustainable development means developing in a way that meets the immediate needs without compromising those of future generations. The Brundtland report is widely accepted as defining sustainable development [16]. Sustainable development is also perceived as a process rather than a result, such as supporting decision-making, investment direction, and technological reorientation. During the 1992 UN Conference on Environment and Development, three dimensions of sustainable development were established (economic, social, and ecological), along with a fourth dimension (institutional. There is a possibility of overlap and interdependency between these dimensions.

Due to the fact that many outdoor lighting systems are installed along roads, street lighting is closely related to sustainable transportation. As a result of sustainable transportation, air pollution, climate change, noise pollution, habitat loss and biodiversity loss, hydrological impacts, and the depletion of non-renewable resources are all problems associated with the industry [17]. In addition to traffic congestion and infrastructure costs, there are costs associated with consumer goods, mobility barriers, and accidents. A number of social factors affect human well-being, including equity, health, accessibility barriers, aesthetics, and community cohesion [18, 19].

Social factors influence human well-being in a number of ways, including equity, accessibility, aesthetics, cohesion, and liveability [20, 21]. ALAN's global expansion has increased the intensity and color of night light, which is now spreading into previously unlit areas [22]. In an effort to detect light pollution, studies conducted over the last two decades have demonstrated that ALAN is an unintentional form of anthropogenic pollution [23] and that ELP has increased [24]. In some studies, it has been linked to improper management and application of artificial light sources and urbanization, leading to unneeded and unwanted emissions of ALAN [24, 25].

19.3 Transdisciplinary Framework for Urban Lighting Research: Actors, Framework, and Four Steps

For sustainable lighting applications to be achieved, four main domains need to be integrated: research, practice, production inputs, and policy initiatives. As a result of collaboration between these four pillars, data can be understood as knowledge. That knowledge can then be applied as standards, which will serve as guidelines for lighting equipment design and methodologies for designing night-time lighting schemes. To gain a better understanding of ALAN as a pollutant caused by applied lighting technology, the ULRTF envisions four domains and a variety of actors. There are various strategies and techniques associated with lighting, which makes it not a uniform domain. An illuminating engineer (e.g., a street light engineer) and a lighting designer (e.g., an architectural lighting designer or an urban lighting designer) will approach lighting applications differently. Using visible luminaires, a street lighting engineer creates homogeneous horizontal illumination for pedestrian paths and vehicular circulation, as defined by strict lighting specifications. Rather than focusing on luminaires, lighting design focuses on light on a surface. As well lighting designers combine a variety of expertise (architecture, urban planning, industrial design, landscape design, and perception psychology) to create light that shapes space, accentuates materials, creates gradients, and reveals shadows within a defined space and context as shown in Figure 19.1. They also improve the accessibility and comfort of night-time spaces.

The ALAN conference will bring together researchers and experts in lighting applications, with scientific findings assessing how light can be used to protect species and ecosystems sensitive to light [26]. Lighting equipment suppliers will participate in the conference. As part of the program, data on light pollution is collected, lighting technologies are developed, testing tools are developed to assess the effectiveness of lighting, and assistance with its application is provided [27–29]. The role of transdisciplinary teams cannot be ignored if we ignore the users, such as urban planners, landscape designers, policymakers, planning officers, sustainability consultants, and environmental lawyers. ALAN is applied to evaluate the balance between security, safety, and nocturnality across nightscapes while regulating fundamental aspects such as urban lighting masterplans [30–32].

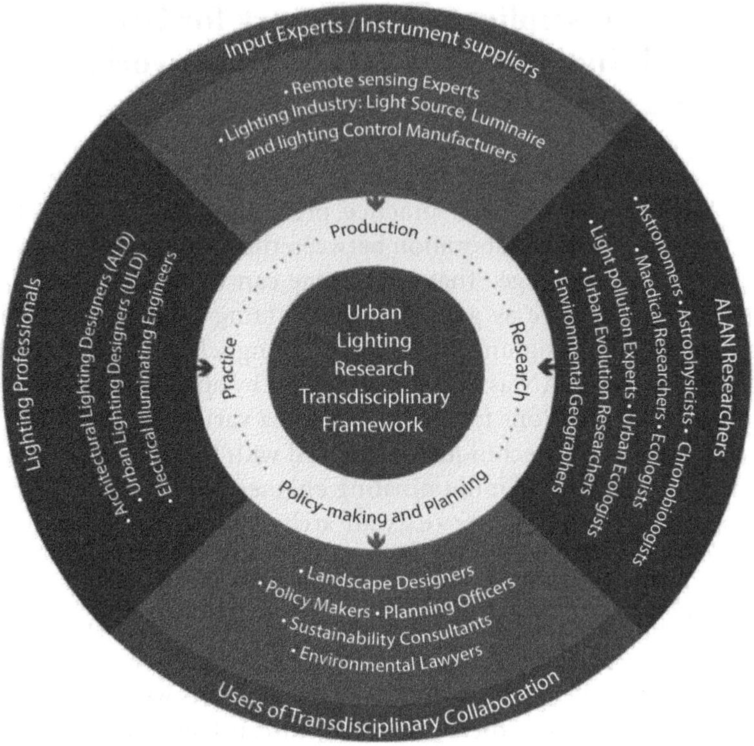

Figure 19.1 Research framework for urban lighting.

By participating in these venues, you can encounter a foreign field of expertise for the first time, acquire knowledge and jargon, establish preliminary social contacts at an acceptable level, form ULRTFs, and establish new platforms (such as conferences and technical committees). On these platforms, ULRTF would enhance communication and collaboration in an effort to promote lateral thinking strategies and problem-solving techniques rather than monologues that focus on linear and vertical thinking [33]. A collaborative framework can only be part of a process for aligning, structuring, and developing long- and short-term objectives that will be sustainable. By collaboratively developing crucial processes and steps, diverse knowledge and expertise can be spread across domains. Our goal is to facilitate the diffusion of diverse expertise and knowledge by creating cross-disciplinary connections between emerging approaches and procedures that are based on scientific practice curricula, as well as to create a platform for transdisciplinary professionals who wish to incorporate scientific knowledge into lighting practices.

Table 19.1 Lighting guides to urban lighting.

Steps	Category	Description
Step 1	Problem Definition	Establish a clear definition of the problem, define the topic(s) involved, and describe existing solutions;
		Identify and compare lighting practice, scientific, and scientific literature related to the defined problems;
		Research questions and hypotheses should be derived from problem-driven topics.
Step 2	Research Design Development	To conduct this research, you should determine what artificial lighting properties will be most appropriate to examine;
		Find out how artificial lighting is researched;
		Defining and evaluating the lighting parameters for future research is crucial;
		Knowledge should be shared and disseminated across disciplines;
		A description of the study's limitations should be provided.
Step 3	Conducting Research (Collecting and Analyzing Data)	For the identification and exchange of current standards, regulations, and guidelines, the collection, exchange, and interpretation of data, and the application of lighting control methods, lighting sources, and luminaires, there must be collaboration across organizational boundaries.
Step 4	Take Actions (Reporting Research Findings)	Transform research results into specific lighting publications and share them with lighting practitioners;
		Assist other team members in the writing of scientific papers;
		Conferences and seminars on lighting;
		Using the results of the research studies, develop guidelines and recommendations to improve existing lighting approaches.

By enhancing public safety, connected lighting systems reduce crime rates and create a sense of security. A dynamic lighting system that responds to pedestrian and vehicular movements improves public spaces' safety and accessibility. A pleasant urban ambience can also be created by adaptable lighting so that light is softened in residential zones and intensified in commercial zones. As a result of this dynamic adjustment, light pollution is reduced, residents' quality of life is improved, and their natural circadian rhythms are preserved. Lighting professionals are expected to be involved in urban lighting research in four stages, as shown in Table 19.1.

19.4 Understanding Connected Lighting Systems

The term "connected lighting" refers to lighting systems that incorporate sensors, controls, and network communication capabilities, enabling lights to adapt to real-time data and respond to specific environmental conditions. Lighting systems with connected technology are capable of optimizing lighting levels, reducing waste, and integrating with other smart urban infrastructure systems [34].

Sensors could be used to control the brightness of a streetlight based on pedestrian presence or daylight levels, for example, as well as integrating connected lighting systems into a digital network, data can be collected, analyzed, and used for urban infrastructure improvements [24].

19.5 Energy Efficiency and Reduced Carbon Emissions

It can drastically reduce energy consumption with connected lighting, which is a significant advantage. In traditional street lighting, no matter whether you need full illumination, the output is fixed. It is feasible to dim or brighten connected lighting systems in response to real-time conditions, effectively reducing electric consumption [35].

Los Angeles, for instance, installed smart LED streetlights that reduce energy usage by as much as 63%. There is enormous potential for energy savings across global cities due to street lighting, which accounts for 40% of a city's energy use. In addition, connected lighting systems reduce carbon emissions by reducing energy usage, an essential step in slowing global warming [36].

a. Integrating with Other Urban Systems

In smart cities, connected lighting is often a fundamental component, enabling it to work synergistically with traffic control, environmental monitoring, and public transportation systems. Streetlights can be integrated with traffic sensors, allowing pedestrians to be warned when vehicles are approaching, increasing safety [37]. Connected lighting systems can also provide city planners with information on peak pedestrian hours, pollution levels, and other patterns, making urban planning more effective. Cities like Barcelona have integrated connected lighting into traffic management and emergency response systems to improve efficiency and reduce delays [38].

b. Economic and Environmental Benefits

It is not only possible to save energy now, but it is also possible to save money and the environment in the long run. These systems, for instance, can predict when repairs are required rather than relying on reactive maintenance schedules to reduce maintenance costs. As a result, labor costs are minimized, operational costs are reduced, and energy waste caused by faulty lighting is prevented [39].

From an environmental standpoint, connected lighting minimizes light pollution and reduces greenhouse gas emissions. Light pollution disrupts local ecosystems and negatively impacts wildlife, particularly nocturnal animals and migratory birds, in urban areas. Connected lighting can be implemented to preserve urban biodiversity and reduce the ecological footprint of cities [40].

c. Overcoming Barriers and Implementing Solutions

Cities face challenges in implementing connected lighting systems, including high initial costs, privacy concerns, and integrating them with existing infrastructure [41]. Installing such a system requires a significant amount of capital, though the long-term savings are substantial. Governments and local municipalities must explore financing mechanisms, such as public-private partnerships and grants, to offset the initial costs. Data security and privacy are another challenge, particularly since connected lighting integrates with public and private networks. For these systems to be trusted by the public, cities must implement strong data protection measures [42]. The city can also adopt a phased approach to connected lighting, beginning with high-priority zones or pilot programs in a limited number of neighborhoods. As a result of this gradual implementation, cities can adapt to challenges, refine their systems, and gradually reap the benefits of connected lighting.

Throughout the 20th century, artificial lighting has been used by cities and towns across the world to light up the night. Illuminating the nightscape from a human-centric perspective promotes security and safety, underscores past and present architecture and history, enables a 24/7 lifestyle, and boosts tourism and the economy [21, 43, 44]. The use of artificial light at night (ALAN) improves visibility by illuminating pedestrian and traffic routes, buildings, landscape elements, networks, and infrastructure so that our society and users can function efficiently. Our study addressed the following challenges by analyzing 32 open-source and proprietary applications. Our findings have been organized into categories based on mitigation strategies, ICT functionalities, and innovation mechanisms. As a result of our analysis, we identified the underlying trends in smart city applications, as well as the gaps in research and development. It is important to note that the findings can benefit a variety of stakeholders, including users, application developers, and policymakers. It has been observed that a number of factors have contributed to the fragmentation of the urban smart and sustainable landscape, both technologically and policy-wise. The "zero vision" offers a host of opportunities that have yet to be explored, including smart, sustainable development. To maximize these opportunities, however, different and innovative approaches need to be taken.

19.6 Conclusion

A connected lighting system can reduce energy consumption, carbon emissions, and light pollution in urban areas, which contributes to sustainable cities. Data privacy concerns and high initial costs are among the challenges associated with implementing these systems. A phased deployment approach, partnerships between public and private organizations, and robust data security are necessary to address these barriers. Establishing sustainability indicators will support the integration of connected lighting systems with ecological, economic, and social sustainability goals. Using this framework, lighting solutions can be tailored to meet current needs as well as future sustainability requirements. With this integrated approach to lighting, cities can create a sustainable future while simultaneously improving their quality of life.

References

1. Jägerbrand, A., New Framework of Sustainable Indicators for Outdoor LED (Light Emitting Diodes) Lighting and SSL (Solid State Lighting). *Sustainability*, 7, 1, 1028–1063, Jan. 2015, doi: 10.3390/su7011028.
2. Fiaschi, D., Bandinell, R., Conti, S., A case study for energy issues of public buildings and utilities in a small municipality: Investigation of possible improvements and integration with renewables. *Appl. Energy*, 97, 101–114, Sep. 2012, doi: 10.1016/j.apenergy.2012.03.008.
3. Gaston, K.J., A green light for efficiency. *Nature*, 497, 7451, 560–561, May 2013, doi: 10.1038/497560a.
4. De Almeida, A., Santos, B., Paolo, B., Quicheron, M., Solid state lighting review – Potential and challenges in Europe. *Renew. Sustain. Energy Rev.*, 34, 30–48, Jun. 2014, doi: 10.1016/j.rser.2014.02.029.
5. Miyairi, K., Endo, K., Matsuda, S., Kokuryo, J., Study of CO_2 Emission in Relation to Economic Efficiency in LED Street Lighting. *J. Illum. Engng. Inst. Jpn.*, 97, 2, 57–64, 2013, doi: 10.2150/jieij.97.57.
6. Gong, Y. and Han, P., Research on Energy-Saving Scheme Based on LED Street Lamp Management-System. *AMM*, 448–453, 2850–2855, Oct. 2013, doi: 10.4028/www.scientific.net/AMM.448-453.2850.
7. Falchi, F., Cinzano, P., Elvidge, C.D., Keith, D.M., Haim, A., Limiting the impact of light pollution on human health, environment and stellar visibility. *J. Environ. Manage.*, 92, 10, 2714–2722, Oct. 2011, doi: 10.1016/j.jenvman.2011.06.029.
8. Bierman, A., Will switching to LED outdoor lighting increase sky glow? *Light. Res. Technol.*, 44, 4, 449–458, Dec. 2012, doi: 10.1177/1477153512437147.
9. Lyytimäki, J., Nature's nocturnal services: Light pollution as a non-recognised challenge for ecosystem services research and management. *Ecosyst. Serv.*, 3, e44–e48, Mar. 2013, doi: 10.1016/j.ecoser.2012.12.001.
10. Boyce, P.R., *Lighting for Driving: Roads, Vehicles, Signs, and Signals*, CRC Press, Boca Raton, 2008, doi: 10.1201/9781420008159.
11. Fotios, S., LRT Digest 1 Maintaining brightness while saving energy in residential roads. *Light. Res. Technol.*, 45, 1, 7–21, Feb. 2013, doi: 10.1177/1477153512464141.
12. Hall: Understanding and applying the concept of sustainab.. - Google Scholar, Accessed: Nov. 13, 2024. [Online]. Available: https://scholar.google.com/scholar_lookup?title=Understanding+and+Applying+the+Concept+of+-Sustainable+Development+to+Transportation+Planning+and+Decision-Making+in+the+U.S.&author=Hall,+R.P.&publication_year=2006.
13. Gudmundsson, H., Harmer, C., Hewitt, A., Jensen, A.V., Sustainability Definitions for NRAs – Framework Part 1: Report v2.0. ERA-NET Road, Report, Apr. 2013.

14. Rani, P. and Sharma, R., Intelligent transportation system for internet of vehicles based vehicular networks for smart cities. *Comput. Electr. Eng.*, 105, 108543, 2023.
15. Rani, P. and Sharma, R., IMFOCA-IOV: Intelligent Moth Flame Optimization based Clustering Algorithm for Internet of Vehicle, in: *2023 14th International Conference on Computing Communication and Networking Technologies (ICCCNT)*, IEEE, pp. 1–6, 2023, Accessed: Nov. 09, 2024. [Online]. Available: https://ieeexplore.ieee.org/abstract/document/10307646/.
16. Brundtland: Our common future - Google Scholar, Accessed: Nov. 13, 2024. [Online]. Available: https://scholar.google.com/scholar_lookup?title=Our+Common+Future&author=World+Commission+on+Environment+and+Development&publication_year=1987.
17. Rani, S., Talwar, R., Malhotra, J., Ahmed, S.H., Sarkar, M., Song, H., A novel scheme for an energy efficient Internet of Things based on wireless sensor networks. *Sensors*, 15, 11, 28603–28626, 2015.
18. Indicators of environmental sustainability in transport: An interdisciplinary approach to methods, in: *RECHERCHES*, European Commission, Bron, France, 2010.
19. Yadav, S.P., Jindal, M., Rani, P., de Albuquerque, V.H.C., dos Santos Nascimento, C., Kumar, M., An improved deep learning-based optimal object detection system from images. *Multimed. Tools Appl.*, 83, 10, 30045–30072, 2024.
20. Edensor, T., Introduction to geographies of darkness. *Cult. Geographies*, 22, 4, 559–565, Oct. 2015, doi: 10.1177/1474474015604807.
21. Zielinska-Dabkowska: Historic urban settings, LED.. - Google Scholar, Accessed: Nov. 13, 2024. [Online]. Available: https://scholar.google.com/scholar_lookup?title=Historic+Urban+Settings,+LED+Illumination+and+its+Impact+on+Nighttime+Perception,+Visual+Appearance,+and+Cultural+Heritage+Identity&conference=Proceedings+of+the+5th+SGEM+International+Multidisciplinary+Scientific+Conferences+on+Social+Sciences+and+Arts,+SGEM2018&author=Zielinska-Dabkowska,+K.M.&author=Xavia,+K.&publication_year=2008&pages=277%E2%80%93292.
22. Bará, S., Anthropogenic disruption of the night sky darkness in urban and rural areas. *R. Soc. Open Sci.*, 3, 10, 160541, Oct. 2016, doi: 10.1098/rsos.160541.
23. Riegel, K.W., Light Pollution: Outdoor lighting is a growing threat to astronomy. *Science*, 179, 4080, 1285–1291, Mar. 1973, doi: 10.1126/science.179.4080.1285.
24. Longcore, T. and Rich, C., Ecological light pollution. *Front. Ecol. Environ.*, 2, 4, 191–198, May 2004, doi: 10.1890/1540-9295(2004)002[0191:ELP]2.0.CO;2.
25. An Overview of the Cognitive and Biological Effects of City Nighttime Illumination Including a London Case Study. *CCA*, 2018, 1, 1–9, Oct. 2018, doi: 10.33797/CCA18.01.02.

26. Rani, P., Verma, S., Yadav, S.P., Rai, B.K., Naruka, M.S., Kumar, D., Simulation of the lightweight blockchain technique based on privacy and security for healthcare data for the cloud system. *Int. J. E-Health Med. Commun. (IJEHMC)*, 13, 4, 1–15, 2022.
27. Kyba, C., *et al.*, High-Resolution Imagery of Earth at Night: New Sources, Opportunities and Challenges. *Remote Sens.*, 7, 1, 1–23, Dec. 2014, doi: 10.3390/rs70100001.
28. Jechow, A., Kyba, C.C.M., Hölker, F., Mapping the brightness and color of urban to rural skyglow with all-sky photometry. *J. Quant. Spectrosc. Radiat. Transfer*, 250, 106988, Jul. 2020, doi: 10.1016/j.jqsrt.2020.106988.
29. Jechow, A., Kyba, C.C.M., Hölker, F., Beyond All-Sky: Assessing Ecological Light Pollution Using Multi-Spectral Full-Sphere Fisheye Lens Imaging. *J. Imaging*, 5, 4, 46, Apr. 2019, doi: 10.3390/jimaging5040046.
30. Schroer, S., Huggins, B.J., Azam, C., Hölker, F., Working with Inadequate Tools: Legislative Shortcomings in Protection against Ecological Effects of Artificial Light at Night. *Sustainability*, 12, 6, 2551, Mar. 2020, doi: 10.3390/su12062551.
31. Meier, J., By Night. An investigation into practices, policies and perspectives on artificial outdoor lighting, 2019, doi: 10.14279/DEPOSITONCE-8285.
32. Marchant, P., Hale, J.D., Sadler, J.P., Does changing to brighter road lighting improve road safety? Multilevel longitudinal analysis of road traffic collision frequency during the relighting of a UK city. *J. Epidemiol. Community Health*, 74, 5, 467–472, May 2020, doi: 10.1136/jech-2019-212208.
33. Serious creativity - ProQuest, Accessed: Nov. 13, 2024. [Online]. Available: https://www.proquest.com/openview/b867602babfb901628af868f33ea6603/1?pq-origsite=gscholar&cbl=37083.
34. Sutton, P.C., A scale-adjusted measure of 'Urban sprawl' using nighttime satellite imagery. *Remote Sens. Environ.*, 86, 3, 353–369, Aug. 2003, doi: 10.1016/S0034-4257(03)00078-6.
35. Di Stefano, J., Energy efficiency and the environment: the potential for energy efficient lighting to save energy and reduce carbon dioxide emissions at Melbourne University, Australia. *Energy*, 25, 9, 823–839, Sep. 2000, doi: 10.1016/S0360-5442(00)00015-3.
36. Li, J. and Colombier, M., Managing carbon emissions in China through building energy efficiency. *J. Environ. Manage.*, 90, 8, 2436–2447, Jun. 2009, doi: 10.1016/j.jenvman.2008.12.015.
37. Blaschke, T., Hay, G.J., Weng, Q., Resch, B., Collective Sensing: Integrating Geospatial Technologies to Understand Urban Systems—An Overview. *Remote Sens.*, 3, 8, 1743–1776, Aug. 2011, doi: 10.3390/rs3081743.
38. White, R. and Engelen, G., High-resolution integrated modelling of the spatial dynamics of urban and regional systems. *Comput. Environ. Urban Syst.*, 24, 5, 383–400, Sep. 2000, doi: 10.1016/S0198-9715(00)00012-0.
39. Pimentel, D., *et al.*, Economic and Environmental Benefits of Biodiversity. *BioScience*, 47, 11, 747–757, Dec. 1997, doi: 10.2307/1313097.

40. Chertow, M.R. and Lombardi, D.R., Quantifying Economic and Environmental Benefits of Co-Located Firms. *Environ. Sci. Technol.*, 39, 17, 6535–6541, Sep. 2005, doi: 10.1021/es050050+.
41. Satterfield, M.B., *et al.*, Overcoming Nontechnical Barriers to the Implementation of Sustainable Solutions in Industry. *Environ. Sci. Technol.*, 43, 12, 4221–4226, Jun. 2009, doi: 10.1021/es802980j.
42. Haynes-Maslow, L., Osborne, I., Jilcott Pitts, S.B., Best Practices and Innovative Solutions to Overcome Barriers to Delivering Policy, Systems and Environmental Changes in Rural Communities. *Nutrients*, 10, 8, 1012, Aug. 2018, doi: 10.3390/nu10081012.
43. Mansfield, K., Architectural lighting design: A research review over 50 years. *Light. Res. Technol.*, 50, 1, 80–97, Jan. 2018, doi: 10.1177/1477153517736803.
44. Backmatter, in: *Light for Cities*, pp. 158–168, DE GRUYTER, Berlin, Boston, Birkhäuser, 2007, doi: 10.1007/978-3-7643-8263-6_bm.

20

Responsible and Green AI for Environment Sustainability

Kunal Gagneja[1]*, Sapna Singh[2], Zainab. R. Abdulsada[3], Shivani Pant[4] and Rami Riad Hussien[5]

[1]*Department of Computer Science Engineering, Chandigarh Engineering College, Chandigarh Group of Colleges, Jhanjeri, Mohali, Punjab, India*
[2]*Department of Electrical, Electronics & Electric Vehicle Engineering, NIMS Institute of Engineering & Technology, NIMS University Rajasthan, Jaipur, India*
[3]*College of Technical Engineering, National University of Science and Technology, Dhi Qar, Iraq*
[4]*Uttaranchal Institute of Technology, Uttaranchal University, Uttarakhand, India*
[5]*College of Technical Engineering, The Islamic University, Najaf, Iraq*

Abstract
Throughout the chapter, the need for responsible and green implementation of artificial intelligence (AI) is discussed, emphasising the need for AI practices that are both ethical and sustainable. The complex nature of AI models has resulted in increased energy, water, and resource requirements, creating a concern for the environment. E-waste generation, energy consumption, and carbon footprint are some of the key environmental impacts. Throughout the chapter, strategies including optimising algorithms, using renewable energy, and promoting efficient computing infrastructure are highlighted as means of ensuring AI's sustainability both as an input and as an output. This paper discusses how AI can be used to optimise energy consumption, predict climate, improve mobility, and conserve biodiversity. The chapter emphasises the importance of balancing technological advances with ecological responsibilities when it comes to sustainable AI.

Keywords: Artificial intelligence, efficient computing, green AI, mobility improvement, sustainable AI, ecological responsibility

**Corresponding author*: kunal2848.research@cgcjhanjeri.in

Pethuru Raj, Sudesh Yadav, Manas Kumar Mishra, Satya Prakash Yadav and Victor Hugo C. de Albuquerque (eds.) Artificial Intelligence Technologies for Smart and Sustainable Urban Transportation: Integrated Platforms and Use Cases, (337–354) © 2026 Scrivener Publishing LLC

20.1 Introduction

Despite AI's rapid transformation of industries and shaping the future, the exponential growth of AI-driven technologies is also raising concerns about the environment. A complex AI system requires a significant amount of energy to train and operate [1]. AI practices that are responsible and environmentally friendly can assist with sustainable environmental management. A number of key aspects of AI have a negative effect on the environment, which we will discuss, as well as methods as to how AI can be made more sustainable, as well as examples of AI that is designed specifically to protect the environment. A variety of industries have been transformed over the past few years by artificial intelligence and machine learning, enhancing efficiency, accuracy, and productivity in sectors such as healthcare, finance, transportation, education, and entertainment. to improve performance, ML models have become more complex, thus requiring estimates of more parameters [2]. Despite their impressive nature, these advances come at a cost, as training and running these models are very time and resource-intensive. It is estimated that ML training requires enormous quantities of computational power, energy, and water to cool the large data centres that house huge amounts of training data, as shown in the following Figure 20.1 [3, 4].

The number of data demands and hyperparameters has increased exponentially over the past decade, but the accuracy of models has not improved. In spite of this, AI has great potential to facilitate countries'

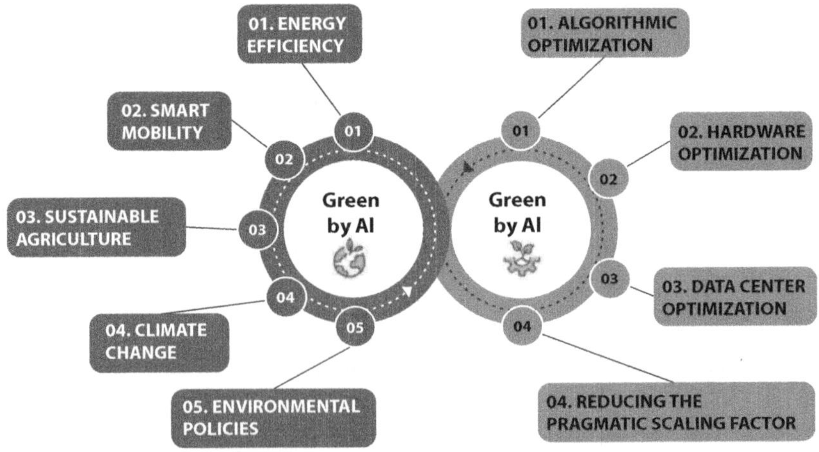

Figure 20.1 An overview of green-by and green-in algorithms.

transition to sustainable practices, thanks to its emphasis on sustainable and efficient solutions. To ensure model results are accurate and transparent and to reflect energy and water usage, sustainability measures must be taken [5]. Approximately 30% of the world's energy consumption will be accounted for by this sector by 2030. This trend is further aggravated by ChatGPT (with GPT-4 as the backbone model), which consumes a great deal of energy [6]. Based on estimates by some authors [3], training GPT-3 on a database of 500 billion words would require 1287 MWh of electricity and 10,000 computer chips, which is approximately the energy needed to run 121 homes a year for a year.

For the USA, powering 121 homes for a year would require 1287 megawatt-hours of electricity and 10,000 chips, according to some researchers [3]. As a result of the training, 550 tons of carbon dioxide was generated, equalling 33 flights between Australia and the United Kingdom. Compared to GPT-3, GPT-4 was trained on 570 times as many parameters, so it required even more energy. There is a cost associated with these systems, both in terms of training and in terms of their use. The GPT-3 estimated that 590 million people had access to electricity in January 2023, which translates to nearly 175,000 individuals using electricity [7]. Additionally, ChatGPT queries consume 260.42 MWh per day, the equivalent amount of energy that a 5 W LED bulb consumes in an hour and a half.

Machine learning (ML) algorithms often require significant data processing and computational power, leading to substantial consumption of energy, water, and the emission of greenhouse gases (GHG). The goal of green AI is to minimise its environmental impacts through the optimisation of algorithms, the improvement of hardware efficiency, and the adoption of sustainable data management practices. Using cloud centres and mobile devices, green AI offers energy-efficient solutions characterised by lower carbon footprints, finer data quality, small models, and minimal computational complexity [8]. As a result, it adds social sustainability to its advantages in providing clear, logical decision-making processes [9].

20.2 AI and the Environment

Machine learning systems, particularly large-scale ones, are computationally intensive. Deep learning models can be trained with more energy than most vehicles will emit in their lifetime. The following areas have a significant impact on the environment:

- **Energy Consumption:** The need for huge data centres makes training large models, like GPT-4, and creating image networks, like DALL-E, very energy intensive. Data centres use processors and cooling systems that consume non-renewable energy, contributing to climate change [10].
- **Carbon Footprint:** As AI models run on fossil fuel-powered infrastructure, they generate CO2 emissions throughout their lifetime, from collecting data and storing it to training and deploying it. As a result of research showing that training a large language model can generate more CO2 than several cars over its lifetime, responsible AI must strive to reduce its carbon footprint [11].
- **E-Waste Generation:** In response to the rapid growth in AI hardware demand, electronic waste (e-waste) is increasing. GPUs, custom AI chips, and other high-performance hardware become obsolete as newer and faster technology emerges. There is often difficulty recycling these components, and they end up in landfills, where they release harmful chemicals [12].

Large companies primarily invest in ML development since they possess the financial, human, and computational resources necessary to do so. As defined by the literature, green algorithms maximise energy efficiency and reduce environmental impact while supporting the implementation of AI to tackle a range of environmental challenges. As a result, two types of algorithms are considered: those that are designed to be green, i.e., energy-efficient in training and usage, and those that are specifically designed to overcome environmental obstacles. Examples of climate change-related policies include the Paris Climate Agreement, the Sustainable Development Goals (SDGs), and the European Green Deal [13, 14]. Artificial intelligence is often distinguished from green-by-AI.

20.2.1 Green-by AI

With AI, the Green Deal can be realised while simultaneously minimising its carbon footprint. It is possible to reduce greenhouse gas emissions by leveraging AI across a broad range of sectors, from energy production to agriculture to land use to biodiversity management to communications to transportation. Furthermore, AI's growing utility makes it possible to predict, adapt, and manage climate change effectively [7]. An analysis of AI's impact on the UN's Sustainable Development Goals revealed that 134 could be enabled, while 59 could be inhibited. Listed below are the main areas where artificial intelligence can enhance sustainability.

a. Energy efficiency

Increasing energy efficiency is vital to becoming a more sustainable society. Smart grids use artificial intelligence to align regional power generation with local demand, thereby reducing overall electricity consumption. A smart grid, on the other hand, allows for electricity flows to be controlled in magnitude and direction, as opposed to traditional grids. Supply and demand are always in balance, which enhances efficiency.

Machine learning techniques are crucial to integrating renewable energy seamlessly into the grid. A growing number of researchers are studying the forecasting of renewable energy due to its variability. Optimising power management and improving service quality require accurate forecasting of solar and wind power. Forecasting wind energy is typically done using neural networks [15], and forecasting solar energy usually involves hybrid approaches consisting of NNs, recurrent neural networks, and short-term memory networks [16, 17]. In the context of real-time data analysis, historical data analysis, and outputs from physical models, artificial intelligence tools play an essential role in successful development. As smart grids grow more complex and uncertain and collect an enormous amount of data in real-time, AI-enabled tools become increasingly important.

It is also possible to make structures more energy-efficient through green AI. Sensor data from smart buildings can be analysed using machine learning models [18, 19] and cities [20, 21] by optimising HVAC, lighting, and energy use patterns, significant energy savings can be achieved, and carbon emissions can be minimised [22, 23].

b. Smart mobility

Artificial intelligence is vital to the development of smart mobility and the improvement of traditional transportation systems' efficiency, responsiveness, and sustainability. There is an escalating problem of congestion and pollution in large cities as the urban population continues to grow. It is possible to alleviate these problems through smart cities, fostering economic growth and improving residents' overall quality of life.

With the help of artificial intelligence, traffic patterns can be predicted, and routes can be optimised in real-time. The prediction of traffic is based on a number of models, including feature-based, Gaussian process, and state-space models, as well as models based on deep learning [24–26]. Dynamic scheduling, powered by artificial intelligence, can improve public transportation as well.

c. Climate change

A wide range of sectors are dependent on AI to mitigate and adapt to climate change [27]. The production of electricity contributes approximately 25% to greenhouse gas emissions caused by human activity, for example, so machine learning techniques are used to forecast electricity demand and supply. Solar and wind power availability were forecasted using a variety of forecasting methods [28, 29]. To analyse this data, supervised machine learning, fuzzy logic, and hybrid physical models are used to combine historical data, outputs of physical models, images, and videos. Quantum computing could speed up the development of large-scale technologies as well, such as solar panels and batteries [30].

20.2.2 Green-in AI

There is a vast amount of potential for innovation in the field of sustainability with AI's multifaceted applications. To make AI more sustainable, it must not consume too much energy; otherwise, it will become counterproductive. Artificial intelligence must be designed with efficiency in mind if it is to function as an energy-saving tool. As a result, we must leverage AI solutions that optimise energy consumption in applications while being naturally low-energy consumers. The key to AI's sustainability is striking the right balance between addressing the problems it aims to solve and not causing them to worsen. A discussion about red versus green artificial intelligence is at the centre of this debate [31]. It emerged from a study of 60 papers presented at the most prestigious conferences that accuracy was more important than efficiency. It led to the concept of red AI, which uses massive amounts of computing power to achieve better results. According to the study, 75% to 90% of papers prioritised accuracy over efficiency. In contrast to Moore's Law, which predicts a double in processing power every two years in large ML models, a 2018 study shows that processing power has doubled every 3.4 months since 2012 [32].

Although ML algorithms consume a significant amount of energy, they shouldn't be viewed as a significant problem or a deterrent to innovation. Ideally, green-in AI reduces or avoids computational costs while producing new results. There are several approaches to reducing computation costs in the literature, not only from a software perspective but also from a hardware perspective. A number of promising approaches are described below.

d. What is Sustainable AI?

There is still a lot to learn about 'sustainable AI' since it is still in its infancy. Sustainable AI is a research area that examines both the technology of AI

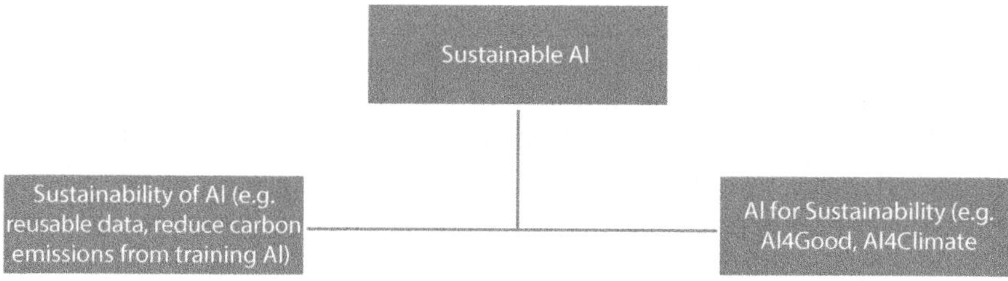

Figure 20.2 The environmental impact of AI.

(the hardware used to power AI, the methods used to train AI, and the data it processes) and its applications while also addressing concerns about AI sustainability and/or sustainable development. Furthermore, sustainable AI should consider not just how AI is implemented or used but how it is designed, trained, developed, validated, re-tuned, and implemented and used.

AI for sustainability versus AI as sustainability (see Figure 20.2) are critical distinctions under this umbrella term. AI for sustainability is led by the well-known non-profit organisation "AI4Good". During this branch, we will examine how artificial intelligence (AI) can be applied to sustainable development in some way, for instance, using machine learning (ML) to achieve the United Nations Sustainable Development Goals (SDGs).

e. Environmental sustainability of AI

As Coeckelbergh points out, we should use AI "to deal with problems related to the environment and climate" [33, 34]. The AI I'm describing here could best be described as AI for sustainability, based on my earlier distinction between sustainable AI and AI for sustainability. I agree with you completely, but I also think AI must be sustainable. There is a fundamental shift in framing since AI for Climate and AI4Good cannot be discussed without also discussing how the development of a particular AI model will affect the environment.

Only a few people are studying and developing artificial intelligence models, and they are already highlighting areas that need further research. It is argued in a 2019 paper [30], that DL models for NLP are both financially and environmentally costly. A financial cost is a result of hardware, electricity, or cloud computing time (which raises ethical questions about who has access to such hardware and other costs.) While a modern tensor processing device has a carbon footprint that contributes to environmental costs. The hardware for training such models consumes a lot of energy since they are trained over some time, such as weeks or months [35].

f. Towards Sustainable AI

As noted by [36], consumers and industry need to change their habits to mitigate climate change. A number of environmental activists, such as Greta Thunburg, are tirelessly fighting for political reform to repair climate change's damage to the planet. It is catching the attention of politicians, and they are taking action. The US Climate Accord Footnote has been re-joined by the European Commission, while the European Commission has incorporated the European Green Deal Footnote. Many policy and industry decisions are based on environmental resilience, so why shouldn't that be the core of every AI ethics discussion? Large tech companies are increasingly using computational power and storing data in data centres across the globe, which requires energy and cooling systems to keep up with their growing demand for data.

In 2021, it will be impossible to deny that any significant global production of products (or innovations) affects the environment in some way. Biodiversity has been demonstrated to be significantly affected by mass farming. Water supplies have been negatively affected by mass clothing production, and electronic waste can leach chemicals and poison into the environment when it is dumped (often in poorer areas). If industry and policymakers are right, AI/ML will have a profound impact across industries. Currently, we see applications for assisting medical practitioners with diagnosis, supporting legal officials in delivering judgements, and assisting human resources with recruitment. There is no end to the list. Therefore, the environmental impact of this technology cannot be ignored. AI must consider its environmental impact.

20.3 Principles of Responsible AI

As part of responsible AI, AI technologies are developed ethically and sustainably while minimising their environmental impact and maximising their social and economic benefits. Key principles include:

- **Transparency:** The goal of transparent AI is to ensure users are aware of how models make decisions by explaining and understanding them. In the context of conservation and environmental monitoring, trust and clarity can contribute to more effective policies and user compliance.
- **Efficiency:** The purpose of efficiency-focused AI is to minimise energy consumption by optimising algorithms and the

use of data. Developing streamlined, non-intensive models can reduce both energy use and carbon emissions.

- **Fairness and Inclusivity:** Climate change can have a big impact on marginalised groups as well, so AI needs to take into account the diverse groups that are impacted by environmental changes. It is possible to create more equitable and sustainable solutions by using responsible AI models that incorporate these perspectives.
- **Accountability:** It is important to hold AI developers responsible for the impact their models have on the environment. Using standardised carbon reporting and lifecycle assessments, we can better gauge AI's true environmental impact.

20.4 Sustainable AI for Human and Planetary Flourishing

A number of human activities have negatively impacted the environment, particularly over the past five decades, including unsustainable population growth, urbanisation, industrialisation, and consumerism [37, 38]. Among the most catastrophic threats humanity faces today are those listed in [39].

- Earth system pollution, including atmospheric and oceanic pollution;
- Ecology is collapsing, and biodiversity is disappearing;
- Natural resources are declining, especially water;
- Changes in climate caused by human activities;
- Overpopulation of humans on Earth;
- Deficiencies in understanding and acting preventatively on these risks on a national and global scale;
- Weapons of mass destruction, including nuclear weapons; pandemics of infectious diseases that are untreatable;
- Lack of nutrition and rising food insecurity;
- Technological advances that is powerful and uncontrollable.

A key threat is being able to understand and address these risks in a preventative manner. Neither governments nor the public are capable of understanding and acting against threats that will precipitate catastrophes or at least lead to them [40, 41]. The government and the public are unable to understand or respond to threats that will lead to catastrophes in the near future [40, 41]. Despite targeting sustainability, AI solutions fail due to their focus on making our cities more efficient and productive rather than addressing anthropogenic global threats [42, 43]. Humanity flourished in the Holocene because the planet's climate and ecosystem were ideal during that time [44]. Anthropocene survival depends on artificial intelligence, according to recent research [45].

a. Green Sensing, Communications and Computing

Our discussions on policy and high-level green AI issues have been followed by a discussion on how digital infrastructure for green AI would be developed. In this discussion, we will provide a comprehensive overview of these infrastructure issues as well as high-level issues so different communities involved in infrastructure and policy research can gain a better understanding of these cross-disciplinary issues and collaborate to find global optimum solutions. A number of technologies have developed and converged since the advent of the Internet of Things (IoT), including miniaturisation of sensors, data-driven methods, AI-driven optimisations, cloud computing, fog computing, and edge computing, resulting in greater integration between sensors, communication, and computing. A collaborative and distributed nature is inherent in smart applications and systems, which is causing a convergence of ecosystems of smart applications and systems.

We must make smart applications and systems more effective at engaging with, sensing, acting on, analysing, and making timely, effective, and sustainable environmental decisions [46]. In conjunction with the increasing number of IoT devices, preprocessing, storing, and processing of data will also increase. With AI, the brain of the smartness is provided; that is, it is responsible for analysing and making decisions. Despite requiring significant computing resources (storage, communication, and processing), artificial intelligence (AI) can improve infrastructure efficiency by analysing and optimising the entire ecosystem.

- **Green AI Practices:** Efforts are being made to minimise the environmental impact of artificial intelligence by optimising the entire lifecycle of an AI system, from the collection of data to the deployment of models. The following are some practices that can be used to make AI more environmentally friendly.
- **Model Efficiency and Optimisation:** One of the main ways to reduce AI's carbon footprint is to design more efficient

algorithms. Techniques like pruning, quantisation, and knowledge distillation allow models to perform complex tasks while consuming less power [47].

- **Model Pruning:** As a result of pruning, the model's size is reduced without performance being compromised by the removal of unnecessary weights. Computing costs are reduced, and energy consumption is reduced with this approach.
- **Quantisation:** By reducing calculation precision, quantisation enables a model to run faster and use less energy. In addition to being compatible with mobile devices and edge computing systems, it is also scalable.
- **Knowledge Distillation:** A smaller, lightweight model can be trained to mimic a larger model through distillation, enabling it to perform similarly while consuming a fraction of the resources.
- **Renewable Energy for Data Centers:** The shift to renewable energy will help AI become more eco-friendly. To power their AI infrastructure, many tech companies turn to solar, wind, and other renewable sources of energy. A good example of this is Google's commitment to run its data centres on carbon-free energy by 2030. Investing in greener technologies across the industry is encouraged by such commitments, which reduce AI's overall carbon footprint [48].
- **Edge Computing:** A cloud-based AI system consumes a significant amount of energy because it constantly transmits and processes data on centralised servers. In contrast, edge computing reduces transmission energy and latency by processing data closer to the source. Using AI at the edge can reduce power consumption and environmental impact by reducing the need for power-hungry data centres [49].
- **Lifecycle Assessment and Carbon Tracking:** AI models are assessed throughout their lifecycle, from the moment they are developed until they are disposed of, to understand their environmental impact. Using carbon tracking, companies can assess and offset their AI-related emissions. To provide transparency, some companies publish carbon reports so that stakeholders and consumers can make informed decisions regarding sustainable technologies [50].

20.5 AI for Environmental Sustainability

Furthermore, AI technologies can actively contribute to environmental sustainability, aside from reducing their environmental impact. Some examples are as follows:

20.5.1 Climate Prediction and Disaster Management

Using artificial intelligence, climate models can be developed, natural disasters can be predicted, and extreme weather can be monitored. Machine learning models can be used to predict events such as hurricanes, floods, and wildfires based on satellite imagery, weather patterns, and historical data. These predictions enable governments and communities to respond proactively, minimising loss of life and environmental damage [51].

20.5.2 Precision Agriculture

AI-driven precision agriculture uses data analytics and machine learning to optimise farming practices, conserving resources and reducing environmental harm. Farmers can apply fertiliser and water more efficiently through the use of drones and sensors powered by artificial intelligence. In this way, higher crop yields are achieved, pesticides are used less, and water is used less.

20.5.3 Wildlife Conservation and Biodiversity

To preserve biodiversity, artificial intelligence monitors endangered species and habitats. By analysing image and audio data from remote cameras and microphones, machine learning algorithms can detect animal species, track population changes, and detect illegal poaching. Conserving species and maintaining ecosystems is made possible by this method [52].

20.5.4 Renewable Energy Optimisation

With AI models, energy grids can be optimised by predicting supply and demand, integrating renewable energy sources, and reducing waste. As a result of machine learning, energy providers are able to predict power usage more accurately, enabling them to distribute electricity more efficiently. Clean energy is therefore promoted, and fossil fuels are reduced as backup power sources.

20.6 Challenges and Future Directions

There are several challenges involved in the implementation of green and responsible AI, including maintaining transparency, maintaining accuracy, and navigating ethical issues. With the advancement of AI technology, the industry must create sustainability-focused standards and frameworks. Future directions may include:

- **Standardised Green Metrics for AI:** Developing metrics for assessing and comparing AI's environmental impact can help companies adopt green practices.
- **Incentives for Green AI Development:** A government or organisation can encourage companies with AI systems that have a minimal carbon footprint to promote sustainable AI.
- **Innovations in AI Hardware:** Increasing demand for energy-efficient AI will lead to the development of hardware, such as neuromorphic computing and quantum computing, which can further reduce energy consumption and environmental impact.

20.7 Conclusion

Responsible AI practices are essential to minimising AI's ecological impact as well as maximising its environmental impact-reduction potential. The priority of green AI in industries can contribute to reducing their carbon footprints and promoting sustainability across all industries. For sustainable AI to achieve its full potential, continuous innovation in model efficiency, adoption of renewable energy, and standardisation of green metrics are essential. AI applications are becoming increasingly prevalent, so it is essential to foster a balanced approach that considers both their benefits and environmental costs.

References

1. Singh, A., *et al.*, Smart Traffic Monitoring Through Real-Time Moving Vehicle Detection Using Deep Learning *via* Aerial Images for Consumer Application. *IEEE Trans. Consum. Electron.*, 70, 4, 7302–7309, 2024. doi: 10.1109/TCE.2024.3445728.

2. Rani, P. and Sharma, R., Intelligent transportation system for internet of vehicles based vehicular networks for smart cities. *Comput. Electr. Eng.*, 105, 108543, 2023.
3. Patterson, D., *et al.*, Carbon Emissions and Large Neural Network Training, in: *arXiv: arXiv:2104.10350*, Apr. 23, 2021, Accessed: Nov. 13, 2024. [Online]. Available: http://arxiv.org/abs/2104.10350.
4. Rani, P. and Sharma, R., An experimental study of IEEE 802.11 n devices for vehicular networks with various propagation loss models, in: *International Conference on Signal Processing and Integrated Networks*, pp. 125–135, Springer, 2022.
5. Lannelongue, L., Grealey, J., Inouye, M., Green Algorithms: Quantifying the Carbon Footprint of Computation. *Adv. Sci.*, 8, 12, 2100707, Jun. 2021. doi: 10.1002/advs.202100707.
6. Rani, P., *et al.*, Federated Learning-Based Misbehaviour Detection for the 5G-Enabled Internet of Vehicles. *IEEE Trans. Consum. Electron.*, 70, 2, 4656–4664, 2023.
7. Vinuesa, R., *et al.*, The role of artificial intelligence in achieving the Sustainable Development Goals. *Nat. Commun.*, 11, 1, 233, Jan. 2020. doi: 10.1038/s41467-019-14108-y.
8. Hussain, N., Rani, P., Chouhan, H., Gaur, U.S., Cyber security and privacy of connected and automated vehicles (CAVs)-based federated learning: challenges, opportunities, and open issues. *Federated Learn. IoT Appl.*, 169–183, 2022.
9. Hussain, N., Rani, P., Kumar, N., Chaudhary, M.G., A deep comprehensive research architecture, characteristics, challenges, issues, and benefits of routing protocol for vehicular ad-hoc networks. *Int. J. Distrib. Syst. Technol. (IJDST)*, 13, 8, 1–23, 2022.
10. Pérez-Lombard, L., Ortiz, J., Pout, C., A review on buildings energy consumption information. *Energy Build.*, 40, 3, 394–398, 2008.
11. Wiedmann, T. and Minx, J., A definition of 'carbon footprint'. *Ecol. Economics Res. Trends*, 1, 2008, 1–11, 2008.
12. Ismail, H. and Hanafiah, M.M., A review of sustainable e-waste generation and management: Present and future perspectives. *J. Environ. Manage.*, 264, 110495, 2020.
13. Wang, H., Li, G., Wang, G., Peng, J., Jiang, H., Liu, Y., Deep learning based ensemble approach for probabilistic wind power forecasting. *Appl. Energy*, 188, 56–70, Feb. 2017. doi: 10.1016/j.apenergy.2016.11.111.
14. Chen, J., Zeng, G.-Q., Zhou, W., Du, W., Lu, K.-D., Wind speed forecasting using nonlinear-learning ensemble of deep learning time series prediction and extremal optimization. *Energy Convers. Manage.*, 165, 681–695, Jun. 2018. doi: 10.1016/j.enconman.2018.03.098.
15. Lin, K.-P., Pai, P.-F., Ting, Y.-J., Deep Belief Networks With Genetic Algorithms in Forecasting Wind Speed. *IEEE Access*, 7, 99244–99253, 2019. doi: 10.1109/ACCESS.2019.2929542.

16. Díaz–Vico, D., Torres–Barrán, A., Omari, A., Dorronsoro, J.R., Deep Neural Networks for Wind and Solar Energy Prediction. *Neural Process. Lett.*, 46, 3, 829–844, Dec. 2017. doi: 10.1007/s11063-017-9613-7.
17. Abdel-Nasser, M. and Mahmoud, K., Accurate photovoltaic power forecasting models using deep LSTM-RNN. *Neural Comput. Applic.*, 31, 7, 2727–2740, Jul. 2019. doi: 10.1007/s00521-017-3225-z.
18. Qolomany, B., *et al.*, Leveraging Machine Learning and Big Data for Smart Buildings: A Comprehensive Survey. *IEEE Access*, 7, 90316–90356, 2019. doi: 10.1109/ACCESS.2019.2926642.
19. Pham, A.-D., Ngo, N.-T., Ha Truong, T.T., Huynh, N.-T., Truong, N.-S., Predicting energy consumption in multiple buildings using machine learning for improving energy efficiency and sustainability. *J. Clean. Prod.*, 260, 121082, Jul. 2020. doi: 10.1016/j.jclepro.2020.121082.
20. Zekić-Sušac, M., Mitrović, S., Has, A., Machine learning based system for managing energy efficiency of public sector as an approach towards smart cities. *Int. J. Inf. Manage.*, 58, 102074, Jun. 2021. doi: 10.1016/j.ijinfomgt.2020.102074.
21. Ghazal, T.M., Hasan, M.K., Ahmad, M., Alzoubi, H.M., Alshurideh, M., Machine Learning Approaches for Sustainable Cities Using Internet of Things, in: *The Effect of Information Technology on Business and Marketing Intelligence Systems, Studies in Computational Intelligence*, vol. 1056, M. Alshurideh, B.H. Al Kurdi, R. Masa'deh, H.M. Alzoubi, S. Salloum (Eds.), pp. 1969–1986, Springer International Publishing, Cham, 2023, doi: 10.1007/978-3-031-12382-5_108.
22. Milojevic-Dupont, N. and Creutzig, F., Machine learning for geographically differentiated climate change mitigation in urban areas. *Sustain. Cities Soc.*, 64, 102526, Jan. 2021. doi: 10.1016/j.scs.2020.102526.
23. Zhang, M., *et al.*, Impact of urban expansion on land surface temperature and carbon emissions using machine learning algorithms in Wuhan, China. *Urban Clim.*, 47, 101347, Jan. 2023. doi: 10.1016/j.uclim.2022.101347.
24. Li, W., *et al.*, A General Framework for Unmet Demand Prediction in On-Demand Transport Services. *IEEE Trans. Intell. Transport. Syst.*, 20, 8, 2820–2830, Aug. 2019. doi: 10.1109/TITS.2018.2873092.
25. Duan, P., Mao, G., Liang, W., Zhang, D., A Unified Spatio-Temporal Model for Short-Term Traffic Flow Prediction. *IEEE Trans. Intell. Transport. Syst.*, 20, 9, 3212–3223, Sep. 2019. doi: 10.1109/TITS.2018.2873137.
26. Yin, X., Wu, G., Wei, J., Shen, Y., Qi, H., Yin, B., Deep Learning on Traffic Prediction: Methods, Analysis, and Future Directions. *IEEE Trans. Intell. Transport. Syst.*, 23, 6, 4927–4943, Jun. 2022. doi: 10.1109/TITS.2021.3054840.
27. Rolnick, D., *et al.*, "Tackling Climate Change with Machine Learning,". *ACM Comput. Surv.*, 55, 2, 1–96, Feb. 2023. doi: 10.1145/3485128.
28. Hong, T. and Fan, S., Probabilistic electric load forecasting: A tutorial review. *Int. J. Forecast.*, 32, 3, 914–938, Jul. 2016. doi: 10.1016/j.ijforecast.2015.11.011.

29. Giebel, G. and Kariniotakis, G., Wind power forecasting—a review of the state of the art, in: *Renewable Energy Forecasting*, pp. 59–109, Elsevier, Woodhead Publishing, 2017, doi: 10.1016/B978-0-08-100504-0.00003-2.
30. Kim, I.H., Liu, Y.-H., Pallister, S., Pol, W., Roberts, S., Lee, E., Fault-tolerant resource estimate for quantum chemical simulations: Case study on Li-ion battery electrolyte molecules. *Phys. Rev. Res.*, 4, 2, 023019, Apr. 2022. doi: 10.1103/PhysRevResearch.4.023019.
31. Dhar, P., The carbon impact of artificial intelligence. *Nat. Mach. Intell.*, 2, 8, 423–425, Aug. 2020. doi: 10.1038/s42256-020-0219-9.
32. Moore, G., Cramming More Components onto Integrated Circuits (1965), in: *Ideas That Created the Future*, H.R. Lewis (Ed.), pp. 261–266, MIT Press, 2021, doi: 10.7551/mitpress/12274.003.0027.
33. Coeckelbergh, M., AI for climate: freedom, justice, and other ethical and political challenges. *AI Ethics*, 1, 1, 67–72, Feb. 2021. doi: 10.1007/s43681-020-00007-2.
34. Basiago, A.D., [No title found]. *Environmentalist*, 19, 2, 145–161, 1998. doi: 10.1023/A:1006697118620.
35. Strubell, E., Ganesh, A., McCallum, A., Energy and Policy Considerations for Deep Learning in NLP, in: *arXiv: arXiv:1906.02243*, Jun. 05, 2019, Accessed: Nov. 13, 2024. [Online]. Available: http://arxiv.org/abs/1906.02243.
36. Klein, N., *On Fire: The (Burning) Case for a Green New Deal*, Simon and Schuster, New York, 2020.
37. Yigitcanlar, T. and Kamruzzaman, M., Planning, Development and Management of Sustainable Cities: A Commentary from the Guest Editors. *Sustainability*, 7, 11, 14677–14688, Nov. 2015. doi: 10.3390/su71114677.
38. Dizdaroglu, D. and Yigitcanlar, T., Integrating urban ecosystem sustainability assessment into policy-making: insights from the Gold Coast City. *J. Environ. Plann. Manage.*, 59, 11, 1982–2006, Nov. 2016. doi: 10.1080/09640568.2015.1103211.
39. Hunter, A. and Hewson, J., There are 10 catastrophic threats facing humans right now, and coronavirus is only one of them, in: *The Conversation*, 2020, http://theconversation.com/there-are-10-catastrophic-threats-facing-humans-right-nowand-coronavirus-is-only-one-of-them-136854.
40. Copland, S., Anti-politics and Global Climate Inaction: The Case of the Australian Carbon Tax. *Crit. Sociol.*, 46, 4–5, 623–641, Jul. 2020. doi: 10.1177/0896920519870230.
41. Hornsey, M.J. and Fielding, K.S., Understanding (and Reducing) Inaction on Climate Change. *Soc. Issues Policy Rev.*, 14, 1, 3–35, Jan. 2020. doi: 10.1111/sipr.12058.
42. Sharma, G.D., Yadav, A., Chopra, R., Artificial intelligence and effective governance: A review, critique and research agenda. *Sustain. Futures*, 2, 100004, 2020. doi: 10.1016/j.sftr.2019.100004.

43. Mortoja, M.G. and Yigitcanlar, T., Local Drivers of Anthropogenic Climate Change: Quantifying the Impact through a Remote Sensing Approach in Brisbane. *Remote Sens.*, 12, 14, 2270, Jul. 2020. doi: 10.3390/rs12142270.
44. Shockley, K., The Great Decoupling: Why Minimizing Humanity's Dependence on the Environment May Not Be Cause for Celebration. *J. Agric. Environ. Ethics*, 31, 4, 429–442, Aug. 2018. doi: 10.1007/s10806-018-9736-5.
45. Stahl, B.C., *et al.*, Artificial intelligence for human flourishing – Beyond principles for machine learning. *J. Bus. Res.*, 124, 374–388, Jan. 2021. doi: 10.1016/j.jbusres.2020.11.030.
46. Alotaibi, S., Mehmood, R., Katib, I., Rana, O., Albeshri, A., Sehaa: A Big Data Analytics Tool for Healthcare Symptoms and Diseases Detection Using Twitter, Apache Spark, and Machine Learning. *Appl. Sci.*, 10, 4, 1398, Feb. 2020. doi: 10.3390/app10041398.
47. Lerlertpakdee, P., Jafarpour, B., Gildin, E., Efficient production optimization with flow-network models. *SPE J.*, 19, 06, 1083–1095, 2014.
48. Zhang, Y., Wang, Y., Wang, X., GreenWare: Greening Cloud-Scale Data Centers to Maximize the Use of Renewable Energy, in: *Middleware 2011, Lecture Notes in Computer Science*, vol. 7049, F. Kon and M. Kermarrec (Eds.), pp. 143–164, Springer Berlin Heidelberg, Berlin, Heidelberg, 2011, doi: 10.1007/978-3-642-25821-3_8.
49. Cao, K., Liu, Y., Meng, G., Sun, Q., An overview on edge computing research. *IEEE Access*, 8, 85714–85728, 2020.
50. Wang, J., Shen, G.Q., Xue, F., Carbon tracking in the building sector: A 'cabbage'framework, in: *The 23rd International Conference on Construction Applications of Virtual Reality*, Florence, Italy, 13/11/2023-16/11/2023, 2023, Accessed: Nov. 13, 2024. [Online]. Available: https://hub.hku.hk/handle/10722/344280.
51. Jaber, M.M., *et al.*, Predicting climate factors based on big data analytics based agricultural disaster management. *Phys. Chem. Earth Parts A/B/C/*, 128, 103243, 2022.
52. Redford, K.H. and Richter, B.D., Conservation of Biodiversity in a World of Use. *Conserv. Biol.*, 13, 6, 1246–1256, Dec. 1999. doi: 10.1046/j.1523-1739.1999.97463.x.

21

Integrating AI into Mobility as a Service (MaaS): The Future of Urban Transportation

Kunal Gagneja[1*], Zainab. R. Abdulsada[2], Sapna Singh[3], Ruby Pant[4] and Ramy Al-Fatlawy[5]

[1]*Department of Computer Science Engineering, Chandigarh Engineering College, Chandigarh Group of Colleges, Jhanjeri, Mohali, Punjab, India*
[2]*College of Technical Engineering, National University of Science and Technology, Dhi Qar, Iraq*
[3]*Department of Electrical, Electronics & Electric Vehicle Engineering, NIMS Institute of Engineering & Technology, NIMS University Rajasthan, Jaipur, India*
[4]*Department of Mechanical Engineering, Uttaranchal Institute of Technology, Uttaranchal University, Dehradun, Uttarakhand, India*
[5]*College of Technical Engineering, The Islamic University, Najaf, Iraq*

Abstract
To revolutionize urban transportation, this chapter examines how Artificial Intelligence (AI) can be integrated with Mobility as a Service (MaaS). As a result of leveraging emerging technologies, MaaS provides seamless, user-centric platforms that are flexible, efficient, and sustainable. Through AI, real-time analysis, predictive modeling, and intelligent decision-making are enabled, addressing challenges such as congestion, inefficiencies in urban mobility, and environmental sustainability. Policymakers, technologists, and urban planners will learn tips and techniques for harnessing AI's potential in creating MaaS ecosystems, such as mobility control enters, adaptive transport systems, and rural accessibility solutions.

Keywords: Artificial intelligence (AI), mobility as a service (MaaS), urban transportation, smart mobility, predictive modeling, real-time analysis, environmental sustainability, adaptive transport systems

**Corresponding author*: kunal2848.research@cgcjhanjeri.in

Pethuru Raj, Sudesh Yadav, Manas Kumar Mishra, Satya Prakash Yadav and Victor Hugo C. de Albuquerque (eds.) Artificial Intelligence Technologies for Smart and Sustainable Urban Transportation: Integrated Platforms and Use Cases, (355–374) © 2026 Scrivener Publishing LLC

21.1 Introduction

A growing number of people around the world have been using shared mobility services in recent years. In shared mobility, vehicles are used to perform a trip in a shared manner. Among these services are bike-share, carshare, and ride-hailing. The first carsharing system was established in Zurich in 1948, and Amsterdam's first bicycle-sharing scheme was established in 1965. Due to advancements in information and communication technologies (ICTs), new mobility solutions have become more convenient and personalized. Additionally, several MaaS platforms have been developed, tested, and implemented in the last five years for the integration of these services.

It is possible to save money, benefit from convenience, and reduce vehicle ownership by using shared mobility services [1]. As such services are introduced in cities, they must be evaluated properly to avoid inefficiency and ineffectiveness. For policymakers to make informed decisions, transport methods that can be evaluated are necessary. Considering the modal split is crucial in this regard since it implies how travellers use shared mobility services, how they differ among them, and how they impact urban mobility and sustainability. To capture this phenomenon, a model is needed. Bicycle-sharing, carsharing, and ride-hailing services haven't been studied in the literature.

21.1.1 The MaaS Concept

Researchers, practitioners, and policymakers have been paying close attention to MaaS because it can enhance our perception of and experience of transportation, as well as improve sustainability and efficiency [2]. The boundaries and characteristics of MaaS have been extensively defined; however, there is no universally accepted definition [3]. An important aspect of MaaS is its integration and interoperability, which are crucial concepts in the mobility paradigm [4, 5]. In the early 1990s, mobility management was developed to increase the efficiency and accessibility of public transportation [6, 7]. Several factors characterized today's society, including urbanization, technological advances, and the increasing penetration of smartphones, have led to an increase in the popularity of shared mobility services (such as bike sharing, e-scooter sharing, car sharing, etc.) and new generation preferences are being influenced by a cultural shift [8]. As a result, MaaS relates closely to new economic paradigms, such as sharing and crowd-sourcing [9–11]. Users can access transportation services on-demand instead of owning their vehicles due to MaaS' emphasis on accessibility.

Additionally, it provides viable and sustainable alternatives to private cars that are as flexible as they are environmentally friendly [7, 12, 13]. The platform allows for the integration and combination of different modes, enabling intermodal travel and allowing consumers to choose from a wide variety of existing mobility options (buses, taxis, metros, trains, sharing and on-demand services, even ferries, or a combination of these). There are many different types of services (for example, parking, charging stations, and so forth) that can be provided within a MaaS service, such as food delivery, logistics, television packages, retail discounts, and so on [12]. It is generally believed that the degree of integration in MaaS schemes differs according to their level of integration; the higher the integration level, the more "MaaS-ness" is expressed [5]. MaaS services are offered *via* apps, just like many other sharing-labeled services. The app provides real-time information on timetables, route prices, vehicle availability, and traffic conditions, as well as centralizing booking and payment processes (one-stop-shop) [5]. This app may also provide weather information, sync with personal activity calendars, show trip history, re-order tickets, display invoices, and provide feedback [14]. A variety of payment options are available, including single trips (pay-as-you-go) and monthly subscription packages that offer discounts on mobility bundles [15]. As a user-centric service, MaaS offers seamless, comfortable, and integrated (or personalized) travel experiences that meet users' needs [16]. Moreover, MaaS fosters a collaborative, cohesive, and interconnected transportation market. A dynamic ecosystem is created when all participants continuously exchange data and interact, resulting in efficient resource utilization [17]. In this way, MaaS is able to operate smoothly through leveraging the collective expertise of all stakeholders [18].

There are countless industries where technology can make a positive impact, such as retail, finance, insurance, healthcare, and even sports. Many of these companies have been able to reduce their operating costs and increase their efficiency as a result. As the transportation industry evolves, modern technology is being applied to overcome obstacles such as traffic congestion, delays, and routing that affect vehicle efficiency as well as system efficiency.

Mobility is changing profoundly as a result of the digital revolution [19]. Using new technologies, mobility can be managed more efficiently and sustainably, balancing efficiency with sustainability [20].

As cooperative-intelligent transportation systems (ITSs) evolve, artificial intelligence (AI) is becoming an important tool for decision-making and continuously adjusting transport supply to meet changing mobility demands through digital platforms and mobility cities. Besides calibrating demand simulation models and interpreting phenomena, combining artificial intelligence with big data collection from video analysis systems and field systems (both at the edge and central levels) can be used to analyze phenomena both under normal conditions and during disruptions [21, 22].

Almost every large city in the world suffers from traffic and logistics problems. Many factors contribute to this, including increased vehicle traffic and an increase in travel by freight and people. An environmentally friendly transportation system can be designed and operated with the help of technology [23]. The use of artificial intelligence (AI) to acquire real-time data from moving cars and leverage Mobility on Demand can, therefore, contribute to traffic management. By integrating AI-based routing and transportation network services, traffic management can be improved by ensuring their security.

A lot of research has been conducted in the area of autonomous vehicles and infrastructure management in transport systems, but artificial intelligence has yet to be applied to these areas. Additionally, AI can be applied very efficiently to plan and operate the MaaS (Mobility as a Service) concept, which focuses on providing adaptive and intermodal services [24].

The use of AI for analyzing information allows effective decision support, which, in combination with process automation and remote ITS management, changes the way mobility management is done [25]. As a result, both costs and errors are reduced, and door-to-door mobility services are customized [26].

In modern metropolitan cities, AI-based Mobility Control Centers (MCCs) are becoming more common to facilitate a digital ecosystem by integrating combinational systems and cooperative elements to manage, regulate, and control urban mobility, as well as to centralize data and control field systems remotely. Figure 21.1 shows these elements and systems:

- Large amounts of heterogeneous data can be acquired using big data solutions to learn about the transport system historically and dynamically, as well as calibrate models using evidence;
- Based on dynamic models, this software simulates and predicts traffic flow using methods and techniques from the field of transport engineering;

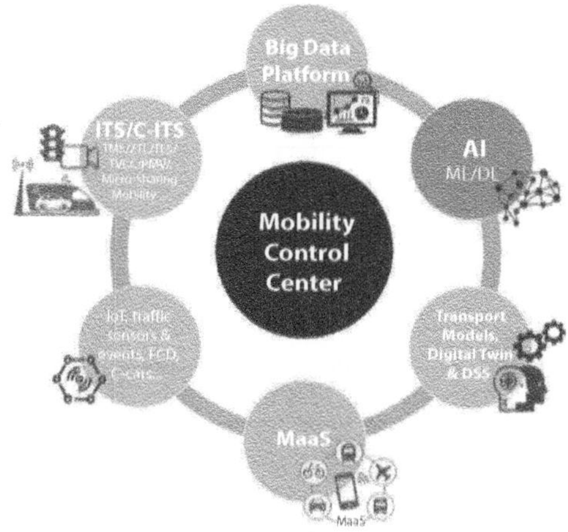

Figure 21.1 A Mobility Control Center (MCC) consists of the following elements.

- In addition to being used for real-time operations management of mobility systems, DSS can also be used for offline transport planning;
- Integrating MaaS with MCCs and communicating "best-in-class" dynamic services to end users will maximize its benefits. In addition, advanced information about the multimodal journey, real-time conditions, segment selections, booking, and payment options can be provided.

21.2 Mobility in Rural Areas is a Problem

Furthermore, compared to more densely populated areas, rural mobility centers are challenged by lower population densities and lower investment levels, as well as changing citizen expectations. ARTS, SMARTA, MAMBA, ITF, and ITF, as well as information gathered from 20 years of experience. There are a variety of intelligent transport systems (ITS) in cities, such as car priority systems, bike rental services, and digital tickets. As a result of the digitalization process, consumers are able to access information more easily, whether they are planning a journey, booking a trip, or paying for a trip. Numerous cities are integrating contactless tickets, using tools like City Mapper to plan their transportation, and increasing the use of new mobility services such as Uber and car clubs, as shown in Figure 21.2.

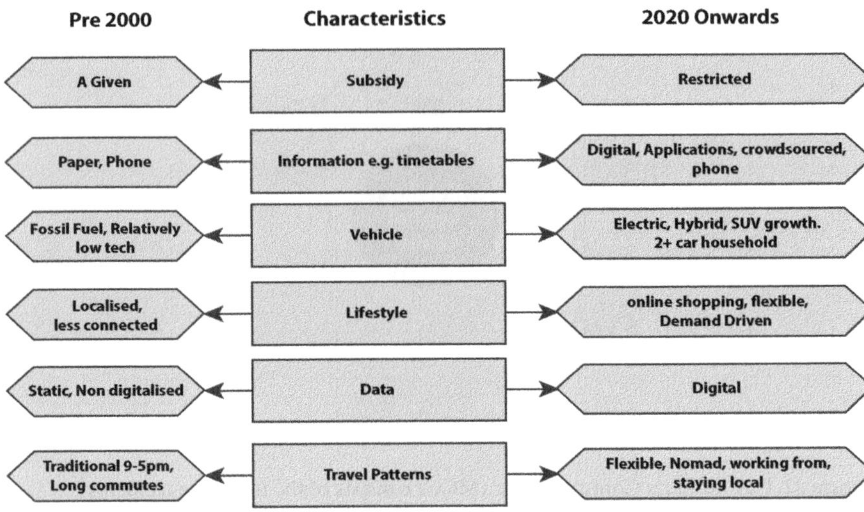

Figure 21.2 Past and present examples of rural mobility.

The role of digital connectivity in enabling consumer choice and offering in rural areas cannot be overstated. In rural areas, an older population may be less digitally literate and more reluctant to engage with technology. Digital competence and comfort cannot be assumed by the whole population [27]. Digital services are delivering solutions in some rural areas, but this globalization presents challenges. A limited number of data services are available in rural areas in the UK, for example [28]. Consequently, rural areas in Finland do not suffer from the "digital divide" [29].

21.3 Transportation Systems and Artificial Intelligence: A Critical Review

21.3.1 Artificial Intelligence-Assisted Smart Cities

Transport-related issues can be solved, and transportation systems can be made more efficient through the use of ITSs, which were discussed in the previous section. In recent decades, smart cities have been gaining concern over ITS, which falls under smart mobility. Smart cities are not defined in any one way, and there are many definitions in the literature.

A smart city is a city that uses ICT to maximize human, social, and relational capital, in addition to achieving environmental sustainability as mention in Table 21.1. In addition to stakeholders' backgrounds and

Table 21.1 The following areas characterize smart cities.

Range	Explanation	Instances
Governing smartly	Enhances public services, improves the efficiency of government, and engages citizens in decision-making through the usage of knowledge.	• Platforms for e-government; • Tools for civic engagement online; • Making policy based on data.
Economic smarts	Aims to foster a dynamic and resilient urban economy through the use of technology to grow, innovate, and sustain.	• A technology park and an innovation hub; • Smart payments are used to, and digital payments; • A data-driven approach to economic development.
A smarter way to travel	Integrates technology to improve accessibility, mobility, and efficiency of transportation solutions.	• Systems that manage traffic intelligently; • Vehicles powered by electricity and autonomy; • System integration of public transportation.
Smart environments	Ensures environmental quality through monitoring, management, and reduction of pollution in urban areas.	• Systems for managing waste intelligently; • Keeping an eye on the air and water quality; • Street lighting consumes less energy.
Living a smarter life	Engaged in improving living conditions for residents by implementing technology in health care, education, and safety.	• An IoT-enabled smart home; • Health care delivered remotely and *via* telehealth; • Innovations in education.
People with smarts	Promotes citizen engagement and education using technology.	• Literacy programs in digital technology; • Platforms for engaging citizens; • Initiatives to develop inclusive communities.

government priorities, defining terms requires a good understanding of what they mean. A smart city, according to academia, is primarily concerned with improving the quality of life. Nevertheless, private companies often prioritize efficiency over everything else [30]. Although smart cities are defined differently, they share three common characteristics or dimensions: they integrate advanced electronic/digital technology into city infrastructure, they improve stakeholders' interests in various aspects of the system, and they improve stakeholder interest. In transportation problems where the parametric relationships among variables are poorly understood, artificial neural networks (ANNs) can be used since they can perform a nonlinear mapping between inputs and outputs by using hidden layers and sufficient training. There is a lot of literature on the use of artificial neural networks for estimating and forecasting states, detecting incidents, monitoring traffic, and controlling infrastructure. Using an SVM, you can identify stages and scenarios based on input data rather than analyze it. In the context of ITS, SVMs are mostly used to detect incidents and predict accidents despite their application to other transport-related problems.

The difference between artificial neural networks and support vector machines is that Bayesian networks take into account probabilities and conditional dependencies as well as data-driven models. A Bayesian network is typically used in ITS literature for traffic forecasting and investigating incidents and accidents. As a technology that enhances and optimizes various aspects of the transportation system, Artificial Intelligence (AI) is used in transportation systems. With artificial intelligence methods such as machine learning, computer vision, natural language processing, and other technologies, transportation networks can be made safer, more efficient, and more sustainable.

There are numerous areas where artificial intelligence can be used in transportation, including:

- **Automobiles that drive themselves**

In autonomous cars, artificial intelligence enables them to navigate, perceive environments, and make decisions without humans.

- **Managing traffic**

My ability to optimize traffic flow and reduce congestion can be improved with the use of traffic sensors, cameras, and historical data. Traffic patterns can be predicted, signal timings can be optimized, and alternative routes can be suggested to make transportation more efficient.

• **Maintaining your systems predictively**
Monitoring and analyzing vehicle and infrastructure data can be used to predict maintenance needs by using artificial intelligence. By preventing unexpected breakdowns and reducing downtime, proactive maintenance can be carried out, resulting in increased reliability and cost savings.

• **Managing supply chains and logistics smartly**
Analysis of large datasets, route optimization, demand prediction, and inventory management can be optimized by artificial intelligence. As a result, costs are reduced, deliveries are made faster, and resources are better utilized.

• **ITS**
Transportation networks are monitored and managed using artificial intelligence technologies. A variety of features are available to enhance traveller safety and efficiency, including real-time incident detection, dynamic routing, adaptive traffic signal control, and traveller information.

• **Experiences of passengers**
The use of artificial intelligence in chatbots or virtual assistants can enhance the passenger experience by providing real-time information and recommendations as well as assistance at various stages of the journey. Artificial intelligence is a key component of transportation systems that can improve safety, efficiency, and environmental sustainability. Additionally, ethical, cybersecurity, and employment issues arise. These factors must be carefully considered to ensure that AI in transportation is implemented responsibly and beneficially.

Mobile applications integrating artificial intelligence have limited urban development. The last few years have seen a lot of attention paid to mobile applications, smart cities, and artificial intelligence. The majority of them are detailed about a single country rather than the whole world. Generally, existing mobile applications do not undergo SWOT analysis since they can often be updated or adapted to meet changing conditions. Others focus solely on multi-service apps while ignoring single-service apps.

21.3.2 AI Applications Currently in Use

AI has revolutionized mobile applications by revolutionizing the way humans engage with them. Mobile applications integrating artificial intelligence have made personal assistants more useful in this environment.

Table 21.2 Smart cities' smart mobility AI implementations.

Explanation	**Account**
ITS	Real-time traffic flow optimization and congestion reduction can be achieved by using artificial intelligence (AI).
Maintaining vehicles in a predictive manner	Smart mobility systems are being enhanced by AI algorithms that are used to predict vehicle maintenance requirements and reduce downtime.
Vehicles that operate autonomously	The use of artificial intelligence (AI) in autonomous vehicles, such as machine learning and computer vision, is automating urban navigation and operations.
Optimizing ride-sharing services	Using AI algorithms, ride-sharing services can be optimized, routes can be improved, travel times can be shortened, and overall efficiency can be improved.
Parking solutions powered by smart technology	Parking systems based on artificial intelligence reduce traffic congestion and enhance parking experiences by providing real-time parking availability data.
Support for your travels	Smart Mobility provides real-time updates, personalized recommendations, and route planning *via* mobile applications powered by artificial intelligence.
Transport Pricing Dynamic	AI algorithms can be used to optimize resource allocation by dynamically adjusting transportation pricing according to demand and traffic conditions.
Forecasting traffic incidents	Data analysis with AI models allows proactive measures to be taken in smart mobility and improves overall safety by predicting and preventing traffic incidents.
Integrating multiple modes of communication	Using AI-based coordination, designing platforms that seamlessly integrate multiple modes of transportation.

Mobile apps mainly use AI technology to recognize voices and patterns. Mobile and online personas can be integrated with this technology. Technology such as this is mainly useful for mobile information services, but it can also be used to create new business models. As an example, AI-enabled mobile apps can create personal assistants based on user profiles. Among the AI applications for mobile devices are image editing, e-commerce, navigation, social media, gaming, smart home control, and augmented reality as shown in Table 21.2.

21.3.3 Identifying Research Gaps

As smart cities develop, we need to understand better how AI-powered mobile apps impact urban ecosystems. A comprehensive literature review is provided as well, which examines how these technologies can enhance mobility, safety, and sustainability in urban environments. Cross-disciplinary approaches to AI development are emphasized by researchers, advocating for collaboration among policymakers, technologists, and citizens. As a result of this integrated perspective, academic discourse is enriched, but it also offers practical insights that can be applied to smart city initiatives in the real world. AI-powered mobile apps are slow to be adopted by smart cities because of a lack of guidelines and frameworks. Smart cities are characterized by geographic and demographic differences, making it difficult to deploy continuous network coverage. According to the paper, wide and deep learning perspectives are analyzed from a computer science perspective.

21.3.4 Sustainability Implications of Apps

AI-driven mobile apps integrated into KAFD's projects demonstrate the transformational potential of technology for urban sustainability. As these applications streamline city transit and offer a unified mobility platform, they reduce traffic congestion and contribute to the conservation of the environment. Technology and urban planning work together to make cities smarter and more sustainable and to reduce carbon footprints.

In urban areas, artificial intelligence is effective at improving traffic flow and reducing congestion, based on a study by [31]. They provide alternative routes based on real-time data and forecasts of traffic patterns that reduce travel times and emissions. This benefits commuters and contributes to the city's sustainability.

Smart cities have also benefitted from AI-enabled mobile apps. In addition to preventing crimes, police can respond to emergencies using apps such as facial recognition and predictive analytics. Using artificial intelligence, safe city analyzes CCTV footage to detect potential security threats and take proactive steps to protect the community.

Mobile apps powered by artificial intelligence have an enormous environmental impact on smart city ecosystems. Using these apps can reduce energy consumption, reduce waste, and promote sustainable transportation options, thereby mitigating climate change. Eco mobility, for example, encourages users in Barcelona to use eco-friendly transportation methods, such as cycling or public transportation, which reduces greenhouse gas emissions.

21.3.5 The Impact of Urban Development on the Environment

Enhancing urban development with AI and machine learning can be achieved through mobile apps.

In the KAFD case study, we demonstrate how AI can be employed to optimize urban transportation within an ecosystem of smart cities. As a major contributor to pollution and carbon emissions, traffic congestion is a major problem in urban areas. With the use of artificial intelligence algorithms, it is possible to reduce travel times and fuel consumption through the analysis of real-time traffic data and the prediction of congestion patterns using mobile applications [32]. Besides improving commutes for individuals, this also reduces greenhouse gas emissions significantly.

Additionally, smart cities can benefit from AI algorithms that optimize energy use. Data can be collected on building energy usage patterns, vehicles, and other infrastructure by leveraging machine learning techniques in a mobile app, allowing for intelligent energy management at the building, vehicle, and infrastructure levels. The use of artificial intelligence can, for instance, identify energy-saving opportunities from data from smart meters, such as optimizing lighting based on occupant patterns or adjusting temperature settings [33]. The reduction of carbon emissions can be achieved through these energy-efficient practices, which contribute to a more sustainable urban environment.

Along with improving safety and security in cities, AI-powered mobile applications can also help protect the environment. An artificial intelligence-powered surveillance system integrated with a mobile app can detect anomalies, such as bags left unattended, and alert police. Preventing crime and ensuring citizen safety can be achieved by this proactive approach to security [34]. The use of artificial intelligence algorithms for maintaining infrastructure, such as bridges and utility systems, can also help prevent accidents and ensure the resilience of city infrastructure by analyzing sensor data and identifying potential failures before they occur [35].

A robust data privacy and security strategy is essential for successful AI integration into smart city ecosystems. The privacy of users should be protected, and these apps should not be able to collect and analyze large amounts of personal information. For AI-powered mobile apps to be trusted and used responsibly, complying with established data protection regulations is crucial [36].

21.4 Encounters

It is technically challenging to integrate multiple technologies into the same field, so the decision to do so requires a lot of thought and planning. This new concern comes from the fact that these systems manage and use valuable commercial and operational data: data security. Another challenge is the lack of value propositions for citizens in terms of awareness. Ultimately, the successful implementation and development of innovative urban solutions and technologies must persuade the public of their benefits. The most powerful lobbies in a country, such as car makers and major food and beverage companies, may not benefit from smart solutions in the past. These groups face a number of challenges that need to be addressed and resolved carefully. Smart cities are complex concepts, and solution designers and implementers must have a thorough understanding of their relationships.

21.4.1 Challenges in Knowledge

As a result of integration challenges, integrating artificial intelligence ecosystems into smart cities is not easy. Technologies are integrated into smart cities to serve a variety of purposes. As a result of their integration, these systems are more complex and require seamless coordination to function effectively. The process of ensuring interoperability and compatibility between different systems is challenging, requiring standardized interfaces and robust communication protocols [37].

In addition to data security, there are other significant challenges. The use of sensors and citizen data generates and collects enormous amounts of data in smart cities. The information is sensitive and valuable, resulting in a high level of risk if it's misused. Smart city initiatives may suffer from public trust issues caused by data breaches or privacy concerns.

Therefore, to ensure data integrity and confidentiality, it is imperative to implement encryption and access controls as well as to store the data securely [38].

The use of AI technology in smart cities also has a hard time convincing citizens of its benefits. Urban life still has a long way to go before people realize how beneficial these technologies can be. AI-driven solutions can improve traffic management, energy efficiency, and public services by providing citizens with information about their advantages. A public that is skeptical and concerned needs to be engaged and communicated effectively.

Industry and interest groups may also have difficulty implementing AI ecosystems in smart cities. Several industries may be affected by smart solutions, including automobile manufacturing and food and beverage corporations. Collaboration and innovation can be achieved if stakeholders' interests are taken into consideration and a common ground is found where stakeholders' interests can be addressed [39].

Aside from these challenges, technical hurdles must also be overcome. The environment may need different sensors depending on the geographical region of a municipality. Using Wi-Fi, for example, introduces vulnerabilities that require management and maintenance to ensure a reliable connection. Furthermore, the use of cloud-based servers increases security risks, which must be mitigated by encrypting data and managing data securely [40].

21.5 Expected Early Adopter and Users

According to Figure 21.3, the panel expects MaaS to be adopted by different categories of travellers. Only the Full integration or MaaS is considered here. Over 80 per cent of experts view Generation Z as early adopters of MaaS. There is a consensus (98%) among experts that MaaS will be most popular among the millennials. However, most experts (65%) believe generation X is a follower.

Additionally, 81 per cent of respondents viewed the baby boomer as a follower. Furthermore, 54 per cent of experts anticipate the silent generation as the largest non-user group of MaaS. However, 44% consider them to be followers.

According to the panel, the potential MaaS user is categorized according to the mode of transport they currently use (Figure 21.4). MaaS is only considered an early adopter by a small percentage of experts (2%). According to them, most automotive users (63%) are followers or even non-users (35%). A majority of panellists (65%) believe public transport

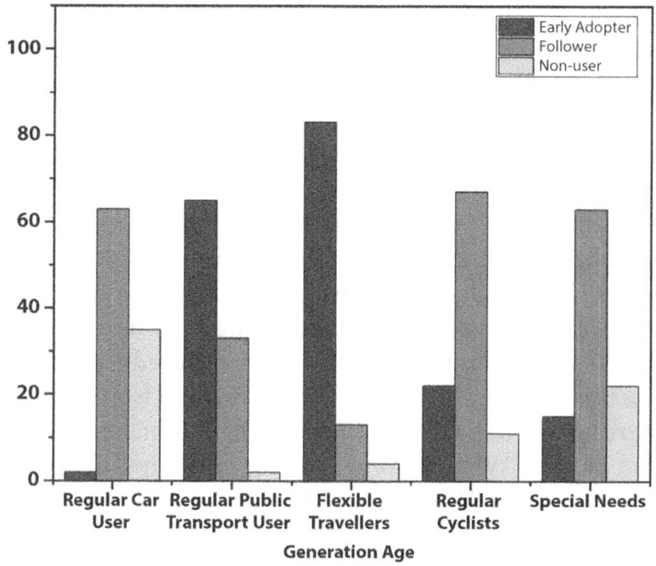

Figure 21.3 Generational breakdown of MaaS users.

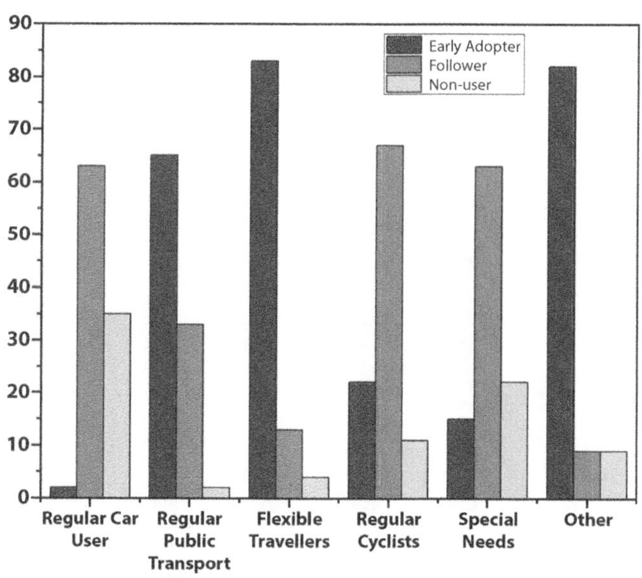

Figure 21.4 Based on the current transport mode, the expected MaaS user.

users will make a switch to MaaS as early adopter, while 33% think they will do so later on in their lives. A mere 2% of respondents don't expect to use MaaS in the near future. Those who combine different modes of transportation to reach their destination are considered early adopters of MaaS by most experts (83%); 13% think they will adopt MaaS later, and 4% don't anticipate using it.

Based on their trip purpose, the panel presents its expectations for potential MaaS users in Figure 21.5. This categorization implies that 63% of commuters will switch to MaaS within the next five years. 65% of business travellers expect an early adoption during their business trip. It seems that the panel's opinion is more evenly split on the educational trip, where 42% believe this category of travellers would benefit from MaaS in the early stages, whereas over half (53%) believe MaaS will make education trips at a later point in time. When it comes to shopping trips, the panel's opinion is more homogeneous. While only 13% predict that these travellers will switch to MaaS at an early stage, the majority (80%) assume that they will do so later. 24% expect MaaS to be adopted early for leisure and social trips, whereas 74% assume it will be adopted later. Among other possible MaaS trip purposes, experts suggest tourism and special events.

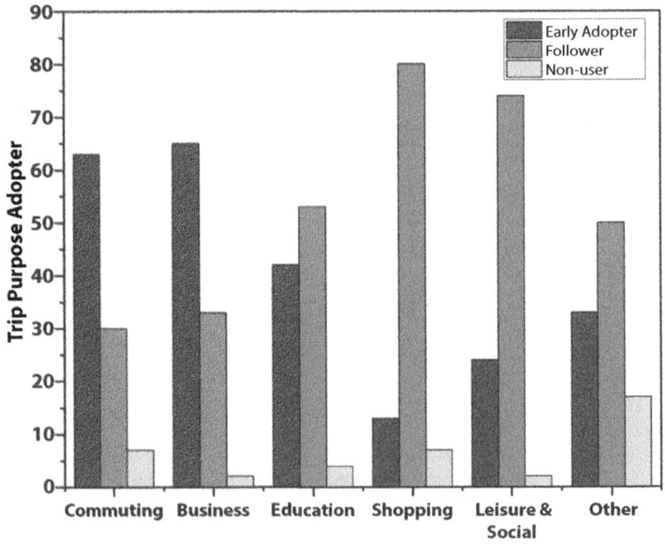

Figure 21.5 Trip purpose-specific MaaS users.

21.6 Conclusion

With AI integrated into mobility as a Service, urban transportation is undergoing a transformative shift, combining technological innovation with sustainability objectives. A MaaS platform powered by AI can enhance mobility experiences, reduce environmental impact, and optimize transportation systems. There are, however, challenges related to implementation, including concerns regarding data privacy, a digital divide, and resistance from traditional stakeholders. Collaboration and robust frameworks are necessary for addressing these challenges. The AI-driven MaaS ecosystem can create sustainable, equitable, and smarter cities by fostering innovation and adaptability.

References

1. Jochem, P., Frankenhauser, D., Ewald, L., Ensslen, A., Fromm, H., Does free-floating carsharing reduce private vehicle ownership? The case of SHARE NOW in European cities. *Transp. Res. Part Policy Pract.*, 141, 373–395, Nov. 2020. doi: 10.1016/j.tra.2020.09.016.
2. *Hensher: Understanding Mobility as a Service (MaaS):.. - Google Scholar*, Elsevier, 2020, Accessed: Nov. 09, 2024. [Online]. Available: https://scholar.google.com/scholar_lookup?title=Understanding%20Mobility%20as%20a%20Service%20(MaaS)%3A%20Past%2C%20Present%20and%20Future&author=D.A.%20Hensher&publication_year=2020.
3. Arias-Molinares, D. and García-Palomares, J.C., The Ws of MaaS: Understanding mobility as a service from a literature review. *IATSS Res.*, 44, 3, 253–263, Oct. 2020. doi: 10.1016/j.iatssr.2020.02.001.
4. Giesecke, R., Surakka, T., Hakonen, M., Conceptualising Mobility as a Service, in: *2016 Eleventh International Conference on Ecological Vehicles and Renewable Energies (EVER)*, Monte Carlo, Monaco, IEEE, pp. 1–11, Apr. 2016, doi: 10.1109/EVER.2016.7476443.
5. Sochor, J., Arby, H., Karlsson, I.C.M., Sarasini, S., A topological approach to Mobility as a Service: A proposed tool for understanding requirements and effects, and for aiding the integration of societal goals. *Res. Transp. Bus. Manag.*, 27, 3–14, Jun. 2018. doi: 10.1016/j.rtbm.2018.12.003.
6. How Mobility as a Service Impacts Public Transport Business Models, in: *International Transport Forum Discussion Papers 2020/17*, Oct. 2020, doi: 10.1787/df75f80e-en.
7. Lynn, L.A., Artificial intelligence systems for complex decision-making in acute care medicine: a review. *Patient Saf. Surg.*, 13, 1, 6, Dec. 2019. doi: 10.1186/s13037-019-0188-2.

8. Hensher, D.A. and Xi, H., Mobility as a service (MaaS): are effort and seamlessness the keys to MaaS uptake? *Transp. Rev.*, 42, 3, 269–272, May 2022. doi: 10.1080/01441647.2022.2044590.
9. Lee, Z.W.Y., Chan, T.K.H., Balaji, M.S., Chong, A.Y.-L., Why people participate in the sharing economy: an empirical investigation of Uber. *Internet Res.*, 28, 3, 829–850, Jun. 2018. doi: 10.1108/IntR-01-2017-0037.
10. Geissinger, A., Laurell, C., Sandström, C., Digital Disruption beyond Uber and Airbnb—Tracking the long tail of the sharing economy. *Technol. Forecast. Soc. Change*, 155, 119323, Jun. 2020. doi: 10.1016/j.techfore.2018.06.012.
11. Tseng, Y.-C. and Chan, C.-L., When the Sharing Economy Meets Established Institutions: Uber and Airbnb in Taiwan. *IEEE Trans. Eng. Manag.*, 68, 6, 1895–1906, Dec. 2021. doi: 10.1109/TEM.2019.2934520.
12. *Kamargianni: The MaaS Dictionary - Google Scholar*, MaaSLab, Energy Institute, University College London, 2018. Accessed: Nov. 09, 2024. [Online]. Available: https://scholar.google.com/scholar_lookup?title=The%20MaaS%20dictionary&author=M.%20Kamargianni&publication_year=2018.
13. Fioreze, T., De Gruijter, M., Geurs, K., On the likelihood of using Mobility-as-a-Service: A case study on innovative mobility services among residents in the Netherlands. *Case Stud. Transp. Policy*, 7, 4, 790–801, Dec. 2019. doi: 10.1016/j.cstp.2019.08.002.
14. Cruz, C.O. and Sarmento, J.M., 'Mobility as a Service' Platforms: A Critical Path towards Increasing the Sustainability of Transportation Systems. *Sustainability*, 12, 16, 6368, Aug. 2020. doi: 10.3390/su12166368.
15. Reck, D.J., Hensher, D.A., Ho, C.Q., MaaS bundle design. *Transp. Res. Part Policy Pract.*, 141, 485–501, Nov. 2020. doi: 10.1016/j.tra.2020.09.021.
16. Jittrapirom, P., Caiati, V., Feneri, A.-M., Ebrahimigharehbaghi, S., González, M.J.A., Narayan, J., Mobility as a Service: A Critical Review of Definitions, Assessments of Schemes, and Key Challenges. *Urban Plan.*, 2, 2, 13–25, Jun. 2017. doi: 10.17645/up.v2i2.931.
17. Kandanaarachchi, T., Nelson, J., Ho, C., Building Trust and Collaboration Among the Stakeholders in a Mobility as a Service Ecosystem–Insights from Two Maas Case Studies, in: *Available SSRN 4253442*, 2022, Accessed: Nov. 10, 2024. [Online]. Available: https://papers.ssrn.com/sol3/papers.cfm?abstract_id=4253442.
18. Zhang, Z. and Zhang, N., A Novel Development Scheme of Mobility as a Service: Can It Provide a Sustainable Environment for China? *Sustainability*, 13, 8, 4233, Apr. 2021. doi: 10.3390/su13084233.
19. Schroten, A., *et al.*, Research for TRAN Committee–The impact of emerging technologies on the transport system. European Parliament, Policy Department for Structural and Cohesion Policies, Brussels, in: *IOP Conference Series: Materials Science and Engineering*, p. 012048, 2020.
20. Comi, A. and Russo, F., Emerging information and communication technologies: the challenges for the dynamic freight management in city logistics. *Front. Future Transp.*, 3, 887307, 2022.

21. Comi, A. and Polimeni, A., Estimating path choice models through floating car data. *Forecasting*, 4, 2, 525–537, 2022.
22. Comi, A., Rossolov, A., Polimeni, A., Nuzzolo, A., Private car OD flow estimation based on automated vehicle monitoring data: Theoretical issues and empirical evidence. *Information*, 12, 12, 493, 2021.
23. Iyer, L.S., AI-enabled applications towards intelligent transportation. *Transp. Eng.*, 5, 100083, Sep. 2021. doi: 10.1016/j.treng.2021.100083.
24. Hensher, D.A., The reason MaaS is such a challenge: A note. *Transp. Policy*, 129, 137–139, 2022.
25. Tirachini, A., Godachevich, J., Cats, O., Muñoz, J.C., Soza-Parra, J., Headway variability in public transport: a review of metrics, determinants, effects for quality of service and control strategies. *Transp. Rev.*, 42, 3, 337–361, May 2022. doi: 10.1080/01441647.2021.1977415.
26. Nuzzolo, A., Crisalli, U., Comi, A., Rosati, L., Individual behavioural models for personal transit pre-trip planners. *Transp. Res. Procedia*, 5, 30–43, 2015.
27. Mounce, R., Beecroft, M., Nelson, J.D., On the role of frameworks and smart mobility in addressing the rural mobility problem. *Res. Transp. Econ.*, 83, 100956, Nov. 2020. doi: 10.1016/j.retrec.2020.100956.
28. Philip, B.V., Alpcan, T., Jin, J., Palaniswami, M., Distributed Real-Time IoT for Autonomous Vehicles. *IEEE Trans. Ind. Inform.*, 15, 2, 1131–1140, Feb. 2019. doi: 10.1109/TII.2018.2877217.
29. Velaga, N.R., Beecroft, M., Nelson, J.D., Corsar, D., Edwards, P., Transport poverty meets the digital divide: accessibility and connectivity in rural communities. *J. Transp. Geogr.*, 21, 102–112, Mar. 2012. doi: 10.1016/j.jtrangeo.2011.12.005.
30. Bharatiya, J., Machine Learning in Cybersecurity: Techniques and Challenges. *Eur. J. Technol.*, 7, 2, 1–14, Jun. 2023. doi: 10.47672/ejt.1486.
31. Wolniak, R. and Stecuła, K., Artificial Intelligence in Smart Cities—Applications, Barriers, and Future Directions: A Review. *Smart Cities*, 7, 3, 1346–1389, 2024.
32. Elias, M., Timber Traceability and Sustainable Transportation Management: A Review of Technologies and Procedures. *Bull. Transilv. Univ. Brasov Ser. II For. Wood Ind. Agric. Food Eng.*, 17, 66, 1, 11–52, 2024.
33. Bayashot, Z.A., The Contribution of AI-Powered Mobile Apps to Smart City Ecosystems. *J. Softw. Eng. Appl.*, 17, 3, 143–154, 2024.
34. Al-Fuqaha, A., Guizani, M., Mohammadi, M., Aledhari, M., Ayyash, M., Internet of things: A survey on enabling technologies, protocols, and applications. *IEEE Commun. Surv. Tutor.*, 17, 4, 2347–2376, 2015.
35. Dou, X., Chen, W., Zhu, L., Bai, Y., Li, Y., Wu, X., Machine Learning for Smart Cities: A Comprehensive Review of Applications and Opportunities. *Int. J. Adv. Comput. Sci. Appl.*, 14, 9, 261–266, 2023. Accessed: Dec. 26, 2024. [Online]. Available: https://search.proquest.com/openview/2b773d51151376afc6e6ec6d7ea3cac1/1?pq-origsite=gscholar&cbl=5444811.

36. Bao, H. and Lee, E.W.J., Examining the antecedents and health outcomes of health apps and wearables use: an integration of the technology acceptance model and communication inequality. *Behav. Inf. Technol.*, 43, 4, 695–716, Mar. 2024. doi: 10.1080/0144929X.2023.2183062.
37. Bibri, S.E. and Krogstie, J., Smart, sustainable cities of the future: An extensive interdisciplinary literature review. *Sustain. Cities Soc.*, 31, 183–212, 2017.
38. Yao, A., *et al.*, A privacy-preserving location data collection framework for intelligent systems in edge computing. *Ad Hoc Netw.*, 161, 103532, 2024.
39. Deakin, M. and Al Waer, H., From intelligent to smart cities. *Intell. Build. Int.*, 3, 3, 140–152, Jul. 2011. doi: 10.1080/17508975.2011.586671.
40. Mavrogiorgos, K., Kiourtis, A., Mavrogiorgou, A., Menychtas, A., Kyriazis, D., Bias in Machine Learning: A Literature Review. *Appl. Sci.*, 14, 19, 8860, 2024.

Index

Printed and bound by CPI Group (UK) Ltd, Croydon, CR0 4YY
17/09/2026
14964265-0002